Candida and Candidamycosis

FEDERATION OF EUROPEAN MICROBIOLOGICAL SOCIETIES SYMPOSIUM SERIES

Recent FEMS Symposium volumes published by Plenum Press

1990 • MOLECULAR BIOLOGY OF MEMBRANE-BOUND COMPLEXES IN PHOTOTROPHIC BACTERIA
Edited by Gerhart Drews and Edwin A. Dawes
(FEMS Symposium No. 53)

1990 • MICROBIOLOGY AND BIOCHEMISTRY OF STRICT ANAEROBES INVOLVED IN INTERSPECIES HYDROGEN TRANSFER
Edited by Jean-Pierre Bélaich, Mireille Bruschi, and Jean-Louis Garcia
(FEMS Symposium No. 54)

1990 • DENITRIFICATION IN SOIL AND SEDIMENT
Edited by Niels Peter Revsbech and Jan Sørensen
(FEMS Symposium No. 56)

1991 • *CANDIDA* AND CANDIDAMYCOSIS
Edited by Emel Tümbay, Heinz P. R. Seeliger, and Özdem Ang̃
(FEMS Symposium No. 50)

Candida and Candidamycosis

Edited by
Emel Tümbay

Ege University
Izmir, Turkey

Heinz P. R. Seeliger

University of Würzburg
Würzburg, Germany

and

Özdem Anǧ

Istanbul University
Istanbul, Turkey

PLENUM PRESS • NEW YORK AND LONDON

Library of Congress Cataloging-in-Publication Data

Candida and candidamycosis / edited by Emel Tümbay, Heinz P.R.
  Seeliger, and Özdem Anğ.
       p.   cm. -- (FEMS symposium ; no. 50)
    Proceedings of the FEMS-Symposium on Candida and Candidamycosis,
  held Apr. 24-28, 1989 in Antalya, Turkey.
    Includes bibliographical references and index.
    ISBN-13: 978-1-4684-5912-8   e-ISBN-13: 978-1-4684-5910-4
    DOI: 10.1007/978-1-4684-5910-4
    1. Candidiasis--Congresses.   I. Tümbay, Emel.   II. Seeliger,
  Heinz P. R. (Heinz Paul Richard) III. Anğ, Özdem.   IV. Federation
  of European Microbiological Societies.   V. FEMS-Symposium on Candida
  and Candidamycosis (1989 : Antalya, Turkey)   VI. Series.
  QR201.C27C36 1991
    616.9'69--dc20   91-2588
                                                                   CIP

Proceedings of a Symposium held under the auspices of the
Federation of European Microbiological Societies, April 24–28, 1989,
in Antalya, Turkey

ISBN-13: 978-1-4684-5912-8

© 1991 Plenum Press, New York
Softcover reprint of the hardcover 1st edition 1991
A Division of Plenum Publishing Corporation
233 Spring Street, New York, N.Y. 10013

PREFACE

    Scientists   from   24   countries who participated in the <u>FEMS-</u>
<u>Symposium on Candida and Candidamycosis, 24-28 April 1989, Antalya,</u>
Turkey, have made important contributions to a better understanding
of <u>Candida</u> and its infections - by their presence, presentations and
discussions in the meeting. The Turkish Microbiological Society,
organizing the Symposium in the name of the Federation of European
Microbiological Societies   (FEMS), thanks   all participants   for
realising this important meeting on such an important subject.

    In this book the full manuscripts of invited and free papers
of the meeting are presented.

    The Editors extend their sincere thanks to all contributors of
papers; to FEMS for encouraging and undertaking the arrangements for
publication; to Ege University Publishing House for printing; to
Plenum Publishing Company for publishing; and to all colleagues and
friends who with their help contributed to this book.

    The Editors wish that this volume would contribute to a better
understanding of <u>Candida</u> and its infections by colleagues working
and/or interested in mycology and mycotic diseases.

                                            The Editors

CONTENTS

Free Papers

INTRODUCING REMARKS AT THE OPENING SESSION OF THE FEMS-SYMPOSIUM ON

CANDIDA AND CANDIDAMYCOSIS

H.P.R. Seeliger

The Honorary President of the Symposium

The Turkish Microbiological Society, organizing this second meeting on a mycological subject within 3 years under the auspices of FEMS in Turkey, has suggested that I say a few introductory words at the opening session. I have accepted this honour with some reservation and with joy at the same time, because it gives me the unique chance to recall the memorable day when after some previous talks under the guidance of Prof. André Lwoff from Paris a group of national delegates met in the CIBA house in London, where I had the honour to preside the official founding of FEMS in the seventies.

There is no European microbiological inspired organization which, in a relatively short span of less than 20 years of life, has comparably succeeded to bring at frequent intervals together over-seeable groups of engaged microbiologists between 50 and 300 capita in several member countries. And this unique and most commendable series of scientific events in microbiology was to a great part made possible by the financial ability of the organization, its journals and its staff.

Many of the speakers, if not all, here and at previous meetings, would share my feelings of gratitude to FEMS for making their participation and presentation of their papers possible, and this quite frequently in countries of the continent which are or have been handicapped in developing microbiology quickly in demands of modern medicine.

With this in mind, I welcome particularly the selection of a rapidly emerging place in South Eastern Turkey, next to what we call Near Orient, thus giving new groups of interested workers in medicine and sciences a unique opportunity to get acquainted with up-to-date progress. Who would ever have believed in a scientific symposium on "Candida and Candidamycosis" in the Eastern Mediterranean Region 20 years ago?

Let us now look forward to the meeting with its presentations, discussions and exchanges of views among colleagues and friends from many ethnical and cultural backgrounds, but with one thing in common: their professional interest and their common endeavour to improve human health and to control those microbial opportunists which can make life and disease quite miserable.

OPENING ADDRESS

E. Drouhet

The President of the Symposium

It is a great honour and pleasure to chair this symposium and also to thank the organizers, Professor Özdem Anğ, President of the Turkish Microbiological Society, and Professor Emel Tümbay, General Secretary of the Symposium, for giving us the privilege to discuss "Candida and Candidamycosis" in this beautiful Mediterranean site, Alanya.

Three years ago, Professor E. Tümbay organized another symposium of the Federation of European Microbiological Societies (FEMS), on "Dermatophytes and Dermatophytoses in Man and Animals" in Izmir, Turkey, which was very successful.

I am sure that this second FEMS Symposium will also be a great success and will constitute an important contribution to a better understanding of Candida and its infections.The most important problems of Candida and candidamycosis will be presented by the best of our colleagues, who are coming not only from Europe and the Mediterranean countries but also from U.S.A., New Zealand, Taiwan, Kuwait, Iran and Iraq.

Thrush has been known from antiquity and was described by Hippocrates (IVth century B.C.) who taught not far from the Turkish coast, in the Mediterranean island of Kos. It represents a tremendously versatile fungal infection which is in constant evolution. F. Odds in his recent book "Candida and Candidosis" analysed no fewer than six thousand references on the subject.

The new aspects of the molecular biology and genetics of Candida (Odds; Nombela), the new developments on morphogenesis (Gow), cell wall physiology (Shepherd; Sentandreu), adhesion (Senet and Robert; Douglas), secreted enzymes (Sullivan and Emerson) and proteinases (Rüchel), treated in this symposium, will represent the most recent research progress on Candida cells. The host-parasite mechanisms (Seeliger); the pathological, epidemiological and clinical aspects of candidosis in heroin addicts, in AIDS and other immunosuppressed conditions (Drouhet and Dupont) will demonstrate developments which have appeared in recent years in this disease. Candidamycosis has also an important place in veterinary medicine (Böhm).

New immunological and sophisticated methods for detecting Candida antigen and antibodies (Poulain; Price and Gentry) and new techniques for identification of Candida species (de Vroey) will lead to an improved diagnosis of candidamycosis.

Finally, this symposium will consider the most recent contributions concerning the mode of action of antifungal agents (Vanden Bossche) and the chemotherapy and prevention of <u>Candida</u> infections (Van Cutsem; Bremm; Polak; Plempel; Troke; Drouhet; Dupont). Numerous free papers and posters will complete the rich programme and will assure the success of this symposium.

The organizing committee is to be congratulated for proposing such exciting topics on the oldest but still the most important current mycosis.

WHY CANDIDA AND CANDIDAMYCOSIS

H.P.R. Seeliger

Institute of Hygiene and Microbiology, University of
Würzburg, 8750 Würzburg, The Federal Republic of Germany

When the etiology of thrush was established some 150 years ago
(BERG, 1839; ROBIN, 1853) as an infection caused by yeast-like
microorganism, the disease was considered frequently as an almost
normal event occurring among newborn and being amenable to local
treatment with boric acid etc., applied by the nurse or midwife. The
general use of vernacular names, such as "thrush" (English), "muguet"
(French) or "Soor" (German) by both the general public and the
medical profession, would indicate a wide-spread occurrence, but also
that the clinical appearance was known since a long time and
apparently did not cause any major concern.

What a change, if one recalls that in 1975 your speaker had a
full auditorium at the University of London, when he delivered his
address on "Opportunistic Fungal Infections with Particular Reference
to Candida albicans"!

This lecture dealt primarily with the same yeast observed in the
early days of medical mycology, and its characters had not changed a
bit since CASTELLANI, subsequent to his work in Ceylon, had published
in 1916 the differential diagnosis of Candida albicans - as the
fungus had been named by ROBIN (1853) - and 5 related species based
on their stable biochemical patterns of sugar fermentation. As a
matter of fact, only very little has been added to the knowledge of
the frequency pattern exhibited by these 6-8 Candida species as
observed in man by this fundamental contribution of Sir ALDO (Fig.1).

Of course, very occasionally a severe or generalized granulomatous or septic Candida infection was seen, and was usually published
as rarity in a first class journal, but nobody would ever have
believed that such mildly pathogenic yeasts could become an essential
agent of secondary infection and a contributor to death in many
individuals, acting as a true opportunist.

Moreover, Candida albicans, the main representative of this group
of imperfect, i.e. anascosporogenous budding fungi, was found to be a
commensal on mucous membranes and in the intestinal tract of many
individuals shownig no signs of an infectious process. Thus, it could
easily be taken as a normal dweller together with other microbes
which make up the normal flora of man. When, however, the normal
condition on mucous membranes and in the intestine was changed as a
result of sprue or the administration of antibacterial agents, such

Fig. 1. Sir Aldo Castellani (1874-1971).

as the broadspectrum antibiotics which left the usually small nembers of <u>Candida</u> cells unaffected, these would then multiply rapidly on the newly created terrain with entirely unforeseen results, thrush-like lesions being only one expression of the disturbed and well equilibrated microbial balance: i.e. balance among the microbial population itself, but also balance between the total of the flora and the host-usually impaired already by reasons stated before.

To these must be added quite a range of other agents: remedies, powerful drugs, but also individual diseases like malignant tumors, metabolic disorders, immunodeficiencies of various nature, and - last but not least- human aberrations, such as alcoholism or drug addiction and finally, the world-wide use of pregnancy inhibiting estrogens i.e. the pill, not to forget age, lack of oral hygiene and food habits as additional contributing factors. They all have one important character in common: they pave the way for the multipli- cation of the <u>Candida</u> cells and to their eventual transition from a relatively harmless commensal to a noxious opportunist which becomes invasive, aggressive and eventually fatal to its victim.

Since a good decade <u>Candida albicans</u> and related species have thus become a powerful and dangerous challange to man with a wide range of clinical syndromes, sometimes hard to recognize as of fungal nature. But medical science has not been asleep. As a matter of fact, various research groups are adding stone to stone for a mosaic which should explain the intrinsic mechanisms, how Candida cells adhere to the endo- and epithelial layers, how they colonize them and how they develop their pseudomycelia to penetrate and to invade the host with subsequent metastatic foci in the internal organs, on the endocardium and in the brain.

At the same time potent anti-Candida drugs were developed which - by various ways of administration- not only were able to cope with overt or suspected clinical infections. Perhaps even more important is the use some of these substances to monitor and even prevent Candida infections in the millions of endangered human hosts.

It cannot be the task of the indroductory presentation to anticipate any of the subsequent communications. But the speaker may be permitted to draw attention at least to the semantics. The present terminology of Candida infections is - in his opinion - somewhat misleading and unclear. If bacteriologists speak of salmonellosis, brucellosis or tuberculosis, they always mean "infection" because none of respective pathogens is a commensal in the human body with the debatable exception of a carrier state with its dangerous implication to other non-immune individuals. The frequently used medical terms "candidosis", "candidiasis" etc. do unfortunately often not differentiate between the most fequently observed commensal nature of Candida cells at this time absolutely harmless to their hosts under normal conditions (presence), and a true infection resulting from its opportunistic exploitation of a set of circumstances with colonization and penetration leading to "Candidamycosis", a designation which leaves no doubt as to its meaning.

The speaker is not very hopeful that this simple and clear terminology will be accepted by the "anti-Candida community" of mycologists and clinicians, but he is more than optimistic that this FEMS meeting will become a great step forward in the battle against a potentially noxious microbial companion of man and in its urgently needed control.

PERSPECTIVE OF THE APPLICATION OF MOLECULAR BIOLOGY TECHNIQUES IN THE

STUDY OF CANDIDA

F.C. Odds

Department of Microbiology, University of Leicester
Leicester LE1 7RH, The United Kingdom

INTRODUCTION

Candida albicans has proved to be a difficult organism to study
in terms of its genetics and molecular biology; which is a pity,
since the power of molecular genetic methods can greatly enhance the
scientific quality of many experiments with the fungus. Because
C.albicans is diploid and has no natural sexual cycle, traditional
approaches to mutation and genetic mapping have yielded few insights
into the properties of its DNA. However, this position is at last
beginning to change rapidly. Genes that code for products of
importance in the study of C.albicans morphogenesis, responses to
antifungal chemotherapy and the pathogenesis of candidosis are now
being successfully cloned, and methods of DNA typing have been
applied to epidemiological studies of candidosis. The future for
molecular genetic approaches to Candida-related problems is a bright
one. This brief overview will amount to a summary of the present
state of the molecular genetic art in the context of Candida and
Candida infections, with consideration of the problems where
DNA-manipulative experimentation can be most usefully applied.

LANDMARKS IN THE DEVELOPMENT OF C. ALBICANS MOLECULAR GENETICS

For many years the genetic of C.albicans lagged far behind that
of other important micro-organisms. It proved difficult to produce a
useful range of C.albicans mutants for elementary genetic work,
transformation of C.albicans genes appeared to be impossible, and
even determination of karyotype could not be achieved because
C.albicans chromosomes, like those of other yeasts, cannot be seen
by light microscopy, nor can they be easily visualized by electron
microscopy.

The major genetic problem with C.albicans is its diploid genome
and the absence of any natural sexual cycle. The relevation of
diploidy in C.albicans came from evidence on three different fronts:
the demonstration of natural heterozygosity in the fungus by Whelan
et al.[1], the analysis of the DNA content per C.albicans cell by
Riggsby et al.[2], and the analysis of recombinants from protoplast
fusion experiments-parasexual cycle analysis - by Poulter et al[3].

Candida and Candidamycosis, Edited by E. Tümbay *et al.*
Plenum Press, New York, 1991

Table 1. <u>Candida albicans</u> genes that have been cloned.

Year	Gene	Gene codes for	Basis of cloning*	Reference
1984	URA3	orotidine 5' phosphate decarboxylase	A	Gillum et al.[4]
1985	HIS3	imidazoleglycerol-phosphate dehydrogenase	A	Rosenbluh et al.[33]
	TRP1	phosphoribosyl-anthranilate isomerase	A	Rosenbluh et al.[33]
1986	ADE2	phosphoribosyl-aminoimi-dazole carboxylase	B	Kurtz et al.[6]
1987	ACT1	actin	C	Mason et al.[16]
	DFR1	dihydrofolate reductase	D	Kurtz et al.[7]
	STA1	glucoamylase	A	Cohen[b]
	CHS1	chitin synthase	A	Au-Yourg[b]
1988	LEU2	β-isopropylmalate dehydrogenase	D	Jenkinson et al.[35]
	L1A1	lanosterol-14α-demethylase	C	Kirsch et al.[30]
	ADE1	phosphoribosyl-aminoimi-dazole-succinocarboxamide synthetase	A	Magee et al.[11]
	BenR	benomyl resistance	E	Magee et al.[11]
	GAL1	galactokinase	A	Magee et al.[11]
	LYS2	2-aminoadipate-semial-dehyde dehydrogenase	A	Magee et al.[11]
	MGL1	growth on α-methyl-D-glucoside	E	Magee et al.[11]
	MtxR	methotrexate resistance	E	Magee et al.[11]
	SOR2	sorbitol dehydrogenase	E	Magee et al.[11]
	SOR9	sorbitol dehydrogenase	E	Magee et al.[11]
	TUB2	β-tubulin	C	Magee et al.[11]
1989	PrA	aspartyl proteinase	C	Lott et al.[28]
	TS	thymidylate synthase	A	Singer et al.[32]

* A, complementation in <u>S.cerevisiae</u>; B, transformation of <u>C.albicans</u> ade2; C, probe from <u>S.cerevisiae</u> gene; D, complementation in <u>E.coli</u>; E, conferred property on <u>S.cerevisiae</u> , [b]see Kurtz et al.[34]

Diploidy does not prevent the isolation and cloning of <u>C. albicans</u> DNA sequences in other recipient organisms and, since 1984, when Gillum et al[4]. published the first report of successful cloning of a <u>C. albicans</u> gene, the number of papers on <u>C. albicans</u> gene cloning has grown rapidly (Table 1). Several <u>C. albicans</u> DNA libraries are now readily available to investigators[5], and the need for <u>de novo</u> creation of new libraries is diminishing.

It can be seen from Table 1 that two main approaches have been used to clone C.albicans genes : complementation of known genes or expression of new genes in appropriate hosts (most commonly Saccharomyces cerevisiae) and probing of C. albicans DNA with known genes from other organisms. It should be possible also to identify C. albicans genes with synthetic oligonucleotide probes based on known amino acid sequences of C. albicans proteins and by recognition of cloned Candida gene products with monoclonal antibodies. Examples of such approaches to cloning are likely to be published in the near future.

Alongside the progress in cloning fragments of C. albicans DNA have come impressive improvements in other methods for molecular genetic experimentation with the fungus. Kurtz and her colleagues[6] achieved transformation in C. albicans in 1986, and developed autonomously replicating plasmids that were published the following year[7]. Cannon et al. have recently constructed a plasmid capable of replicating autonomously in both C. albicans and S. cerevisiae, which should therefore be useful as a shuttle vector between the two yeasts (Shepherd, personal communication). Kelly et al [8,9], have devised a method for site-directed gene disruption in C.albicans, which should be adaptable to new problems. Gene disruption is followed by ultraviolet light treatment to generate homozygous recombinants with specific lesions in both members of the paired diploid chromosomes.

Substantial progress has also been achieved in electrophoretic karyotyping of C. albicans. Snell & Wilkins[10] were the first to apply the technique of gel electrophoresis with slow periods of electric field reversal to separate chromosomes from C. albicans. Several authors have now published results achieved with various refinements of the field alternation principle; most recently Magee et al [11]. have begun the process of assignment of specific genes to individuals among the seven C. albicans chromosomes by hybridization of blots of electrophoresed chromosomes with cloned genes.

APPLICATIONS OF MOLECULAR GENETICS TO PROBLEMS IN THE CANDIDA FIELD

DNA typing of Candida species and strains

For the study of the epidemiology of candidosis, reliable methods for differentation of Candida species and for typing of individual strains within each species are imperative. Molecular genetic approaches offer the basis for determination of species and strain identity at the DNA level. For species differentiation, several DNA probes have been developed. Some of these probes react with only a single Candida species. These include Ca3 and Ca7[12], 27A[13] and certain C.albicans repeat sequences[14], which all react exclusively with isolates of C. albicans, and Ct13-8[15], which is C. tropicalis-specific. Such probes open the possibility for rapid identification of C.albicans or C. tropicalis. Their value would lie mainly in rapid detection and identification of Candida species directly in clinical samples; for identification of C. albicans in culture the traditional germ tube test still offers advantages of economy and ease of performance over the DNA technology.

The acting gene from C.albicans[16] and a repeat sequence from C.albicans rDNA[17,18] are capable of differentiating clinically important Candida species when they are used as probes of Southern blots from restriction endonuclease digests of DNA. This approach to species identification offers a theoretical advantage of precision over methods that rely on phenotypic properties : it is likely to be some time before the method becomes sufficiently simple and economic for routine use.

Some of the probes already mentioned (Ca3, Ca7, 27A and rDNA repeats) plus certain fragments of C.albicans mtDNA[19,20] allow for differentiation of strains of C.albicans in terms of differences in restriction digest/Southern blot patterns. Fox et al.[21] tested several clinical isolates of C.albicans with the mtDNA probe and a biotiny-lated version of probe 27A : they considered the mtDNA probe was insufficiently discriminatory for epidemiological purposes but probe 27A differentiated 60 C.albicans DNA types in isolates from 63 patients and therefore held great promise as an epidemiological tool.

Gene-level discrimination of C.albicans strains can be achieved by electrophoresis of EcoR1 digests of DNA extracted from the fungus, without the use of subsequent blotting and probing procedures. Such typing by restriction fragment length polymorphisms (RFLPs) was described in 1987 by Scherer & Stevens[22], and has been recently applied epidemiologically by Matthews & Burnie[23]. RFLP typing of DNA is inherently more vulnerable to artefacts of technique than typing based on well characterized DNA probes, and there is already evidence that this approach may give different results in different hands. Matthews & Burnie[23] reported that all the phenotypically identical C.albicans strains from one outbreak of disseminated candidosis gave identical RLFPs, but Stevens, Odds & Scherer (submitted) found five different RFLP types among these isolates. It would be helpful if the cause of such differences could be resolved and the tests standardized, since the conclusions to be drawn from the two studies about the relation between genotype and phenotype in serious Candida infections are markedly different.

Electrophoretic karyotyping also reveals differences between strains of C. albicans[18,24,25,26] and these differences have been used to study the distribution of C.albicnas strains among hospital patients[27]. Karyotyping methods, like RFLPs, seem at present to present problems of standardization between laboratories; Magee et al[11]. state that there are problems of reproducibility even between different electrophoretic devices within a single laboratory. These difficulties need to be addressed before the DNA typing methods can be regarded as universally applicable.

Identification of virulence factors in C. albicans

One of the most powerful applications of molecular genetic technology is the use of specific gene probes and specific gene disruptions to identify the function of gene products in vivo. In the context of the pathogenesis of microbial infection, this means that genes coding for putative virulence factors can be altered selectively, with results that can be measured in animal models of infection in vivo.

It is likely that this approach will first be successfully applied to the proteinase enzyme secreted by C. albicans, which has been recognized as a likely virulence factor for several years. Lott et al.[28] recently published the base sequence of a cloned aspartyl proteinase from C. albicans. They used the PrA gene from S.cerevisiae as a hybridization probe to detect the C. albicans gene. It is not yet clear that the product of the C. albicans gene cloned by Lott et al.[28] is the same secreted enzyme that has been implicated as a virulence factor. Since antibody to the C. albicans secreted enzyme does not react with S. cerevisiae proteinase A (Rüchel, personal communication), and since the N-terminal amino acid sequence of the secreted enzyme differs from that reported by Lott et al.[28] (Turver, Boulnois & Odds, unpublished data) it is possible that the cloned gene codes for a different aspartyl proteinase in C. albicans.

Morphogenesis in C. albicans

The pathological significance of hypha formation by C. albicans is a controversial topic, but the consesus view remains that formation of hypae endows on the fungus an enhanced ability to penetrate host tissue in vivo. In any case, the processes of dimorphism in C. albicans remain of great interest as a model of cellular differentiation. Three genes pertinent to morphogenetic development, ACT1, coding for actin, TUB2, coding for β-tubulin, and CHS1, coding for chitin synthase, have now been cloned (see Table 1) so that information on their regulation and its significance for morphogenesis is likely to be forthcoming in future.

Russell et al. [29] found higher levels of methylated DNA in C. albicans hyphae than in the yeast form, and suggested this means that gene activity is greater in hyphae than in yeasts.

Development and mode of action of antifungal drugs

Cloning of genes for known antifungal drug targets in Candida may facilitate studies of the mechanisms of action of antifungals and could lead to development of new compounds. The gene for the cytochrome P450 (lanosterol 14α-demethylase), which is the target for azole antifungals, has been cloned[30] and sequenced[31]. Similarly, the gene coding for thymidylate synthase, another potential target for antifungals, has been cloned and sequenced, and the enzyme expressed in E. coli[32]. A major advantage of cloning and expressing genes of potential drug target proteins is that it greatly increases the amount of target material available for study and facilitates its purification.

CONCLUDING REMARKS

This short review serves to show that molecular genetic approaches to the study of C. albicans are represented increasingly in the published literature and that they possess the potential to provide novel answers to many questions pertinent to Candida infections. There are weaknesses as well as strengths of experimentation with "DNA high technology" : such experimentation often proves frustratingly difficult in practice, it is not invulnerable to charges of poor reproducibility, and it is costly to develop.

The work that has been done with C. albicans so far has been related almost entirely to gene structure. It is in the field of gene regulation and expression and their relation to behaviour of whole cells that the most exciting developments lie. Cells and tissues are, after all, organized at a level much higher than can be easily comprehended by knowledge of base sequences per se (what DNA sequence is involved when one says the words "gene expression"?). There is now much evidence that C. albicans is highly variable in that its expressed phenotype can adapt rapidly and subtly with changes in its environment. Thus, elucidation of the interaction between mammalian host microenvironments and regulation of phenotype in <u>C. albicans</u> is ultimately the means of obtaining insights into the complex transition between commensal and pathogen that characterizes this problematic fungus.

REFERENCES

1. W.L. Whelan and P.T. Magee,Natural heterozygosity in <u>Candida albicans</u>, <u>J. Bacteriol</u>., 145 : 896 (1981).
2. W.S. Riggsby, L.J. Torres-Bauza, J.W. Wills, T.M. Townes, DNA content, kinetic complexity, and the ploidy question in <u>Candida albicans</u>, <u>Mol. Cell Biol</u>, 2 : 853 (1982).
3. R. Poulter, V. Hanrahan, K. Jeffery, D. Markie, M.G. Shepherd, P.A. Sullivan, Recombination analysis of naturally diploid <u>Candida albicans</u>, <u>J.Bacteriol</u>., 152 : 969 (1982).
4. A.M.Gillum, E.Y.H. Tsay, D.R. Kirsch, Isolation of the <u>Candida albicans</u> gene for orotidine-5'-phosphate decarboxylase by complementation of <u>S. cerevisiae</u> ura3 and <u>E.coli</u> <u>pyrF</u> mutations, <u>Mol.Gen.Genet</u>., 198 : 179 (1984).
5. P.T.Magee, E.H.A. Rikkerink, B.B.Magee, Review methods for the genetics and molecular biology of <u>Candida albicans</u>, <u>Anal. Biochem</u>., 175 : 361 (1988).
6. M.B.Kurtz, M.W. Cortelyou, D.R. Kirsch, Integrative transformation of <u>Candida albicans</u>, using a cloned <u>Candida ADE2</u> gene, <u>Mol. Cell Biol</u>., 6 : 142 (1986).
7. M.B. Kurtz, M.W. Cortelyou, S.M. Miller, M. Lai, D.R. Kirsch, Development of autonomously replicating plasmids for <u>Candida albicans</u>, <u>Mol. Cell Biol</u>., 7 : 209 (1987).
8. R.Kelly, S.M.Miller, M.B. Kurtz, D.R. Kirsch, Directed mutagenesis in <u>Candida albicans</u> : one-step gene disruption to isolate ura3 mutants, <u>Mol. Cell Biol</u>., 7 : 199 (1987).
9. R. Kelly, S.M. Miller, M.B. Kurtz, One-step gene disruption by contransformation to isolate double auxotrophs in <u>Candida albicans</u>, <u>Mol. Gen. Genet</u>., 214 : 24 (1988).
10. R.G.Snell, R.J. Wilkins, Separation of chromosomal DNA molecules from <u>C. albicans</u> in pulsed field gel electrophoresis, <u>Nucl. Acid Res</u>., 14 : 4401 (1986).
11. B.B. Magee, Y. Koltin, J.A. Gorman, P.T. Magee, Assigment of cloned genes to the seven electrophoretically separated <u>Candida albicans</u> chromosomes, <u>Mol.Cell Biol</u>., 8 : 4721 (1988).
12. D.R.Soll, C.J.Langtimm, J.McDowell, J.Hicks, R.Galask, High-frequency switching in <u>Candida</u> strains isolated from vaginitis patients, <u>J.Clin. Microbiol</u>., 25 : 1611 (1987).
13. S.Scherer, D.A. Stevens, A <u>Candida albicans</u> dispersed, repeated gene family and its epidemiologic applications, <u>Proc.Nat.Acad.Sci</u>. (USA), 85 : 1452 (1988).
14. J.E.Culter, P.M.Glee, H.L.Horn, <u>Candida albicans</u> and <u>Candida stellatoidea</u>-specific DNA fragment, <u>J. Clin. Microbiol</u>., 26: 1720 (1988).
15. D.R. Solly, M. Staebell, C. Longtimm, M. Pfaller, J. Hicks, T.V. Gopala Rao, Multiple <u>Candida</u> strains in the course of a single systemic infection, <u>J. Clin. Microbiol</u>., 26 : 1448 (1988).
16. M.M.Mason, B.A.Lasker, W.S.Riggsby, Molecular probe for identification of medically important <u>Candida</u> species and <u>Torulopsis glabrata</u>, <u>J. Clin.Microbiol</u>., 25 : 563 (1987).
17. B.B.Magee, T.M. D'Souza, P.T. Magee, Strain and species identification by restriction fragment length polymorphisms in the ribosomal DNA repeat of <u>Candida</u> species, <u>J.Bacteriol</u>., 169 : 1639 (1987).

18. B.B. Magee and P.T. Magee, Electrophoretic karyotypes and chromosome number in <u>Candida</u> species, <u>J.Gen.Microbiol.</u>, 133 : 425 (1987).

19. P.D.Olivo, E.J. McManus, W.S. Riggsby, J.M. Jones, Mitochondrial DNA polymorphism in <u>Candida albicans</u>, <u>J.Infect.Dis.</u>, 156 : 214 (1987).

20. K.J. Kwon-Chung, W.S. Riggsby, R.A. Uphoff, J.B. Hicks, W.L. Whelan, E.Reiss, B.B.Magee, B.L. Wickes, Genetic differences between type I and type II <u>Candida stellatoidea</u>, <u>Infect. Immun.</u>, 57 : 527 (1989).

21. B.C.Fox, H.L.T. Mobley, J.C. Wade, The use of a DNA probe for epidemiological studies of candidiasis in immunocompromised hosts. <u>J.Infect. Dis.</u>, 159 : 488 (1989).

22. S.Scherer and D.A. Stevens, Application of DNA typing methods to the epidemiology and taxonomy of <u>Candida</u> species, <u>J.Clin. Microbiol.</u>, 25 : 675 (1987).

23. R.Matthews and J. Barnie, Assessment of DNA fingerprinting for rapid identification of outbreaks of systemic candidiasis, <u>Br.Med. J.</u>, 298 : 354 (1989).

24. T.J.Lott, P. Boiron, E. Reiss, An electrophoretic karyotype for <u>Candida albicans</u> reveals large chromosomes in multiples, <u>Mol.Gen. Genet.</u>, 209 : 170 (1987).

25. R.G.Snell, I.F. Hermans, R.J.Wilkins, B.E.Corner, Chromosomal variations in <u>Candida albicans</u>, <u>Nucl. Acid Res.</u>, 15 : 3625 (1987).

26. K.J. Kwon-Chung, B.L. Wickes, W.G. Merz, Association of electro-phoretic karyotype of <u>Candida stellatoidea</u> with virulence for mice, <u>Infect. Immun.</u>, 56 : 1814 (1988).

27. W.G. Merz, C.Connelly, P. Hieter, Variation of electrophoretic karyotypes among clinical isolates of <u>Candida albicans</u>, <u>J. Clin. Microbiol.</u>, 26 : 842 (1988).

28. T.J. Lott, L.S. Page, P. Boiron, J. Benson, E. Reiss, Nucleotide sequence of the <u>Candida albicans</u> aspartyl proteinase gene, <u>Nucl. Acid Res.</u>, 17 : 1779 (1989).

29. P.J. Russell, J.A. Welsch, E.M. Rachlin, J.A. McCloskey, Different levels of DNA methylation in yeast and mycelial forms of <u>Candida albicans</u>, <u>J.Bacteriol.</u>, 169 : 4393 (1987).

30. D.R.Kirsch, M.H. Lai, J. O'Sullivan, Isolation of the gene for cytochrome P450 L1A1 (lanosterol 14α-demethylase) from <u>Candida albicans</u>, <u>Gene</u>, 68 : 229 (1988).

31. M.H.Lai and D.D.Kirsch, Nucleotide sequence of cytochrome P450 L1A1 (lanosterol 14α-demthylase) from <u>Candida albicans</u>, <u>Nucl. Acid. Res.</u>, 17 : 804 (1989).

32. S.C. Singer, C.A. Richards, R. Ferone, D. Benedict, P. Ray, Cloning, purification and properties of <u>Candida albicans</u> thymidylate synthase, <u>J.Bacteriol.</u>, 171 : 1372 (1989).

33. A. Rosenbluh, M. Mevarech, Y. Koltin, J.A. Gorman, Isolation of genes from <u>Candida albicans</u> by complementation in <u>Saccharomyces cerevisiae</u>, <u>Mol. Gen. Genet.</u>, 200 : 500 (1985).

34. M.B. Kurtz, D.R. Kirsch, R. Kelly, The molecular genetics of <u>Candida albicans</u>, <u>Microbiol. Sci.</u>, 5 : 58 (1988).

35. H.F. Jenkinson, G.P. Schep, M.G. Shepherd, Cloning and expression of the 3-isopropylmalate dehydrogenase gene from <u>Candida albicans</u>, <u>FEMS Microbiol. Lett.</u>, 49 : 285 (1988).

POLARIZED MORPHOGENESIS OF <u>CANDIDA ALBICANS</u>:

CYTOLOGY, INDUCTION AND CONTROL

N.A.R. Gow   T. Crombie   W. Gooday

Department of Genetics and Microbiology, Marischal
College, University of Aberdeen, Aberdeen AB9 1AS,
Scotland, The United Kingdom

INTRODUCTION

There can be few systems of cellular morphogenesis which have attracted such interest, yet have proved to be so recalcitrant to investigation as yeast-mould dimorphism in <u>Candida albicans</u>. Until very recently morphological mutants have not been available and the use of molecular genetics is in this organism still in its infancy. Since genetic techniques have not so far provided the means to explore this fascinating phenomenon, we have attempted instead to produce a cytological and physiological description of the yeast to hyphal transition and some of the early events which are associated with this process. In particular we have been motivated by work with other eukaryotic organisms such as the slime moulds and the true filamentous fungi which has shown that temporal and spatial fluxes of inorganic ions are often tightly coupled to cellular differentiation[1,2]. Based on this type of approach our studies on <u>C. albicans</u> suggest that germ tube formation occurs in response to nutrient poor conditions, especially when the supply of nitrogen is limiting. The pattern of cell growth reviewed below seems well adapted to these conditions. We find that there is a good correlation between an early rise in the cytoplasmic pH of the cell and the induction of germ tube growth. Our recent work studying the effects of applied electrical fields on germ tube formation and budding in <u>C. albicans</u> suggests that transmembrane fluxes of calcium ions are also of importance to germ tube formation. This article provides a brief description of these studies and their implication to the control of dimorphism in this organism.

CYTOLOGY OF DIMORPHISM

When the yeast cell of <u>C. albicans</u> forms a germ tube, the cytoplasm within it flows into the hypha and leaves behind a vacuolated space[3,4]. Continued extension of the germ tube is supported by a more or less constant volume of biosynthetically active cytoplasm which leaves in its wake an increasingly vacuolated hypha. Regular septation delimits intercalary compartments which are highly vacuolated but all have single nuclei in central position[5]. The formation of secondary germ tubes or lateral branches is preceded

Candida and Candidamycosis, Edited by E. Tümbay *et al.*
Plenum Press, New York, 1991

by regeneration of the protoplasm which replaces the vacuolar space. This phenomenon has been shown for both laboratory strains and clinical isolates of C. albicans in media containing serum, proline or N-acetyl glucosamine and in defined media based on glucose, salts and various amino acids [6]. This type of growth is typical of other non-dimorphic fungi, such as species of Basidiobolus, Uromyces and Puccinia, and has been likened to the migration of "plasmodia moving about in a system of tubes" [7]. From a teliological viewpoint this mode of growth would seem to maximise the opportunity of moving away from a region of adversity or nutrient depletion, since it maximises extension growth for the minimum cost to protein biosynthesis, a feature which exemplifies the ecological specialisation of filamentous fungi in general. This model provides a basis for understanding other cytological and physiological characteristics of germ tube growth in this organism.

(a) It explains why germ tube extension of C. albicans exhibits linear kinetics which is unlike the exponential kinetics typical of the germination of fungal spores [5,6]. This relates to the fact that the extension of the apex of the germ tube in C. albicans is supported by a constant cytoplasic volume and not an increasing cytoplasmic volume as in spore germination. Growth in this case is therefore not autocatalytic.

(b) Branch formation after septation may be delayed for a short or for an extended time. For example, many cells grown in vivo have very few branches. Since branching is the mycelial equivalent of budding, it might be expected to be regular and coupled tightly to each septation event. However, vacuolated intercalary compartments are biosyntheticaly quiscent and branching is delayed until cytoplasm can be regenerated.

(c) The observation that germinated chlamydospores are empty [8] suggests that outgrowth by protoplasmic migration also occurs during this process.

(d) The observed reduction of the carbon to nitrogen ratio during germ tube formation in N-acetylglucosamine-containing medium [9] reflects the production of new cell wall, but minimal protein biosynthesis.

(e) Cell division of yeast and pseudohyphal cells can be shown to be initiated at a constant cell size if the substantial vacuolar volume is first subtracted from the total cell volume of the pseudohyphal cells [10].

(f) In scanning electron micrographs of yeast and hyphal cells, the vacuolated mother cell and subapical compartments may collapse upon critical point drying. Presumably chemical fixation fails to provide sufficient structural support in the aqueous vacuolated regions of the germlings to prevent collapsing during dehydration [11].

(g) The hyphal form has a higher specific protease activity than the yeast form. Fungal vacuoles are normally compartments which store proteolytic enzymes. Because of this, enzyme preparations of chitin synthase prepared from cell homogenates are less stable than those made from yeast cell cultures (Mc Dougall, Gooday, Gow & Inall, unpublished).

(h) Recent work using the mitochondrial stain DASPMI

(dimethylaminostyrl-1-mthylpyridinium iodide) also reinforces the vacuolation model (Aoki et al, personal communication). This dye fluoresces with an intensity which reflects the degree of energization of the mitochondrion. This is known because petite mutants which lack mitochondrial cristae or strains which lack components of the electron transport chain only fluoresce very weakly. Using this dye these authors show that each apical cell within a mycelium of C. albicans fluoresces brightly, but in subapical compartments fluorescence is weak and the mitochondria may be fragmented and restricted to the periphery of the vacuolated compartments. Regeneration of cytoplasm prior to branch formation is associated with the recovery of bright fluorescence. Again these results emphasise that it is only the tips of a mycelium which are metabolically and biosynthetically active.

INDUCTION OF DIMORPHISM

C. albicans can be stimulated to form germ tubes by a wide range of growth conditions [12,13]. However, ultimately there must be a common control mechanism responsible for conversion from the yeast to the hyphal form. This mechanism may be expressed to a greater or lesser extent in different media or in different strains so that pseudohyphal cells, with a morphology intermediate between buds and hyphae, or true hyphae without constrictions may result. Control of the basic process of filamentous growth, however, like that of budding growth, is unlikely to vary greatly.

It is also evident that this mechanism responds to a range of convergent signal pathways which probably involve a series of membrane receptors which pass environmental information on the cell interior [13]. After signal reception the central processing mechanism presumably then transmits a secondary message which has the effect of activating those processes which bring about collectively the transition from delocalized cell expansion (cell budding) to polarized tip growth (germ tube formation).

There are at least three candidates for the secondary messenger which responds to the signalling pathways and relays information to the morphological regulatory processes; (i) a calcium signal, involving changes in intracellular calcium concentration and the activities of calmodulin, (ii) a cyclic mucleotide signal involving cAMP, cGMP, etc. and the activities of genes and enzymes which respond these effectors, and (iii) a signal based on changes in intracellular pH and consequent changes in enzyme activity or gene expression. Examples of each of these have been described in the control of differentiation in other organisms. We have examined the latter in yeast-mould dimophism in C. albicans and conclude from the results that changes in internal pH accompany the dimorphic transition and may be involved in its active control [2,13,14]. The evidence for this is as follows.

(a) Internal pH rose from a value of 6.8 for stationary phase cells to around 8.0 in cells which were destined to form germ tubes and to only 7.5 or less in cells that were destined to form buds [2]. These data were obtained using DMO (5,5' - dimethyl [2-^{14}C] oxazolidine-2,4-dione) and benzonate as weak acid probes, using ^{31}P nuclear magnetic resonance spectroscopy, and by measurement of the pH of homogenates of cells at different stages of the dimorphic transition. The rise in internal pH was followed by a return to a more neutral pH value prior to cell evagination. The pH changes therefore preceded cell differentiation and may therefore be part of the decision making process.

(b) Mechanisticaly it is suggested that the cytoplasmic alkalinization is brought about by activation of the proton-pumping ATPase in the cell membrane. The proton pump inhibitor diethylstilboestrol inhibited both the pH changes and germ tube formation without affecting the short-term viability of the cells. Proton efflux is probably offset by potassium ion influx which rises in proportion to the extent of cytoplasmic alkalinization (Stewart & Gow, unpublished)(Fig. 1).

(c) The temporal changes in internal pH occurred in germ tube-inducing growth media containing either amino acids, glucose and salts or N-acetyl glucosamine and were not therefore restricted to a unique set of environmental conditions.

(d) These conditions did not induce similar changes in the internal pH of cells of _Saccharomyces cerevisiae_ which in general had a lower and more stable internal pH than _C. albicans_ over a range of external pH's.

(e) Natural isolates of _C. albicans_ which formed few germ tubes and strains which were selected on the basis of a decreased capacity to form germ tubes in comparison to the wild type did not exhibit the extensive cytoplasmic alkalinization seen in germ tube positive stains cultivated under germ tube inducing conditions.

(f) Weak bases such as procaine and nicotine increased internala pH and led to a partial induction of germ tube formation. Possible control over dimorphism mediated by calmodulin and cAMP has been reviewed elsewhere but it is worth pointing out that the activities of calmodulin and the enzymes that make and destroy cAMP are pH

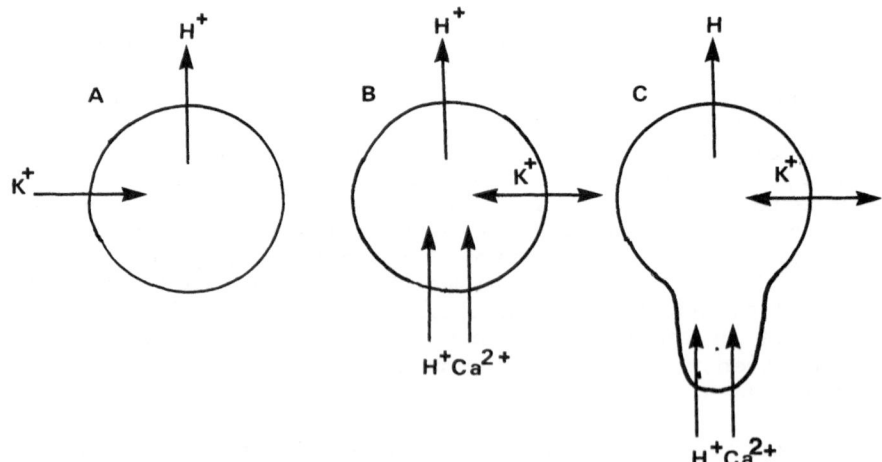

Fig. 1. Inferred ionic fluxes during germ tube formation in _C. albicans_. (a) Proton pumping with charge balance maintained by potassium ion uptake. (b) Transcellular proton current and calcium ion influx, with uptake of the cations correlation with site, selection and germ tube evagination (c).

sensitive (see discussion in refence 2). There may therefore be an interplay between these putative secondary messengers. In the discussion below a case is made for the importance of local fluxes of protons and calcium ions in the selection of the site of germ tube evagination and the maintenance of cell polarity during filamentous growth.

CONTROL OF POLARITY DURING DIMORPHIC GROWTH

In order to form a germ tube wall, biosynthesis must be first localized, then confined to a narrow region at the developing apex of the cell. Growth by budding is less highly polarized. It is difficult to study cell properties such as polarity since there is no obvious quantifiable assay and few ways of inhibiting polarized growth without inhibiting concomitantly growth and cell viability.

Work on various filamentous fungi and plants has shown that tipgrowth is accompanied by the production of circulating currents of inorganic ions by the cell [1]. These currents can be carried by a variety of ions in differnt organisms including protons, calcium ions, potassium ions and others. In two fungal hyphae that have been examined in detail the major ions are protons and the proton current flows for the main into the tip region of growing hypae [15,16]. In one of these fungi, Neurospora crassa, calcium ions are also required for the tip to extend [16]. Because the ions which are streaming into the tip carry electrical charge, the ion current also represents an electrical current. This can be measured directly with an ultrasensitive microelectrode called a vibrating probe [17]. So far the narrow hyphae of C. albicans have proved to be just too small to be examined with this instrument. However, we have recently made some progress in studying the elecrophysiology of dimorphism in this organism using a different approach. This approach has in effect provided us with a quantifiable assay for the polarity of growth of C. albicans and the means to study the effects of compounds which affect polarity without inhibiting cell growth per se.

The experiments involve immobilising cells on a polylysine-coated slide, exposing the cells to an artificial electrical field, then observing the effect of the field on the directional growth of the cells. The protocol was modified from that used to study other filamentous fungi [18]. We reasoned that tip growth is normally associated with the creation of a natural electrical field and so applied voltage gradients may also affect the mechanism(s) bringing about apical extension. Our results have shown that cell polarity is affected markedly by electrical fields and that exogenous calcium ions may have a critical role to play in the selection of the sites at which germ tube are formed. The following findings have been made.

(a) When exposed to elecric fields of up to 2mV. cell diameter^{-1} buds and germ tubes were formed at the cathode-facing side of the cell (Fig. 2). By taking the average cosine that the site of evagination made to the anode-cathode axis and multiplying this value by 100 to give the percentage we have quantified this response at all field-strengths. Maximal polarization was 82% for buds and 92% for germ tubes. The threshold field capable of causing polarity was around 2mV. cell diameter^{-1}. These results show unequivocally that the site of evagination is not fixed and can occur at any position on the cell surface.

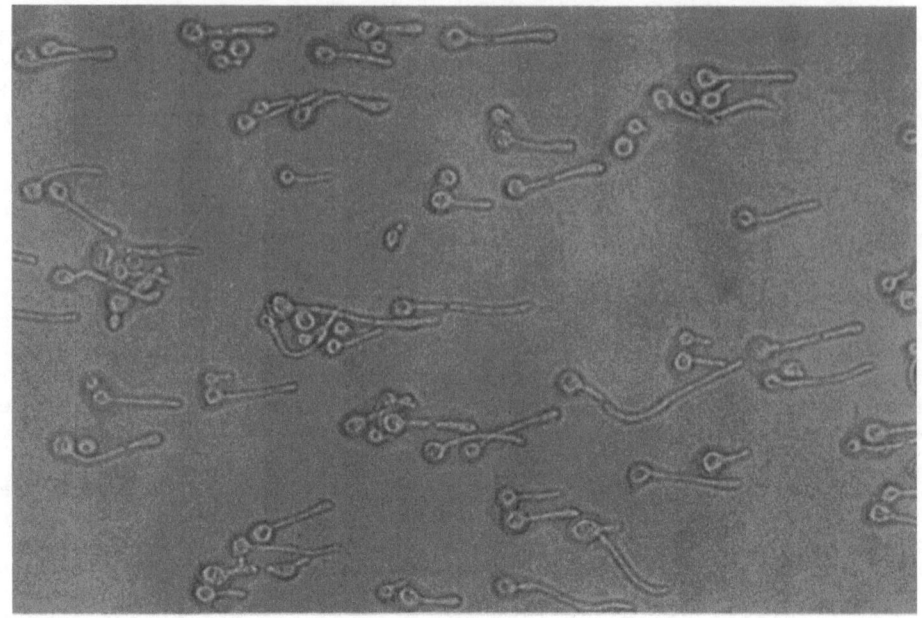

Fig. 2. Germ tube formation and growth of <u>C. albicans</u> in an
electric field of 28 mV. cell diameter^{-1}. Growth is
towards the cathode.

(b) The electric field polarized the site of evagination and the
continued extension of the germ tube, but buds were not converted to
germ tubes in these experiments. In other words, conditions which
supported budding maintained this mode of growth despite the
influence of a polarizing electric field. This suggests to us that
there is more to the control of dimorphism than physiological
regulation as has been suggested recently.

(c) An electric field was not required throughout the time
required to form a germ tube. In these experiments germ started to
form after 90 min, but cells exposed to fields of 28mV. cell
diameter^{-1} for only 30 min were polarized when the germ tube
eventually emerged. Cells therefore retained a "memory" of the field.
We are at present investigating how the cells which have been
programmed to form a germ at the cathodic end of the cell have been
changed.

(d) Cells require an exogenous supply of calcium to be able to
make a germ tube tip. Once the tip has been formed, however, much
less exogenous calcium is required to maintain tip growth. The
enlarging vacuoles and mitochondria which may serve as internal
calcium stores once growth has been initiated (Fig 1).

(e) Less exogenous calcium is required for the initiation of
budding growth.

It is our contention that artificial electrical fields cause
differential ion transport at the poles of the cell due to induced
changes in the membrane potential. Thus the electric field may induce
proton and calcium ion uptake at the cathode end of the cell which,

in some as yet undefined way, must cause local growth to be favoured. We are at present only able to speculate as to the mechanisms, but the phenomenon is clear and the many of the possibilities are now experimentally accessible.

CONCLUSIONS

Our cytological and electrophysiological experiments suggest that germ tube formation in C. albicans is triggered by nutrient poor conditions and results in unbalanced growth where extension is supported by a fixed, migratory cytoplasmic volume. The decision to form a bud or a germ tube may be dependent on the extent of cytoplasmic alkalinization which occurs in response to a wide range of environmetal stimuli. The site of budding and germ tube formation can be influenced by artificial electrical fields which affect cell polarity without affecting cell growth. Germ tube formation requires an external source of calcium ions which is not required for the initiation of buds. Proton efflux and calcium ion influx may therefore tell the cell when and where to form a germ tube.

REFERENCES

1. N.A.R. Gow, in : "Spatial Organization in Eukaryotic Microbes", pp. 25-41, R.K. Poole and A.P.J. Trinci, eds., IRL Press, Oxford (1987).
2. E.Stewart, N.A.R. Gow and D.V. Bowen, J. Gen. Microbiol., 134: 1079 (1989).
3. N.A.R. Gow and G.W. Gooday, J. Gen. Microbiol., 128:2195 (1982).
4. N.A.R. Gow and G.W. Gooday, Sabouraudia:J. Med. Vet. Mycol., 22:137 (1984).
5. N.A.R. Gow and G.W. Gooday, J. Gen Microbiol., 128:2187 (1982).
6. N.A.R. Gow, G. Henderson and G.W. Gooday, Microbios, 47:97(1986).
7. P.H. Gregory, in: "The Fungus Spore", pp.1-14, M.F. Madelin, ed., Butterworth, London (1966).
8. A.Cassone, N.Simonetti and V. Stripoli, in: "Spores VI", pp.172-178,P. Gerhardt, R. N. Costilow and H.L. Sadoff, eds., ASM Press, Washington (1975).
9. E. Mattia, Carruba, Angionlella, A. Cassone, J. Bacteriol., 152: 555, (1982).
10. K. Yokoyama and K. Takeo, Arch. Microbiol., 134:251 (1983).
11. D.L. Brawner and J.E. Cutler, Sabouradia: J. Med. Vet. Mycol., 23:389 (1985).
12. F.C.Odds, Crit. Rev. Microbiol., 12: 45 (1985).
13. N.A.R. Gow, in: "Proceedings of the Xth International Congress of the Society for Human and Animal Mycology", pp. 73-77, J.M. Torres-Rodriguez, ed., J.R. Prous Science, Barcelona (1988).
14. E. Stewart, S. Hawser and N.A.R. Gow, Arch. Microbiol., 151:149 (1989).
15. D.L. Kropf, J.H. Caldwell, N.A.R. Gow and F.M. Harold, J.Cell. Biol., 99: 486 (1984).
16. A.M. McGillivray and N.A.R. Gow, J. Gen. Microbiol., 133:2875 (1987).
17. L.F. Jaffe and R. Nuccitelli, J. Cell Biol., 63:614 (1974).
18. A.M. McGillivray and N.A.R. Gow, J. Gen. Microbiol., 132:2551 (1986).

CANDIDA ALBICANS: CELL WALL PHYSIOLOGY AND METABOLISM

M.G.Shepherd   P.K.Gopal

Experimental Oral Biology Unit, Faculty of Dentistry
Box 647, Dunedin, New Zealand

INTRODUCTION

The cell wall is of considerable importance for a variety of reasons: it forms the contact point for adhesion of the fungus to host cells; it is the rigid structure which confers mechanical stability on the cell and maintains its characteristic shape; it acts as a protective barrier; it contains components that act as immuno-genic determinants and immunomodulators and it is intimately involved in the regulation of secreted hydrolytic enzymes. The cell wall has also attracted interest as a potential target for novel antifungal agents. This is because the glucans and chitin of the wall are found in the pathogen but not the host and, therefore, these compounds and particularly the enzymes involved in their biosynthesis are potentially safe targets for antifungal agents. Finally, in order for us to understand fungal morphogenesis we need to know the mechanisms of wall synthesis and the assembly of the wall leading to its final architecture because it is the temporal and spatial arrangement of the wall components that dictate the final shape of the cell.

The cell envelope of Candida albicans is defined as the plasma membrane, the periplasmic space of the cells, the cell wall, and the fibrous layer associated with the outer region of the wall. The plasma membrane forms a permeability barrier between the cytosol of the cell and the external environment. The periplasmic space is the region bounded by the cell wall and the plasma membrane that includes space created by membrane evagination. The material external to the periplasmic space is the cell wall which is made up of a mosaic of components which are discussed in some detail below. The cell envelope and cell wall of Candida albicans have recently been reviewed by Shepherd [1], Shepherd et al., [2] Odds [3] and Reiss [4].

MATERIALS and METHODS

Isolation of mannoproteins

Candida albicans ATCC 10 261 was grown to stationary phase in a 18L fermentor in glucose-salts-biotin medium at 28°C[5].

Mannan was extracted from harvested cells by autoclaving for 90 min in 20 nM citrate buffer pH 7.0. After centrifugation, the super-

Candida and Candidamycosis, Edited by E. Tümbay et al.
Plenum Press, New York, 1991

21

natant was saved and the paste was re-extracted in the same manner. The extracts were combined and the crude mannans precipitated by addition of sodium acetate and three vols of ethanol. The precipitate was collected, redissolved in water, dialysed and lyophilised; the yield was approximately 3 gm/100 gm wet wt of cells.

The crude mannan (4g) was dissolved in 100 ml of water and a Cetavlon solution (4g in 50 ml water) added. After standing the mixture at room temperature for six hours, the precipitate was isolated by centrifugation and washed in 50 ml of water. The supernatant and washings were combined, and 100 ml of 1 % boric aid was added. The solution was stirred and the pH adjusted to 8.8 with 2N NaOH. The precipitate formed after standing the solution for an hour was collected, washed with 0.5 % sodium borate, pH 8.8 and dissolved in 50 ml of 2 % acetic acid. To this solution 1 gm sodium acetate and 3 vol of ethanol were added to reprecipitate the mannoproteins. This precipitate was isolated by centrifugation, washed with 2 % acetic acid in ethanol and then redissolved in water. The resulting solution was neutralised, dialysed against water and lyophilised. The final purification of the mannoprotein was with ion exchange chromatography. Mannoprotein (Igm) was dissolved in water and applied to a column of DEAE-Sephadex A-50 (4x80 cm) equilibriated with 50mM potasium phosphate buffer, pH 6.8. The column was eluted with a linear gradient of 0 to 500 mM NaCl. The major carbohydrate peak was pooled, dialysed and lyophilised.

Elimination reaction

A 100 mg portion of the purified mannoprotein was dissolved in 0.1 N NaOH containing 0.8 M sodium borohydride and incubated at 40°C for 24 hrs. The excess borohydride was decomposed by dropwise addition of acetic acid; the solution was then neutralised (NaOH), concentrated to 2 ml and applied to a Biogel P2 column.

Proton - NMR

High resolution ^{1}H NMR was performed on a Bruker AM 500 MHZ spectrometer at the NMR facility, Department of Chemistry, University of California, Berkeley. A pulse width of 90°C was used with a sweep width of 2500 HZ. Chemical shifts are described relative to an internal acetone standard.

Solid State NMR

High resolution solid state ^{13}C NMR spectra of total isoluble fraction (acid insolube/alkali insoluble) of cell wall glucan was recorded at 75.43 MHZ on Varian VXR 300 spectrometer equipped with a CP-MAS probe. Solid samples were contained in a rotor which was spun at 9000 rpm with compressed air. Contact time was 500 usec with a repitition time of 2s. Spectra were accumulated for 1000-2000 times.

RESULTS AND DISCUSSION

The predominant components of the wall are mannoproteins (20-30 %) and β-glucans (50-60 %) and small quantities of chitin (0.6-2.7 %), protein (5-15 %) and lipid (2-5 %) [6-10]. The glucan and manno-protein content is similar in both yeast and hyphal cells although the hyphal cells do contain at least three times as much chitin as yeast cells [7,9,11]. Transmission electronmicroscopy show the wall to be composed of a number of different layers and the thickness and number of these layers varies in germ-tube formation and the cell cycle [12-16]. The striated appearance would indicate that there are layers of different chemical composition, but it is likely that the changes in electron density reflect quantitative rather than complete qualitative changes. A schematic diagram showing the organisation of the cell envelope of <u>Candida albicans</u> is given in Figure 1 and the different layers are to be regarded as zones of enrichment rather than areas restricted to the molecule named. For example, it is known that the mannoprotein forms the outermost layer and is distributed throughout the entire wall from lectin binding studies and extraction of the cell wall with different solvents, [17,18] cytochemical staining [19] and the cross reaction of ferritin-conjugated antibodies.

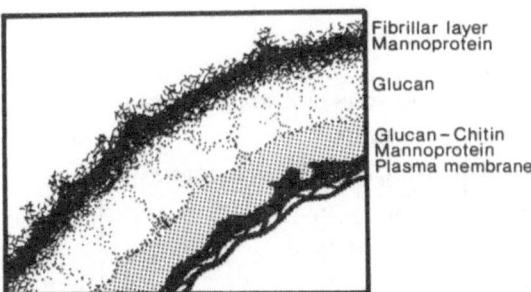

Fig. 1. Proposed architecture for the cell envelope of <u>Candida albicans</u>.

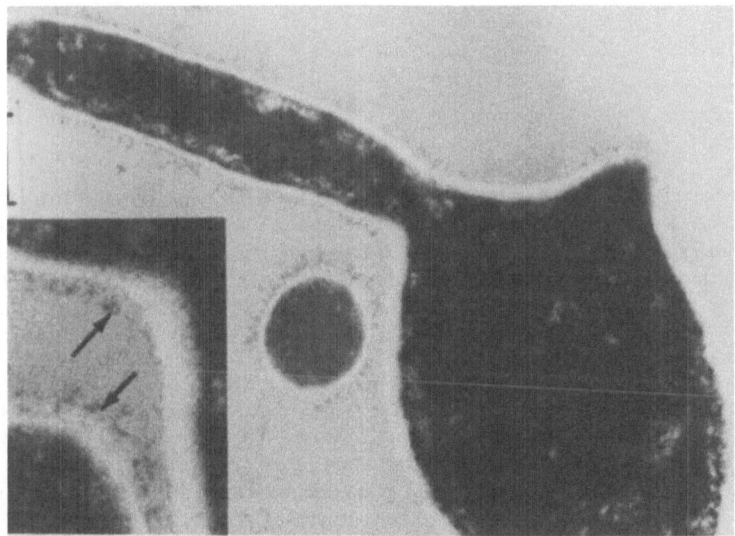

Fig. 2. Transmission electron micrograph of developing germ-tube of <u>C. albicans</u>. An electron-dense outer layer of the germ tube wall (arrows in inset). In the cross sectioned germ tube the fibrils are seen to be arranged radially and perpendicular to the cell wall. Reprinted from Hubbard et al [14].

Tronchin et al. [21] used wheat agglutinin to show that one layer of chitin was distributed in the inner wall layer near the plasma membrane and there was a smaller amount found on the outer cell wall coat variously called slime layer, mucous layer or fuzzy layer, [13,14,22,23] and this layer is believed to play an important role in pathogenicity, phagocytosis and adherence. It is also known that there are considerable alterations in the number of layers and the thickness of the wall in the different modes of growth [12,14,16].

In Figure 2 it can be seen the developing germ-tubes have distinct fibrils arranged radially and perpendicular to the cell but these are absent from the mother cell. Details of the components of these cell wall layers are given below.

MANNOPROTEIN OF *CANDIDA ALBICANS*

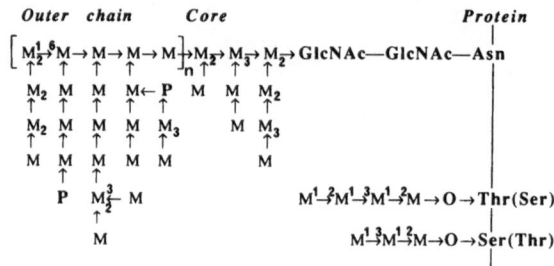

Fig. 3. Structure of C. albicans cell wall mannoprotein.

Mannoproteins

The mannoproteins of the C. albicans cell wall account for some 20-30 % of the weight and they are the major antigenic components of the intact yeast cells. In Figure 3 the major structural features of C. albicans mannoprotein is outlined. The mannan glycoprotein is characterized by three domains: an outer chain containing the major antigenic sites, an inner core and the base-labile oligosaccharides attached through serine and threonine. Detailed information is not available on the fine structure of either the protein or the carbohydrate moieties. However, it is known that the mannoproteins of Candida are similar to those found in other yeasts. Gopal (P.K. Gopal and M.G. Shepherd, unpublished results) has recently elucidated the structure of the alkali labile oligosaccharides. The elution profile of the oligosaccharides released by controlled B elimination of the cell wall mannoprotein of Candida albicans ATCC 10261 is shown in Figure 4. The three fractions isolated on this sizing column [(Biogel P_2 (400 mesh) 2x195 cm] were identified as mannobiose, mannotriose and mannotetraose. The chemical structure of these oligosaccharides including the nature of the glycosidic linkages and the monosac-

24

charide sequence was determined by high resolution ^{1}H NMR (Figure 4). The structure of the tetrasaccharide was 0-M-^{2}M-^{3}M-^{2}M as determined by the anomeric proton signals of 5.04, 5.12, 5.18 and 5.36 shown in Figure 4. The signal assignment was as described by Cohen and Ballou [24] and Tsai et al [25]. Similarly, structures of trisaccharide and disaccharide were assigned as 0-M-^{2}M-^{3}M and 0-M^1-^{2}M, respectively. These data are the first report on the linkage analysis of 0-linked oligosaccharides in the mannoprotein of <u>Candida albicans</u>. Although the structure of the disaccharide and trisaccharide of <u>Candida</u> mannoproteins are identical to those found in S<u>accharomyces cerevisiae</u>, the structure of the tetrasaccharides are different. The 0-linked oligosaccharides make up about 12 % of the total mannosyl residues [26].

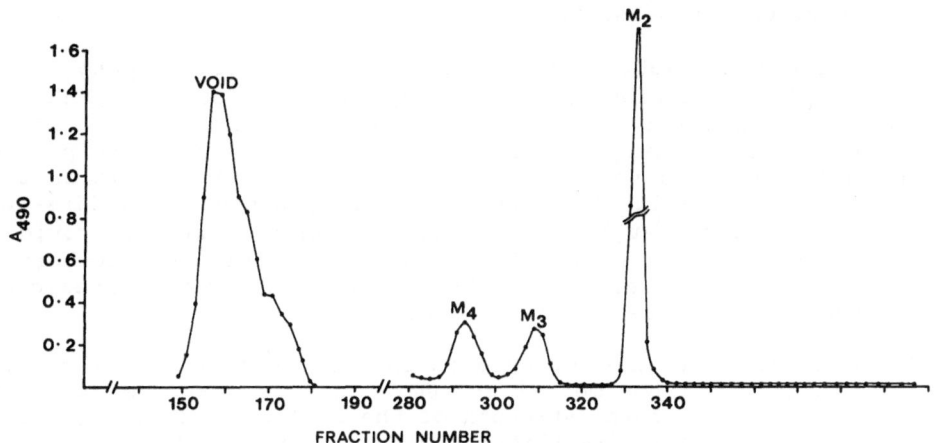

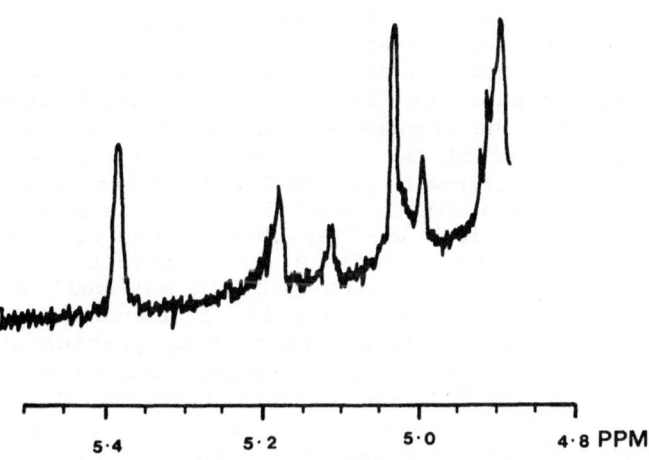

Fig. 4. The 0-linked oligosacchariades of <u>C. albicans</u> wall mannoprotein: (a) elution profile on Biogel P2 of oligosaccharides produced by β elimination; (b) ^{1}H NMR spectrum of tetrasaccharide.

The major portion of the polysaccharide is an a-1,6 linked poly-mannose joined to the protein through a chitobiose bridge at aspa-ragine. In the outer chain of C. albicans, 55 % of the mannose units are substituted at the 2 position with oligosaccharides of mannose joined by 1,2 linkages and occasionally 1,3 linkages [27]. There are two serotypes of C. albicans, serotype A and serotype B [28,29] and these two serotypes are distinguished by the side chain oligosac-charides attached to the backbone 1,6 linked mannose. In serotype A mannan the immunodominant haptan (Factor 6) is believed to be a linear α (1-2) linked mannohexaose with an α-1,3 linkage at the penultimate sugar from the reducing end [4]. The B serotype appears to contain a branched molecule. It should be noted, however, that the structures postulated do not take account of the important role of phosphate residues in antigenic responses [30] and there is still some debate on the role of B-linked mannose residues the presence of which are indi-cated from proton NMR spectrum. Mannan immunochemistry and immunology has been reviewed by Reiss [4].

The molecular weight of the isolated mannoprotein depends upon the method of preparation. In many of the earlier studies, Fehling solution was used which resulted in alkaline degradation and a mole-cular weight for the mannoprotein of only 40 kDa [31]. In these prepa-rations acid was used to dissociate the copper-mannan complex and therefore the acid labile 1,3 mannobiosyl and phosphodiester linkages were lost. The borate-cetyltrimethyl ammonium bromide method provides higher molecular weight material. If the dissociation of the borate and mannan complex is carried out by strong acid, critical residues such as the 0-phosphates would be lost.

The mannoproteins are embedded in the glucan-chitin wall matrix and at least two experimental systems indicate that they are not required for the structural strength of the cell wall. Firstly, cells become osmotically sensitive throught degradation by β-1,3 glucanases alone and secondly when protoplasts are regenerated in the stabilising medium of magnesium chloride containing glucose the resultant osmoti-cally resistant cells have synthesized glucan but no mannoprotein is formed on the surface [32] . A recent study by Molloy et al [33] on vecto-rial iodination of proteins of the cell surface showed that a zymo-lyase degradation of a wall preparation, released material that was not contaminated with membrane and cytosolic proteins but did release 93 % of the iodine with a specific activity 45 fold higher than the original SDS extract. The requirement for β-1,3 glucanase to release the ^{125}I labelled material is consistent with either a covalent attachment or tight entrapment of the labelled proteins in the cell wall glucans. The major iodinated material moved as a diffuse band at 260 kDa but it is not clear whether this is one mannoprotein or a mixture. By analogy with the study of Frevert and Ballou [34] on S. cerevisiae this 260 kDa mannoprotein may well be the major structured glycoprotein in the cell wall. It contains 1.5 % protein and 98.4 % hexose and amalysis of the peptide portion showed it to contain 36-38% of serine and theonine residues typical of mannoproteins containing 0-linked oligosaccharides. Elorza et al. [26] found C. albicans mannoproteins to contain 7 % protein, 92 % carbohydrate. The carbohydrate was mostly mannose but a small amount of glucose was also present.

Glucans

As with the mannoproteins the difficulty in determining the structure of the glucans from C. albicans has been in obtaining pure material that has not undergone extensive degradation. One scheme that has been used in both S. cerevisiae and C. albicans is to separate the different glucans on the basis of their solubility in acid and alkali. Purified fractions are then classified as alkali soluble glucan, acetic acid soluble glucan and the acid/alkali insoluble glucan. It is known that the β-glucans are the major structural elements of the wall since osmotically sensitive cells are generated after degradation with a β-1,3 glucanase [32]. Methylation and 13C - NMR analyses of the glucan from yeast and hyphal cells of C. albicans showed the β-1,6 glucan to be the major polymer (50-70 %) which was highly branched. There was also a highly branched β-1,3 glucan and a mixed β- 1,3 1,6 glucan [35]. Although the nature and amounts of glucan were similar in yeast and hyphal cells the alkali insoluble glucan from germ-tube forming cells contained considerably more β-1,3 linkages (67 %) than the fraction in yeast and hyphal [36]. If the β-1,3 to β-1,6 ratio in the mother cell is taken into account it would appear that during the early stages of germ-tube formation there is almost exclusive synthesis of β-1,3 linked glucans and this is analogous to the situation observed with regenerating spheroplasts [35,37] which preferentially synthesize β-1,3 glucans. The development of the secondary wall, which may include β-1,6 glucan synthesis, is believed to be important in developing the final shape of the cell [1]. Surarit et al. [38] have shown that there is a formal chitin-glucan glycosidic linkage between the sixth position of GlcNAc on chitin and the one position of a β-1,6 glucan. Chitin is known to be dispersed around the entire wall [21] and its association with glucan would provide a structural scaffold to which the remaining components could attach. There are also reports of glucan mannoprotein complexes [39], but the nature of these linkages has not been established.

The microfibrils of glucan and glucan-chitin endow rigidity on the wall and these β-glucans probably form helical structures [40]. It is well recognised that high resolution solid state ^{13}C NMR can reveal a great deal of information not only on the nature of glycosidic linkages of the polysaccharides but also on the tertiary structure of the polymer. In a recent review Saito [41] has described the usefulness of this technique as a means of determining the conformational characteristics of the polymer. A major advantage of ^{13}C NMR spectroscopy over other methods is that conformational analysis can be performed on noncrystaline as well as crystalline molecules. Here we report the application of this technique on the analysis of the total insoluble glucan from the cell wall of Candida albicans. As shown in Figure 5, the spectrum depicts typical features of a predominantly β-1,3 linked glucan; the anomeric peak of C-1 occurs at 103.7 ppm and the peaks due to C-2 and C-5 at 74.2 ppm. The most noteworthy feature of the spectrum is the signal due to C-3; 3A at 85.7 ppm and 3B at 83.5 ppm (Figure 5). The displacement of the C-3 signal is about 2.2 ppm. Saito et al. carried out a systematic ^{13}C CP-MAS NMR study of various B glucans and from their findings concluded that at least four types of conformation for glucans can be distinguished on the basis of characteristic displacement of NMR signals at the C-1 and C-3 positions. The two C-3 signals of 85.7 ppm and 83.5 ppm observed in Candida glucan, correspond to the laminaran type and paramylon type conformation, respectively [42]. From the relative proportion of the two C-3 signals i.e. it is further possible to conclude that the degree of polymerisation of this glucan is < 380. Furthermore, by analogy with the structures of paramylon and cellulose, it is possible that the heterogeniety of conformation observed in this glucan provides a degree of crystallinity which would fit its proposed role as a structural component of the wall [35].

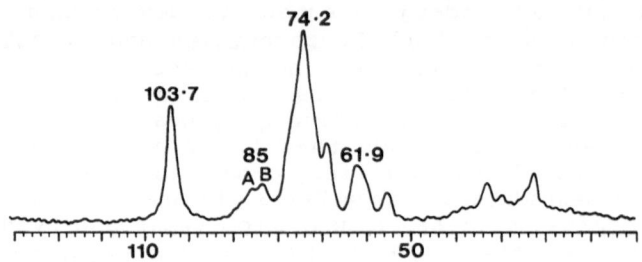

Fig. 5. The 75.43 MHZ 13C– CP–MAS NMR spectra of acid/alkali insoluble glucan obtained from cell wall of <u>Candida albicans</u> ATCC 10261. Contact time was 500 usec.

The biosynthesis of β-1,3 glucan occurs through a transmembrane enzyme (β-1,3 glucansynthetase) that catalyses the vectorial synthesis of the polymer. The substrate UDP glucose is added on the cytosolic side and the product emerges from the outer cell side. These are linear molecules and many questions remain unanswered such as how the β-glucans are: (i) branched (ii) crosslinked with chitin (iii) complexed with mannoproteins. One further and intriguing question critical to our understanding of the biology of yeasts and fungi, is how these extruded molecules undergo intussesception into the existing wall material (e.g. with bud formation or in the branching of hyphae). The β-glucanases are the most abundant cell wall hydrolysases which have β-glucans as their structural component, and it is proposed that the glucanases clip the existing glucan allowing addition glucan molecules to be incorporated as reviewed by Nombela et al [43]. It is also likely that chitinase is involved in a similar process. This is summarized in Figure 6. The structural polysaccharides of glucan and chitin are synthesized from UDP glucose and UDP GlcNAc, respectively, whereas the glycoproteins are formed and transported through the complex pathway involving the endoplasmic reticulum and the Golgi and finally transported to the plasma membrane in a secretory vesicle in the process we call exocytosis [44]. Since it is unlikely that there are independent sources of energy such as ATP existing in the extracellular wall matrix it is most likely that the branching that occurs in the glucan structures is through transglycosylation either from the same molecule as is found in branching of glycogen or between different molecules as is found with the transglycosylation characteristic of the extra cellular matrix of <u>Streptococci</u>. The final event in the wall architecture appears to be crosslinking between the polymers, particularly chitin and glucan. As is shown in Figure 7, it is proposed that after individual glucan and chitin microfibrils are formed at the point of apical growth, there is then secondary wall synthesis which includes both crosslinking of polymers and crosslinking and complexing of glucan and chitin. In this context the timing of secondary wall formation provides a model which accounts for the difference between the yeast and hyphal modes of growth. For example, if the glucan-chitin complex were formed almost immediately behind the apical tip this would give a rigid structure resulting in a hyphal element. If, however, secondary wall formation with its crosslinks were delayed, a more plastic wall would result allowing the formation of a spherical cell.

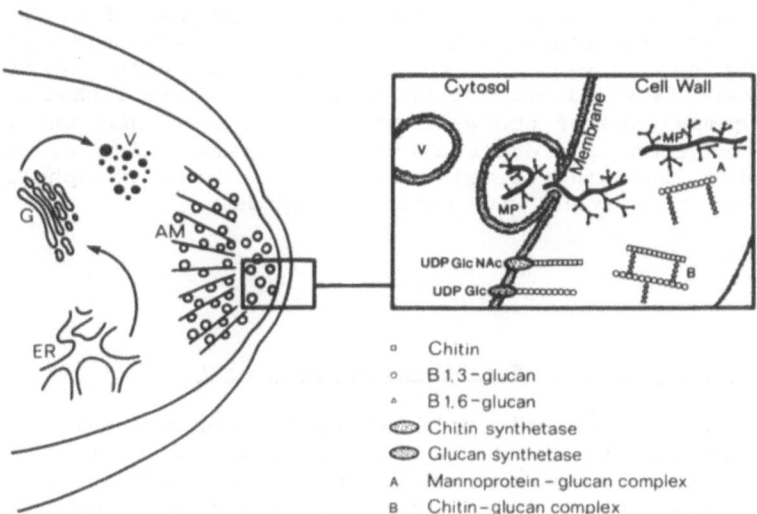

- ▫ Chitin
- ○ B 1,3 – glucan
- △ B 1,6 – glucan
- ⬭ Chitin synthetase
- ⬭ Glucan synthetase
- A Mannoprotein – glucan complex
- B Chitin – glucan complex

Fig. 6. Cellular and biochemical mechanisms involved in C. albicans wall formation.

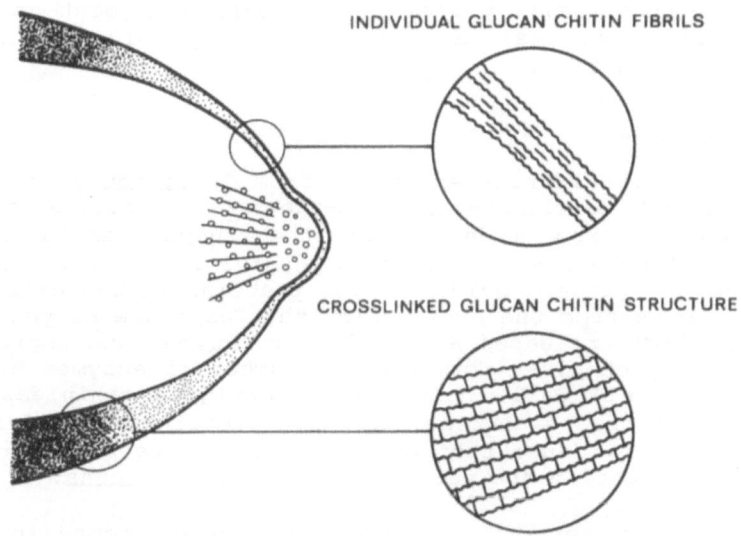

INDIVIDUAL GLUCAN CHITIN FIBRILS

CROSSLINKED GLUCAN CHITIN STRUCTURE

Fig. 7. Secondary wall development during growth of Candida albicans. It is proposed that individual β 1,3 – and β 1,6 – glucan and chitin microfibrils are formed at the point of apical growth. Subsequently, there is secondary wall synthesis which includes additional polymer synthesis and cross linking of glucan and chitin.

Lipids

There are reports that lipids make up between 1 to 5 % of the cell wall [7,9] and indeed that the lipid content and composition changes during the yeast to mycelial transformation. Despite these reports there is still some doubt among yeast workers that lipids are integral components of the wall and many believe that the lipids in the wall fractions are derived from the plasma membrane. Indeed the major lipids found in the wall are triglycerides, phospholipids and sterolesters consistent with membrane lipids.

Proteins

(i) Number and quantity of proteins in cell wall

There are considerable differences between the various reports on the percentage contribution of proteins to the cell wall [7,9,26,47]. Leading on from the variation in protein content are the studies showing a large number of proteins, in some reports more than 40, that move as discrete bands on electrophoresis [26,48]. Molloy et al. [33] have shown by vectorial iodination that walls prepared by mechanical breakage or by SDS extraction are highly contaminated with intracellular material. Zymolyase degradation of a wall preparation released material that did not contain these contaminating bands but did release 93 % of the iodine with a specific activity 45 fold higher than in the SDS extract. From this study it was concluded that there were no more than twelve proteins in C. albicans yeast cells that resolved by electrophoresis and autoradiography and were accessible to vectorial iodination.

(ii) Other proteins

The thread-like fibrils extending from C. albicans to the epithelial cells and the clumping of germ-tubes is indicitive of proteinaceous material and indeed germ-tube clumping is inhibited by digestion with proteinases or with sulphydryl reagents but not with amannosidase [49]. Further evidence for proteins in the cell wall of C. albicans comes from the observation that exposure of yeast cells to dithithreitol releases extracellular enzymes and improves the efficiency of protoplast formation. A number of enzymes have been located in the cell wall including N-acetylglucosaminidase, acid phosphatase, glucanases and the aspartate proteinase as reviewed by Odds [3]. There is evidence that the proteinase acts as an aggressin and is a determinant in the pathogenicity of C. albicans.

Recently a receptor for iC3b has been found in Candida albicans [50,51]. In mammalian cells the iC3b interacts with the CR3 receptor on phagocytic cells to promote phagocytosis and stimulate the release of intracellular microbiocidal agents such as superoxide or myeloperoxidase. The presence of the receptor on Candida allows the Candida cells to bind iC3b noncovalently and avert recognition by neutrophil CR3 and in this way the recognition site for neutrophil CR3 is masked. Therefore, the surface receptors for C3 fragment iC3b present on Candida albicans promote pathogenicity by inhibiting phagocytosis. The complement receptors on C. albicans and CR3 are antigenically related but are not identical [52].

REFERENCES

1. M.G.Shepherd, Cell envelope of _Candida albicans_, CRC _Crit.Rev._ _Microbiol._, 15: 7 (1987).
2. M.G.Shepherd, R.T.M.Poulter, P.A.Sullivan, _Candida albicans_: Biolog, genetics, and pathogenicity, _Ann.Rev.Microbiol._, 39: 579 (1985).
3. F.C.Odds, "_Candida_ and Candidosis. A Review and Bibliography", 2nd Edition, Bailliere Tindall, London (1988).
4. E.Reiss, "Molecular Immunology of Mycotic and Actinomycotic Infection", Elsevier, New York (1986).
5. M.G.Shepherd, Y.Y.Chiew, S.P.Ram, P.A.Sullivan, Germ tube induction in _Candida albicans_, _Can.J.Microbiol._, 26: 21 (1980).
6. C.T.Bishop, F.Blank, P.E.Gardner, The cell wall polysaccharides of _Candida albicans_: glucan, mannan and chitin, _Can.J.Chem._, 38: 869 (1960).
7. F.W.Chattaway, M.R.Holmes, A.J.E.Barlow, Cell wall composition of the mycelial and blastospore forms of _Candida albicans_, _J.Gen.Microbiol._, 51: 367 (1968).
8. H.F.Hasenclever and W.O.Mitchell, A study of yeast surface antigens by agglutination inhibition, _Sabouraudia_, 3: 288 (1964).
9. P.A.Sullivan, Y.Y.Chiew, C.Molloy, M.D.Templeton, M.G.Shepherd, An analysis of the metabolism and cell wall composition of _Candida albicans_ during germ-tube formation, _Can.J.Microbiol._, 29: 1514 (1983).
10. R.J.Yu, C.T.Bishop, F.P.Cooper, H.F.Hasenclever, F.Blank, Structural studies of mannans from _Candida albicans_ (serotype A and B) _Candida parapsilosis_, _Candida stellatoidea_ and _Candida tropicalis_, _Can.J.Chem._, 45: 2205 (1967).
11. P.C.Braun and R.A.Calderone, Chitin synthesis in _Candida albicans_: comparison of yeast and hyphal forms, _J.Bacteriol._, 133: 1472 (1978).
12. A.Cassone, N.Simonetti, V.Stippoli, Ultrastructural changes in the wall during germ-tube formation from blastospores of _Candida albicans_, _J.Gen.Microbiol._, 77: 417 (1973).
13. W.Djaczenko, A.Cassone, Visualization of new ultrastructural components in the cell wall of _Candida albicans_ with fixatives containing TAPO, _J.Cell Biol._, 52: 186 (1971).
14. M.J.Hubbard, P.A.Sullivan, M.G.Shepherd, Morphological studies of N-acetylglucosamine induced germ tube formation by _Candida albicans_, _Can.J.Microbiol._, 31: 696 (1985).
15. D.Poulain, G.Tronchin, J.F.Dubremetz, J.Biguet, Ultrastructure of the cell wall of _Candida albicans_ blastospores: study of its constitutive layers by the use of a cytochemical technique revealing polysaccharides, _Ann.Microbiol._, 129A: 141 (1978).
16. C.Scherwitz, R.Martin, H.Ueberberg, Ultrastructural investigation of the formation of _Candida albicans_ germ tubes and septa, _Sabouraudia_, 16: 115 (1978).
17. A.Cassone, E.Mattia, L.Boldrini, Agglutination of blastospores of _Candida albicans_ by concanavalin A and its relationship with the distribution of mannan polymers and the ultrastructure of the cell wall, _J.Gen.Microbiol._, 105: 263 (1978).
18. G.Tronchin, D.Poulain, J.Biguet, Cytochemical and ultrastructural studies of the cell wall of _Candida albicans_. I.Localization of mannan by means of concanavalin A on ultrathin sections, _Arch.Microbiol._, 123: 245 (1979).
19. A.Cassone, D.Kerridge, E.F.Gale, Ultrastructural changes in the cell wall of _Candida albicans_ following cessation of growth and their possible relationship to the development of polyene resistance, _J.Gen.Microbiol._, 110: 339 (1979).

20. R.A.Venezia and R.C.Lachapelle, The use of ferritin-conjugated antibodies in the study of cell wall components of _Candida albicans_, _Can.J.Microbiol._, 19: 1445 (1973).

21. G.Tronchin, D.Poulain, J.Herbaut, J.Biguet, Localization of chitin in the cell wall of _Candida albicans_ by means of wheat germ agglutinin, _Eur.J.Cell Biol._, 26: 121 (1981a).

22. M.Pesti, E.K.Novak, L.Ferenczy, A.Svoboda, Freeze fracture electron microscopical investigation of _Candida albicans_ cells sensitive and resistant to nystatin, _Sabouraudia_, 19: 17 (1981).

23. G.Tronchin, D.Poulain, J.Herbaut, J.Biguet, Cytochemical and ultrastructural studies of _Candida albicans_. II. Evidence for a cell wall coat using Concanavalin A, _J.Ultrastruct.Res._, 75: 50 (1981b).

24. R.E.Cohen and C.E.Ballou, Linkage and sequence analysis of mannose-rich glycoprotein core oligosaccharides by proton nuclear magnetic resonance spectroscopy, _Biochemistry_, 19: 4345 (1980).

25. P.K.Tsai, J.Frevert, C.E.Ballou, Carbohydrate structure of _Saccharomyces cerevisiae_ mannoprotein, _J.Biol.Chem._, 259: 3805 (1984).

26. M.V.Elorza, A.Marcilla, R.Sentandreu, Wall mannoproteins of the yeast and mycelial cells of _Candida albicans_: Nature of the glycosidic bonds and polydispersity of their mannan moieties, _J.Gen.Microbiol._, 134: 2393 (1988).

27. E.Reiss, D.G.Patterson, L.W.Yert, J.S.Holler, B.K.Ibrahim, Structural analysis of mannans from _Candida albicans_ serotypes A and B and from _Torulopsis glabrata_ by methylation gas chromatography, mass spectrometry and exo-a-mannanase, _Biomed.Mass Sprectrom._, 8: 252 (1981).

28. H.F.Hasenclever and W.O.Mitchell, Antigenic studies of _Candida_. I.Observation of two antigenic groups in _Candida albicans_, _J.Bacteriol._, 82: 570 (1961).

29. D.F.Summers, A.P.Grollman, H.F.Hasenclever, Polysaccharide antigens of the _Candida_ cell wall, _J.Immunol._, 92: 491 (1964).

30. Y.Okubo, Y.Honma, S.Suzuki, Relationship between phosphate content and serological activities of the mannans of _Candida albicans_ strains NIH A-207, NIH B-792, and J-1012, _J.Bacteriol._, 137: 677 (1979).

31. T.Nakajima and C.E.Ballou, Characterization of the carbohydrate fragments obtained from _Saccharomyces cerevisiae_ mannan by alkaline degradation, _J.Biol.Chem._, 249: 7679 (1974).

32. P.Gopal, P.A.Sullivan, M.G.Shepherd, Metabolism of [14C] glucose by regenerating spheroplasts of _Candida albicans_, _J.Gen. Microbiol._, 130: 325 (1984 a).

33. C.E.Molloy, M.G.Shepherd, P.A.Sullivan, Identification of envelope proteins of _Candida albicans_ by vectorial iodination, _Microbios_, 1989 (in press).

34. J.Frevert and C.E.Ballou, _Saccharomyces cerevisiae_ structural cell wall mannoprotein, _Biochemistry_, 24: 753 (1985).

35. P.K.Gopal, P.A.Sullivan, M.G.Shepherd, Isolation and structure of glucan from regenerating spheroplasts of _Candida albicans_, _J.Gen.Microbiol._, 130: 1217 (1984b).

36. P.K.Gopal, M.G.Shepherd, P.A.Sullivan, Analysis of wall glucans from yeast, hyphal and germ-tube forming cells of _Candida albicans_, _J.Gen.Microbiol._, 130: 3295 (1984c).

37. D.R.Kreger and M.Kopecka, On the nature and formation of the fibrillar nets produced by protoplasts of _Saccharomyces cerevisiae_ in liquid media: an electronmicroscopic, X-ray diffraction and chemical study, _J.Gen.Microbiol._, 92: 207, (1975).

38. R.Surarit, P.K.Gopal, M.G.Shepherd, Evidence for a glycosidic linkage between chitin and glucan in the cell wall of Candida albicans, J.Gen.Microbiol., 134: 1723 (1988).

39. J.Friis and P.Ottolenghi, The genetically determined binding of Alcian blue by a minor fraction of yeast cell walls, Comp.Rend. Trav.Lab.Carlsberg., 37: 327 (1970).

40. R.H.Marchessault and Y.Deslandes, Texture and crystal structure of fungal polysaccharides, in : "Fungal Polysaccharides", P.S. Sandford and K.Matsuda, eds., American Chemical Society Symposium Series No. 126, pp. 221-250, American Chemical Society, Washington DC (1980).

41. H.Saito, Conformation-dependent ^{13}C chemical shifts: A new means of conformational characterization as obtained by high-resolution solid-state ^{13}C NMR, Magnetic Resonance in Chemistry, 24: 835 (1986).

42. H.Saito, R.Tabeta, T.Sasaki, Y.Yoshioka, A high-resolution solid-state 13C NMR study of (1-3)-B-D-glucans from various sources. Conformational characterisation as viewed from the conformation-dependent 13C chemical shifts and its consequence to gelation property, Bull.Chem.Soc.Jpn., 59: 2093 (1986).

43. C.Nombela, M.Molina, R.Cenamor, M.Sanchez, Yeast B-glucanases: a complex system of secreted enzymes, Microbiol.Sci., 5: 328 (1988).

44. R.Scheckman, Protein localization and membrane traffic in yeast. Ann.Rev.Cell Biol., 1: 115 (1985).

45. D.E.Bianchi, The lipid content of cell walls obtained from juvenile, yeast-like and filamentous cells of Candida albicans, Ant. van Leeuwenhoek J., 33: 324 (1967).

46. M.A.Ghannoum, G.Janini, L.Khamis and S.S.Radwan, Dimorphism-associated variations in the lipid composition of Candida albicans, J.Gen.Microbiol., 132: 2367 (1986).

47. H.Yamaguchi, Effect of biotin insufficiency on composition and structure of cell wall of Candida albicans in relation to its mycelial morphogenesis, J.Gen.Appl.Microbiol., 20: 217 (1974).

48. W.L.Chaffin and D.M. Stocco, Cell wall proteins of Candida albicans, Can.J.Microbiol., 29: 1438 (1983).

49. L.Rahary, R.Bonaly, J.Lematre, D.Poulain, Aggregation and disaggregation of Candida albicans germ-tubes, FEMS Microbiol., Lett., 30: 383 (1985).

50. B.J.Gilmore, E.M.Retsinas, J.S.Lorenz, M.K.Hostetter, An iC3b receptor on Candida albicans: Structure, function, and correlates for pathogenicity, J.Infect.Dis., 157: 38 (1988).

51. F.Heidenreich and M.P.Dierich, Candida albicans and Candida stellatoidea, in contrast to other Candida species, bind iC3b and C3d but not C3b, Infect.Immun., 50: 598 (1985).

52. A.Eigentler, T.F.Schulz, C.Larcher, E.Breitwieser, B.L.Myones, A.L.Petzer, M.P.Dierich, C3bi-binding protein on Candida albicans: temperature-dependent expression and relationship to human complement receptor Type 3, Infect.Immun., 57: 616 (1989).

TRANSGLUCOSYLATION CATALYSED BY THE EXO-β-GLUCANASE

OF CANDIDA ALBICANS

P.A.Sullivan   G.W.Emerson   M.J.Broughton   H.J.Stubbs

Department of Biochemistry, University of Otago
P.O. Box 56, Dunedin, New Zealand

INTRODUCTION

The β-1,3/1,6-glucan that constitutes 70% of the cell wall of Candida albicans [1] is qualitatively similar to the glucan of Saccharomyces cerevisiae but it contains higher proportions of branch points (7%) and β-1,6 linkages (43-50%)[2]. Cabib at al. [3] have reviewed the current understanding of the synthesis of the call wall. Linear chains of β-1,3 glucan are synthesized by a UDP-glucose synthetase located in the plasma membrane but nothing is known about the formation of β-1,6 oligosaccharide chains, branching or cross linking to other polymers. In this paper we describe a rapid transglucosylation reaction catalyzed by an exo-β-glucanase from C.albicans.

MATERIAL AND METHOD

Exo-β-glucanase was purified to homogeneity from the culture medium (yeast extract, casein acid-hydrolysate, and glucose: 2,10,20 g/l) of C.albicans ATCC 10261 (unpublished). The enzyme assay, substrates, units of enzyme and other methods were as described previously by Ram et al.[4,5] The assays with laminara-oligosaccharides (Seikagaku Kogo, Tokyo) were for 6 h at 30°C. Each assay (0.1 ml) contained: substrate (0.2-8 mg), enzyme 4 μg and 2 μmol Na acetate buffer, pH 5.6. At various times samples were analysed by HPLC with a HPX 42A column (Biorad, Richmond, Ca.) operated at 85°C with H_2O as the solvent (0.4 ml/min) and detection by refractive index.

RESULTS

The exo-β-glucanase from the culture medium is clearly the same or very similar to the enzyme from cell extracts [5,6]: M_r 4 x 10^4, specific activity with laminaran 65 U/mg, pH optimum 5.6, V_{max} 240 μmol/min/mg and K_m 3.7 mg/ml. It is also similar in many respects to the exo-β-glucanases from other yeasts and fungi [7]. The enzyme hydrolysed pustulan (V_{max} 25 μmol/min/mg; K_m 3.3 mg/ml), p-nitrophenyl-β-glucoside (V_{max} 8.5 μmol/min/mg; K_m 1.6 mg/ml), all disaccharides of the formula n-0-(β-D glucopyranosyl)-D glucopyranoside where n=3,2,4 and 6 in order of decreasing reactivity and the series of laminara-oligosaccharides G_2 to G_6. With laminaratriose (G_3) the V_{max} and K_m for the hydrolytic reaction were 83 µmol/min/mg enzyme and 2.5 mg/ml. HPLC with a column which separates oligosaccharide on the basis of size showed that the enzyme catalysed a transferase reaction (Fig.1). At 80 mg/ml the disappearance of G_3 (5400 µmol glucosyl units/min/mg) was much faster than the rate of formation of glucose (52 µmol/min/mg). Glucosyl units however were transferred to G_3 (and subsequently to higher oligomers) at 3700 µmol glucosyl units/min/mg. A scheme for this process is:

(i) $G_n + E \rightarrow G_{(n-1)} + EG$

(ii) $EG + G_n \rightarrow E + G_{(n+1)}$

(iii) $EG + H_2O \rightarrow E + G$

where G is glucose and oligomers thereof, E is enzyme, EG is enzyme-glucosyl complex and in (ii) and (iii) G_n and H_2O are competing acceptors.

The transferase reaction (ii) did not show saturation kinetics and the rate increased with substrate concentration up to 80 mg/ml. At 80 mg/ml the yield of transferase products (G_3 + >G_3) was more than 35% of the original glucosyl units and these were stable under the reaction conditions for at least 6h. The products of the transferase reaction are presently being analysed by NMR spectroscopy and gc/ms.

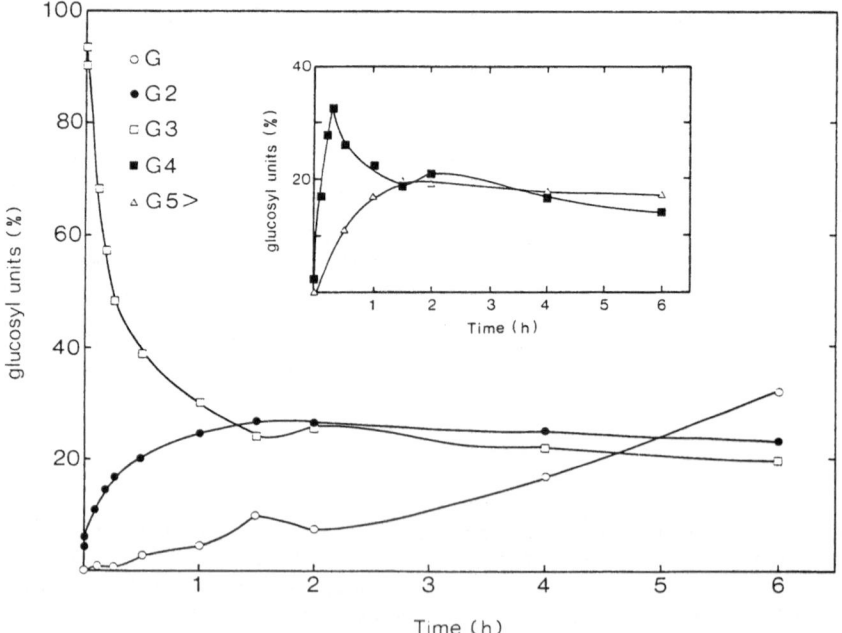

Fig. 1. HPLC analysis of the products from laminaratriose. Laminaratriose (80 mg/ml) was incubated with exoglucanase as described in Method and aliquots were analyzed at the times indicated.

DISCUSSION

Exoglucanases readily hydrolyze soluble β-glucans and glucosides _in vitro_ and it is quite likely that together with endo-β-glucanases they catalyze the localized hydrolysis of the existing wall glucan to initiate a new phase of growth: e.g. budding, germ-tube formation, spore germination. It is noteworthy however that this exoglucanase and many other glucanases are secreted not only at the initiation of new growth but also constitutively during growth. Cabib et al.[3] have suggested that cell wall expansion may occur by "intussusception" thus obviating the requirement for continuous degradation during active growth. Our results lead us to speculate that, in the region of new wall growth, glucanases may function as transferases that rearrange the initial linear (1,3) product. Three possible types of reaction are illustrated below: branching, elongation and conversion of β-1,3 to β-1,6 linkages where (a) is a transferase reaction predicted for an endo-enzyme, (b) for an exo-enzyme and (c) would require an as yet uncharacterized 1,3/1,6 transferase. These reactions could occur on the 1,3 polymer being extruded by the glucan synthetase.

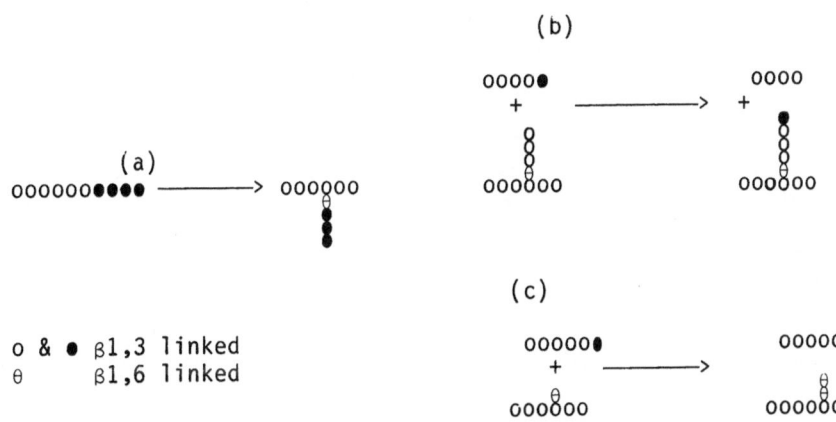

The localized conditions where the newly synthesized glucan is extruded may favour rearrangement rather than hydrolysis because of a high concentration of glucan oligosaccharides and a lowered activity of H_2O. Rearrangement could also be favoured by the relative rates of the forward and backward reactions, product inhibition and finally the formation of the insoluble glucan complex. In the mature wall most of the glucanase activity is latent because wall entrapped β-glucanases are tightly bound to cell wall glucan [8]. Furthermore, it is well established that glucosidases catalyze transglycosylation reactions [9,10]. Confirmation of the model outlined in this paper will require detailed studies but it could provide another area for the systematic development of antifungal agents.

ACKNOWLEDGEMENT

This work was supported by grants from the Medical Research Council of New Zealand, the N.Z. Universities Grants Committee and the University of Otago.

REFERENCES

1. P.A.Sullivan, Y.Y.Chiew, C.Molloy, M.D.Templeton and M.G.Shepherd, Canad.J.Microbiol., 29: 1514 (1983).
2. P.K.Gopal, M.G.Shepherd and P.A.Sullivan, J. Gen. Microbiol., 130: 3295 (1984).
3. E.Cabib, B.Bowers, A.Sburleti and S.J.Silverman, Microbiol. Sci. 5: 370 (1988).
4. S.P. Ram, L.K. Romana, M.G. Shepherd and P.A. Sullivan, J. Gen. Microbiol., 130: 1227 (1984).
5. S.P.Ram, K.H.Hynes, S.P.Romana, M.G.Shepherd and P.A.Sullivan, Life Science Advances - Biochemistry 7 (in press) (1988).
6. M.Molina, R.Cenamor and C.Nombela, J. Gen. Microbiol., 113: 609 (1987).
7. C.Nombela, M.Molina, R.Cenamor and M.Sanchez, Microbiol. Sci., 5: 328 (1988).
8. A.G.Dickerson and R.C.F.Baker, J.Gen. Microbiol., 112: 67 (1979).
9. K.Nisizawa and Y.Hashimoto, in: "The Carbohydrates: Chemistry and Biochemistry", W.W. Pigman and D. Horton, eds., pp. 241-300, Academic Press, New York (1970).
10. K.G.Nilsson, Tibtech., 6: 256 (1988).

PROTEINASES SECRETED BY CANDIDA SPECIES

R. Rüchel

Department of Medical Microbiology, University of
Göttingen, Göttingen, The Federal Republic of Germany

The extracellular proteolytic activity of C. albicans was
discovered by Staib [1] and various secretory candidal proteinases have
been characterized [2-9]. The enzymes are aspartic proteinases
(E.C.3.4.23.6) related to pepsin, renin, and cathepsin-D; like
these, they are inhibited by pepstatin-A. Most Candida proteinases
are glycoproteins of approximately 45 kD molecular mass; only a
proteinase of C. parapsilosis had a mass of 33 kD [7]. The nucleotide
sequence of a proteinase gene of C. albicans has recently been
published [10]; howover, it is not clear at present, whether this gene
codes for a secretory enzyme.

Proteinases are secreted by virtually all isolates of C.albicans
and C.tropicalis [11,12]. Each fungal strain secretes a single proteinase
which may be type-specific [11]. Extracellular proteolytic activity was
also detected among isolates of C.parapsilosis [7,13] and C.glabrata [14].
The clinical relevance of these enzymes is debatable. Other candidal
species, which are occasionally found in clinical specimens, are
non-proteolytic in vitro and in vivo. Hence, the distribution
of extracellular Candida proteinases largely reflects the order of
virulance of these opportunistic yeast-like fungi. Staib [15] was also the
first to suggest a role of Candida proteinases in the pathogenesis of
mycosis. An association between proteolytic C.albicans and mucosal
disease (denture stomatitis) was first observed by Budtz-Joergensen [16],
and Macdonald and Odds [3] confirmed, by immunofluorescence, the
induction of Candida proteinase during murine candidosis of the
kidney.

There is ample evidence that proteolytic C.albicans and
C.tropicalis secrete their enzymes in humans during mycosis: High
titers of specific antibodies can be monitored in sera of patients
suffering from candidosis [3] and proteinase-related antigens can be
demonstrated in mycotic human tissues [17,18]. Proteinase antigen can be
detected by enzyme immunoassay in the sera of patients, who are
suspected of deep-seated candidosis [19].

Evidence for a role of Candida proteinase in the pathogenesis of
mycosis was briefly reviewed by Rüchel [18]. A correlation between
proteolytic activity of C.albicans and mucosal infection in vitro and
in vivo was shown by Ghannoum and Abu Elteen [20]; and Cassone et al [21].
MacDonald and Odds [22] found that a non-proteolytic mutant strain of
C.albicans was less virulent in mice than the strongly proteolytic

parental strain. This was confirmed by Kwon-Chung et al. [23] who showed also that revertants with restored proteolytic activity regained their virulence.

The role of fungal proteinase as a factor of virulance is supported by additional evidence: Proteolytic candidal cells are highly cytotoxic for human monocyte-like cells (U 937), while non-proteolytic blastospores cause little cytotoxicity [24]. This cytotoxic effect of proteolytic Candida blastospores can be modulated in a dose-dependent fashion by the specific proteinase inhibitor pepstatin-A, whereas a non-proteolytic isolate of C.tropicalis was not at all affected by pepstatin [25].

Pepstatin interferes with the candidal infection of murine epidermis [8], and it largely prohibits the invasion of human oral mucosa by Candida albicans in vitro [26], but the solubilized inhibitor proved ineffective upon intravenous application in mice with established systemic candidosis [27]. This lack of success may be due to the fast clearance of solubilized pepstatin from the blood stream. Intravenously applied pepstatin suspension produces lasting levels of the inhibitor and can indeed modulate the outcome of murine candidosis [28].

A variety of target proteins for Candida proteinases have been proposed. Most fungal proteinases degrade serum albumin into small fragments. This indicates a low substrate specificity, since albumin is stabilized by 17 disulfide bonds. Hence, very few proteins were found to resist degradation by Candida proteinase at acidic pH: lysozyme resists by virtue of its extreme basic charge, and ferritin is broken down only very slowly.

Among the proteins of oral secretions, lactoferrin, lactoperoxidase and mucin are degraded in the same way as the secretory immunoglobulins [29]. Candida proteinases hydrolyse immunoglobulin A2, which resists degradation by the IgA-1 proteinases of various pathogenic bacteria [30]. Degradation of immunoglobulins in vivo by C.albicans was demonstrated by immunoblot of proteins derived from mycotic murine kidneys [31]. Degradation of antibodies by proteolytic Candida may indeed be critical, if humoral immunity plays an important role in the control of the systemic spread of mycosis, as was suggested by Matthews et al [32]. Immunization against purified candidal proteinase caused a significant protective effect in lethally infected mice [25].

Candida proteinase degrades collagens [33] and acts as a keratinase in vitro [5]; hence, it may be involved in fungal invasion through orthokeratinized mucosa.

In submucosal tissue, the fungus encounters blood vessels which are readily invaded. At this stage of infection, certain strains of C.albicans and C.tropicalis may cause blood coagulation, particularly by activation of coagulation factor X in a yet unexplained fashion, and the enzymes may also exert a renin-like effect [34]. The highest activity of this sort was found among proteinases which were non-covalently bound to candidal cell fragments [35]; these enzymes may be vacuolar proteinases which have been described by Farley et al [36].

Enzymic effects of acid Candida proteinases require a lowered pH. Physiological pH at 37°C prohibits activity of Candida proteinases and causes alkaline denaturation. The pH-dependent activity profiles of Candida proteinases, and resistance to alkaline denaturation, are strain-specific. Enzymic activity may begin at pH 6.5 by causing limited hydrolysis. In the proximity of the activity maximum (pH 3-3.5) Candida proteinases act in a rather indiscriminate fashion.

Suitable conditions for proteolytic activity may exist in sites of poor blood circulation (acidosis), in secretions with lowered pH, and within phagolysosomes. Indeed, phagocytosed C.albicans and C.tropicalis were shown to express proteinase antigen [37]. Candida, itself, acidifies its micro-environment by secretion of organic acids [38], thus creating a zone of lowered pH which may allow proteolytic activity. The acidification may be particularly strong in budding yeasts and in the tips of filamentous cells [39].

Secretion of protons by C.albicans depends upon the concentration of available glucose [29,40]. The critical concentration of glucose at which proteinase is induced, varies among the species; it was particularly high in a strain of C.parapsilosis, rendering it comparatively avirulent [7]. In the few cases of fungemia due to C.parapsilosis, which we have observed, neither proteinase antigen nor rise of anti-proteinase antibodies was found [19]. The typical susceptibility of diabetics to candidosis may reflect the induction of fungal proteinase by secretions rich in glucose.

Candida albicans comprises two major serotypes [41]. Both the blastoconidia and filamentous cells of serotype A express the proteinase antigen, while various strains of serotype B carry the proteinase antigen on the blastoconidia only [37]. This pattern reflects the superior virulence of serotype A and its prevalence among clinical isolates.

There is evidence that other factors modulate the activity of candidal proteinases during infection [27,42]. In C.stellatoidea (a sucrose-negative variant of C.albicans), proteolytic activity in vitro did not correlate with virulance in mice [43]. However, the majority of the accumulated data support a crucial role of extracellular acid proteinase in the establishment of mycosis by C.albicans and C.tropicalis. Thus, Ross et al. [9] found a 1000-fold higher virulence for mice of a proteolytic wild type strain as compared with a proteolytic mutant of C.albicans.

Candidal proteinases may be targets for prevention or even the therapy of mycosis. The recent attention given to the related proteinases of pathogenic retroviruses [44] could lead to the development of such inhibitors, which may be modified pepstatins.

REFERENCES

1. F. Staib, Sabouraudia, 4 : 187 (1965).
2. H. Remold, H. Fasold, F. Staib, Biochimica et Biophysica Acta, 167 : 399 (1968).
3. F. MacDonald and F.C. Odds, J. Med. Microbiol., 13 : 423 (1980).
4. R. Rüchel, Biochimica et Biophysica Acta, 659 : 99 (1981).
5. M. Hattori, K. Yoshiura, M. Negi, H. Ogawa, Sabouraudia, 22 : 175 (1984).
6. H. Kaminishi, Y. Hagihara, S. Hayashi, T. Cho, Infect. Immun., 53 : 312 (1986).
7. R. Rüchel, B. Boening, M. Borg, Infect. Immun., 53 : 411 (1986).

8. T.L. Ray and C.D. Payne, _Infect. Immun._, 56 : 1942 (1988).
9. I.K. Ross, F. De Bernardis, G.W. Emerson, A. Cassone, P.A. Sullivan, _J. Gen. Microbiol._, 136 : 687 (1990).
10. T.J. Lott, L.S. Page, P. Boiron, E. Reiss, _Nucleic Acids Res._, 17 : 1779 (1988).
11. R. Rüchel, K. Uhlemann, B. Boening, _Zbl. Bakt. Hyg._, Orig. A, 255 : 537 (1983).
12. F. MacDonald, _Sabouraudia_, 22 : 79 (1984).
13. F. De Bernardis, R. Lorenzini, R. Verticchio, L. Agatensi, A. Cassone _J. Clin. Microbiol._, 27 : 2598 (1989).
14. J. Reinholdt, P. Krogh, P. Holmstrup, _Acta Pathol. Microbiol. Immunol. Scand._ Sect. C, 95 : 265 (1987).
15. F. Staib, _Mycopathol. Mycol. Appl._, 37 : 345 (1966).
16. E. Budtz-Jörgensen, _Sabouraudia_, 12 : 266 (1971).
17. C.F. Zimmermann, DDS Thesis, Faculty of Medicine, University of Göttingen, The Federal Republic of Germany.
18. R. Rüchel, _in_ : "Oral Candidosis", L.P. Samaranayake and T.W. MacFarlane, eds., pp. 47-65, Wright, London (1990).
19. R. Rüchel, B. Böning-Stutzer, A. Mari, _Mycoses_, 31 : 87 (1988).
20. M. Ghannoum and K. Abu Elteen, _J. Med. Vet. Mycol._, 24 : 407 (1986).
21. A. Cassone, F. De Bernardis, F. Mondello, T. Ceddia, L. Agatensi, _J. Infect. Dis._, 156 : 777 (1987).
22. F. MacDonald and F.C. Odds, _J.Gen.Microbiol._, 129 : 431 (1983).
23. K.J. Kwon-Chung, D. Lehman, C. Good, P.T. Magee, _Infect. Immun._, 49 : 571 (1985).
24. M. Borg, D. Kirk, H. Baumgarten, R. Ruechel, _Sabouraudia_, 22:357 (1984).
25. H. Stutzer, MD Thesis, Faculty of Medicine, University of Göttingen, The Federal Republic of Germany.
26. M. Borg and R. Ruechel, _Infect. Immun._, 56 : 626 (1988).
27. A.M. Edison and M. Manning-Zweerink, _Infect. Immun._, 56 : 1388 (1988).
28. R. Rüchel, B. Ritter, M. Schaffrinski, _Zbl. Bakt._, 273 : 391 (1990).
29. R. Rüchel, _Microbiol. Sci._, 3 : 316 (1986).
30. M. Kilian, J. Mestecky, M.W. Russel, _Microbiol. Rev._, 52 : 296 (1988).
31. R. Rüchel, _Zbl. Bakt. Hyg. A_, 257 : 266 (1984).
32. R. Matthews, J. Burnie, D. Smith, I. Clark, J. Midgley, M. Conolly, B. Gazzard, _Lancet_, II : 263 (1988).
33. Y. Hagihara, H. Kaminishi, T. Cho, M. Tanaka, H. Kaita, _Arch. Oral Biol._, 33 : 617 (1988).
34. R. Rüchel, _Zbl. Bakt. Hyg._, I. Abt. Orig. A, 255 : 368 (1983).
35. R. Rüchel, B. Boening, E. Jahn, _Zbl. Bakt. Hyg.A_, 260 : 523 (1985).
36. P.C. Farley, M.G. Shepherd, P.A. Sullivan, _Biochem. J._, 236 : 177 (1986).
37. M. Borg and R. Ruechel, _J. Med. Vet. Mycol._,28 : 3 (1990).
38. L.P. Samaranayake, D.A.M. Geddes, D.A. Weetman, T.W. MacFarlane, _Microbios_, 37 : 105 (1983).
39. G. Turian, _Experientia_, 37 : 1278 (1981).
40. L.P. Samaranayake, A. Hughes, T.W. MacFarlane, _J.Med.Microbiol._, 17 : 13 (1984).
41. H.F. Hasenclever and W.O. Mitchell, _J.Bacteriol._,82 : 570 (1961).
42. Y. Kondoh, K. Shimizu, K. Tanaka, _Microbiol. Immunol._, 31: 1061 (1987).
43. T. Blundell and L. Pearl, _Nature_, 337 : 596 (1989).
44. K.J. Kwon-Chung, B.L. Wickes, W.G. Merz, _Infect. Immun._,56 : 1814 (1988).

ADHESION OF CANDIDA ALBICANS TO HOST SURFACES

L.J. Douglas

Department of Microbiology, University of Glasgow
Glasgow G12 8QQ, The United Kingdom

INTRODUCTION

Colonization of host surfaces by the pathogenic Candida species depends on the ability of the fungi to adhere. Adhesion enables the organisms to resist the flushing action of the fluids that continuously bathe such surfaces. In the case of mucosal surfaces, epithelial cells with adherent microorganisms are repeatedly lost by desquamation and so the progeny of these organisms much attach to newly exposed cells if colonization is to continue. This host defence mechanism provides a selective pressure which may result in the elimination of less adhesive organisms. Adhesion is, therefore, the first step in the process leading to persistent colonization and infection, and the ability to adhere constitutes an important virulence factor.

Increasing recognition of the importance of adhesion has stimulated a great deal of research and Candida adhesion to a variety of surfaces has now been tested in vitro. These include epithelial and endothelial cells, fibrin-platelet matrices, bacteria, neutrophils, an inert materials such as denture acrylic and Teflon [1]. This contribution will focus primarily on Candida adhesion to epithelial cells.

MEASUREMENT OF ADHESION IN VITRO

Adhesion assays have been devised using both exfoliated and cultured epithelial cells [2]. Because of the widespread occurrence of Candida infections of the mouth and vagina, most studies have concentrated on fungal adhesion to exfoliated buccal or vaginal cells. Either type of cell preparation is readily obtained from human volunters by gently swabbing or scraping the mucosal surface. In assays involving Candida ablicans, the yeast form of the fungus is most commonly used both for convenience of counting and because yeasts generaly predominate in the early stages of colonization.

Adhesion to buccal or vaginal cells can be measured very simply using a protocol devised originally by Kimura and Pearsall [3]. This involves preparing and mixing equal volumes of carefully standardized suspensions of yeasts and epithelial cells. After incubation at 37°C, epithelial cells are removed on polycarbonate filters which have a

Candida and Candidamycosis, Edited by E. Tümbay *et al.*
Plenum Press, New York, 1991

pore size (12μm) sufficient to allow passage of unattached yeasts. The epithelial cells with their adherent yeasts can then be fixed, stained and counted directly on these transparent filters under a light microscope. Alternatively, radiolabelling methods can be used.

ADHESION AND VIRULENCE

The relationship between pathogenicity and adhesion was first demonstrated by King et al.[4] in experiments with buccal and vaginal cells. With either cell type, C. albicans -undoubtedly the most virulent Candida species- was also the most adherent, followed by C. tropicalis and C. parapsilosis. The less virulent species, C. guilliermondii, C. krusei and C. tropicalis, showed relatively little adhesion. Similar species differences have been reported for adhesion to epidermal corneocytes [5], intestinal epithelial cells[6], vascular endothelium [7] and fibrin-platelet matrices [8] in vitro. Thus, there is a clear relationship between adhesive ability and pathogenic potential.

MECHANISM OF ADHESION

In vitro assays have been widely used to investigate different factors which can affect adhesion to epithelial cells. These include factors relating to the yeast, the epithelial cells and the assay environment [1]. Much of this work was done as a prelude to studies on the mechanism of adhesion, where the long-term goal is the development of measures to inhibit adhesion in vivo and so prevent infection.

A variety of mechanisms have been suggested for Candida adhesion to epithelial surfaces. They comprise three types:

1. Non-specific adhesion, which depends on van der Waals forces, electrostatic and hyrophobic interactions, and hydrogen bonding.

2. Adhesin-receptor interactions, where binding molecules (adhesin) on the fungal cell surface interact in a stereospecific way with complementary receptor molecules on the host cell membrane.

3. Coadhesion, in which adherent bacteria or other yeasts mediate attachment.

Although a combination of some or all of these mechanisms is likely to operate in vivo, most of the available evidence indicates the primary importance of adhesin-receptor interactions. A number of yeast components have been suggested as possible adhesins. They include mannan or mannoprotein [9,10,11], chitin[12] and lipids [13]. Similarly, various different epithelial receptors have been proposed, including glycosides[14,15], lipids[13] and fibronectin [16].

These apparent contradictions may be partly explained by the ability of C. albicans to engage in more than one type of adhesin-receptor interaction. It seems that, as with many bacterial species, different strains can possess different adhesins [14]. However, a number of points must be experimentally proven before it can be concluded that a specific molecule is involved in adhesion [17]; some of these are listed in Table 1. Whether all of the adhesins proposed for C. albicans fulfill these criteria remains to be demonstrated.

Table 1. Some requirements for the demonstration that a given micro-
         bial component is directly involved in adhesion (adapted
         from Curtis, 1979).

---

1. Isolation and purification of the molecular species involved.
2. Demonstration of the surface location of this molecule.
3. Organisms denuded of the molecule should show greatly reduced adhesion.
4. Enzymic dissection of the molecule should destroy adhesion.

EVIDENCE FOR A MANNOPROTEIN ADHESIN

There is now considerable evidence for a mannoprotein adhesin
which binds via its protein moiety to fucose-containing glycosides on
the surface of buccal and probably also vaginal epithelial cells.
This adhesin appears to be present in a number of C. albicans strains
originally isolated from active infections. Adhesion of these strains
to epithelial cells can be significantly increased by growing the
organisms in defined medium containing a high concentration of
certain sugars as the carbon source [18]. Yeasts grown in medium
containing 500mM galactose, for example, are up to ten times more
adherent than control organisms harvested from medium with a
relatively low concentration (50mM) of glucose. Enhanced adhesion
appears to be due to the production, under these conditions, of a
fibrillar surface layer which can be visualized in thin sections of
yeasts treated with ruthenium red or polycationic ferritin [19].
Similar structures have been demonstrated on the surfaces of yeasts
present in clinical specimens from patients with Candida infections;
they may represent appendages analogous to bacterial fimbriae.

Fibrils appear to be released from the surface of galactose-
grown infective strains relatively readily, especially upon prolonged
incubation at 37°C, and extracellular polymeric material (EP) can be
isolated from culture supernatants. Pretreatment of buccal epithelial
cells with EP results in inhibition of yeast attachment when the
cells are subsequently used in adhesion assays [20]. This suggests that
EP contains an adhesin which can bind to, and block, epithelial cell
receptors. Preliminary results indicate that EP can also have an
inhibitory effect in vivo. For example, mice given vaginal rinses of
EP appear to be more resistant to subsequent vaginal infection with
C. albicans than are untreated control animals [21].

Chromatographic analysis of acid-hydrolysed EP has revealed
mannose, glucose and glucosamine as the only identifiable sugars
present, while quantitative analysis indicates that the overall
composition of different EP preparations is similar, irrespective of
the carbon source or incubation period [20]. They contain 65-82%
carbohydrate (mainly mannose), together with smaller amounts of
protein (7%), phosphorus (0.5%) and glucosamine (1.5%). Serological
studies have demonstrated that EP preparations obtained after growth
of the yeast on different carbon sources are immunologically
identical, but that 500mM galactose-grown yeasts have more antigenic
determinants and presumably a more extensive fibrillar layer than
500mM sucrose-grown organisms, whereas 50mM glucose-grown yeasts have
fewest determinants/fibrils [20].

The high mannose content of EP preparations suggests that the
adhesin is probably mannoprotein in nature. Further evidence to
support this conclusion has come from experiments with tunicamycin,

an antibiotic which, in yeasts, specifically inhibits synthesis of mannoprotein but not that of the other major wall components, glucan and chitin. Addition of tunicamycin to cultures of C.albicans in high-galactose medium at the end of exponential growth inhibited formation of the fibrillar layer. As a result, the yeasts showed a decrease in adhesion to buccal cells of over 60% as compared with untreated organisms. Control yeasts (grown on 50mM glucose), which largely lacked the fibrillar layer, were unaffected by tunicamycin treatment [9].

FURTHER CHARACTERIZATION OF THE ADHESIN

Further characterization of a mannoprotein adhesin obviously necessitates identifying that portion of the complex molecule which participates in the interaction with epithelial cells. Experiments designed to determine the minimum structure of EP required to inhibit adhesion in conventional assays indicate that the protein portion of the adhesin is more important that the carbohydrate moiety in mediating attachment [22]. For example, pretreatment of EP with heat, dithiothreitol or proteolytic enzymes (except papain) either partially or completely destroyed its ability to inhibit adhesion to buccal cells, whereas pretreatment with sodium periodate or α-mannosidase had little or no effect. Moreover, the protein-rich fraction obtained by incubating EP with endoglycosidase H inhibited adhesion to a greater extent than did the carbohydrate-rich fraction [22].

Crude EP preparations obtained from culture supernatants contain several different components; partial purification of adhesin has been achieved by a two-step procedure involving chromatography on concanavalin A-sepharose and DEAE-cellulose. The purified material inhibited yeast adhesion to buccal cells thirty times more efficiently (on a weight basis) than unfractionated EP [22]. More recently, a purification/degradation scheme has been devised which produces a protein fragment with a substantially greater ability to block adhesion (see below).

CHARACTERIZATION OF EPITHELIAL CELL RECEPTORS

Elucidation of adhesion mechanisms involves identification not only of the microbial adhesin but also of the host cell receptor. Participation of the protein portion of yeast mannoprotein in the adhesion process would be analogous to many bacterial adhesion mechanisms in which a proteinaceous adhesin interacts with a glycoside receptor (either glycoprotein or glycolipid) on the host cell surface [23]. Inhibition tests with sugars and lectins have been widely used to characterize such receptors. Tests of this type with C. albicans indicate that a receptor containing the 6-deoxyhexose, L-fucose, may commonly be required. For example, addition of L-fucose to in vitro assays with strain GDH 2346 inhibited yeast adhesion to buccal epidhelial cells. Pretreatment of the buccal cells wih lectin from Lotus tetragonolobus (which is specific for L-fucose), but not other lectins, also blocked adhesion [14].

Of six C. albicans strains tested, five behaved in this way [21]. On the other hand, adhesion of strain GDH 2023 was inhibited by N-acetylglucosamine and wheat-germ agglutinin (a lectin specific for N-actylglucosamine) but not by L-fucose or L-tetragonolobus lectin, indicating that a different type of adhesion mechanism function in this organism. With the sensitive strains, addition of L-fucose to

46

assay mixtures caused only partial inhibition of adhesion, suggesting that the natural mucosal receptor is larger than an L-fucose residue and/or that a particular stereochemical configuration is required. Alternatively, additional adhesion mechanisms may operate.

Recently, these findings have been exploited in devising a purification scheme for the yeast adhesin that binds to fucoside receptors. The protocol involves treating EP with N-glycanase, papain and dilute alkali to cleave the protein and carbohydrate portions of mannoprotein. Fucoside-binding protein fragments are then recovered on an affinity column containing the trisaccharide of the H blood-group antigen which terminates in a residue of L-fucose. The purified material is devoid of carbohydrate and inhibits yeast adhesion to buccal cells 220 times more efficiently, on a protein weight basis, than crude EP (F.D. Tosh and L.J. Douglas, unpublished results). On an overall weight basis, inhibition is some 2000-fold greater because crude EP contains considerable amounts or carbohydrate. Further work on the characterization of this protein fraction is currently in progress.

REFERENCES

1. L.J.Douglas, in: "The Yeasts", Vol. 2, 2nd Edition, A. H. Rose and J.S. Harrison, eds., pp. 239-280, Academic Press, London (1987a).
2. L.J. Douglas, CRC Crit. Rev. Microbiol., 15:27 (1987b).
3. L.H. Kimura and N.N. Pearsall, Infect. Immun., 21: 64 (1978).
4. R.D. King, J.C. Lee and A.L. Morris, Infect.Immun.,27:667 (1980).
5. T.L. Ray, K.B. Digre and C.D. Payne, J.Invest. Dermatol., 83:37 (1984).
6. S.A. Klotz and R.L. Penn, Curr. Microbiol., 16:119 (1987).
7. S.A. Klotz, D.J. Drutz, J.L. Harrison and M. Huppert, Infect. Immun., 42:374 (1983).
8. P.A. Maisch and R.A. Calderone, Infect. Immun., 27:650 (1980).
9. L.J. Douglas and J.McCourtie, FEMS Microbiol.Lett., 16: 199 (1983)
10. J.C. Lee and R.D. King, Infect. Immun., 41:1024 (1983).
11. R.L. Sandin, A.L. Rogers, R.J. Patterson and E.S. Beneke, Infect. Immun., 35:79 (1982).
12. E. Segal, N. Lehrer and I. Ofek, Exp. Cell Biol., 50:13 (1982).
13. M.A. Ghannoum, G.R. Burns, K. Abu Elteen and S.S. Radwan, Infect. Immun.,54:189 (1986).
14. I.A. Critchley and L.J. Douglas, J. Gen. Microbiol., 133:637 (1987b).
15. J.D. Sobel, P.G. Myers, D. Kaye and M.E. Levison, J. Infect. Dis., 143:76 (1981).
16. K.G. Skerl, R.A. Calderone, E. Segal, T. Sreevalsan and W.M. Scheld, Can. J. Microbiol., 30:221 (1984).
17. A.S.G. Curtis, in: "Adhesion of Microorganisms to Surfaces", D.C. Ellwood, J. Melling and P. Rutter, eds., pp. 199-208, Academic Press, London (1979).
18. J. McCourtie and L.J. Douglas, Infect. Immun., 45:6 (1984).
19. J. McCourtie and L.J. Douglas, Infect. Immun., 32:1234 (1981).
20. J. McCourtie and L.J. Douglas, J.Gen. Microbiol., 131:495 (1985).
21. I.A. Critchley, Phll. D. Thesis, University of Glasgow (1986).
22. I.A. Critchley and L.J. Douglas, J. Gen. Microbiol., 133:629 (1987a).
23. G.W. Jones and R.E. Isaacson, CRC Crit. Rev. Microbiol., 10:229 (1983).

FACTORS OF THE PATHOGENICITY OF <u>CANDIDA ALBICAN</u>S - A REVIEW

L. Krempl-Lamprecht

The Clinic and Out-Patient Clinic of Dermatology,
Technical University of Munich, Munich, The Federal
Republic of Germany

<u>Candida albicans</u>, the most common cause of pathological yeast infections, is an <u>opportunistic</u> fungus. That means it has <u>specific attributes of pathogenicity</u>, but these different factors can produce a mycosis only if the <u>host is predisposed und reduced in his antimicrobial defense</u>. Therefore, great differences are seen in the severity of a <u>Candida</u>-mycosis. They are caused by the fungus as well as by the host :

1) Different strains of the yeast possess different degrees of virulence. So a low or a high propensity to cause diseases may be observed.

2) The resistance of the host is varying, too : If there is a limited failure of the defense mechanisms, <u>Candida</u> invades superficially. If the defects are serious and numerous, a deep seated mycosis may arise. In patients with T-cell dysfunction ainly chronical infections occur.

In the following I will give a short review about special properties of <u>C. albicans</u>, about its possible factors of pathogenicity.

The primary condition of an infection is the possibility of the microorganism to adhere onto the epithelial cells of the host so strong that it cannot be swept away by any movement. Therefore, adherence capabilities play an important, likely the major role in the pathogenesis of <u>Candida</u>.

The mechanism which controls this phenomenon is an exceptional contact between surface components of the yeast and surface particles of the host cell. Both act together in a lectin-like manner, where a protein containing component is linked to a polysaccharide component.

Different possibilities of lectin binding are imaginable :

1) <u>Protein-adhesins of the yeast cell</u> bind to the <u>carbohydrates of the host cell</u>, which act as <u>receptors</u>.

2) <u>Polysaccharides</u> of the <u>yeast cell</u> react with the <u>proteins of the human cell</u>.

Candida and Candidamycosis, Edited by E. Tümbay *et al.*
Plenum Press, New York, 1991

3) <u>Both</u>, the yeast cell and the host cell, possess specific <u>gly-coproteins</u> with characteristic moieties of carbohydrates.

4) <u>Other microorganisms</u>, like Gram-negative bacteria with exceptional mannose sensitive fimbriae, act as <u>bridging mediators</u>.

Electronmicrographs have shown such a bridge-like contact of fibrillar-floccular appearance between <u>C.albicans</u> and different epithelial cells of the host. Probably it is built by the protein portion of mannoprotein, a characteristic part of the yeast cell wall, and fucose, a carbohydrate constituent of the host cell surface.

A fungus giving rise to a mycosis must be able to damage tissues by secretion of <u>characteristic enzymes</u>, which are adapted to the temperature and pH of the host. They have to prepare compounds for the nutrition of fungus.

Many investigators suggest the involvement of typical enzymes in the pathogenicity of <u>C.albicans</u>. Two groups are mentioned frequently: <u>proteolytic enzymes and phospholipases</u>. Both are hydrolytic enzymes and catalyse the decomposition of peptides, glycosides and esters.

Among the extracellular proteolytic enzymes a <u>carboxylprotei-nase</u>, active at pH 3.2 (possibly with isoenzymes) is mainly investigated. In different <u>Candida</u> strains a considerable variation in the quantity of the enzyme production is observed : Serotype A produces larger amounts of these enzymes than serotype B.

A clear correlation between the proteolytic activity and the degree of invasiveness is not yet proved, but an argument for the valuation as virulence factor is the reduction of the mouse lethality caused by mutant strains, which are deficient in the secretion of proteinases.

<u>Phospholipases</u> (degrading phospholipids into lysophospholipids) and <u>lysophospholipases</u> are able to lyse different biological membranes. They are located at the cell surface, especially at the hyphal tip and secreted into the environment.

It is supposed that during tissue invasion a path is cleared by yeast cells with a high phospholipase activity. Other cells can follow and proceed to invade and colonize the tissues with the production of hyphae.

A correlation found between the enzymatic activity, the adherence to epithelial cells and the virulence in experimental animals support the suggestion that phospholipases are important factors of pathogenicity. They are foud exclusively in <u>C. albicans</u>, not in other species of <u>Candida</u>.

During the invasion, a morphological alteration takes place and the well known <u>dimorphism</u> of <u>C. albicans</u> is expressed. Blastospores produce germ tubes, later on pseudomycelia and true mycelia instead of a continuous budding. The mycelial forms play a distinct role in the initial processes of tissue invasion. Several investigations have shown that the adherence of germ tubes and hyphae exceed the adherence of blastospores.

In the course of its growth <u>Candida</u> meets other microorganisms like yeasts and bacteria. By mutual interactions a negative or a positive effect for Candida may result. It is caused not only by their competition for nutrients, but also by the active secretion of inhibitory or stimulatory substances; for example, some yeasts are able to produce extracellular glycoproteins called killer toxins. They act by interfering with the cytoplasmic membrane of target cells. As a result of the altered membrane permeability leakage of potassium is promoted.

The name "pool efflux stimulating toxins" indicates furthermore a partial dissipation of the ATP-pool : AMP is accumulated in the medium of sensitive cells.

At first it was presumed that such substances could be factors of the Candida pathogenicity, but later investigations have shown that C. albicans has no killing activity, but a high susceptibility to the killer toxins of other yeasts like Saccharomyces.

Commensal bacteria can play an enhancing or a degrading role in Candida growth.

One mechanism by which a positive effect of Gram-negative bacteria has been reported is again an influence on the adherence. Preincubation of human cells with Escherichia coli-suspensions resulted in a significant increase of candidal adherence. The increased candidal adherence is mediated by the presence of mannose sensitive fimbriae (type 1) on the surface of the bacterial cell. The fimbriae may act as a bridging mechanism between C.albicans and human epithelial cells, thus increasing the yeast adherence.

Another interesting interaction between Candida and Gram-negative bacteria concerns the siderophores, oligopeptides, which transport iron. In tissues free iron, an essential growth substance, is restricted due to its being complexed with iron-binding proteins mostly. Therefore many pathogenic microorganisms have evolved a system, called siderophores, which allows them to compete successfully with the host proteins for iron. While cultured under iron-limiting condition the growth of Candida is suppressed.Addition of supplemental siderophores can reverse the suppression. Hereby the yeast can utilise autologous (=candidal) siderophores as well as heterologous (=bacterial) siderophores like enterochelin from enterobacteria.

Observations have shown that the possession of a siderophore-mediated-iron-transport-system is correlated with increased virulence. The ability of C.albicans to utilise not only candidal siderophores but also non-candidal, p.e.from Escherichia, seems to be an added factor to the pathogenicity.

This review has shown that in the last decade a remarkable number of investigations have elucidated the problem of fungal virulence. On the other hand, the necessity to eliminate opportunistic fungi has yielded new antimycotics with high efficacy and without side effects, such as Fluconazole or Itraconazole. And there is also a new principle of therapy, still in the beginning of its development. As many mycoses are based on deficiencies of the defense systems, immunostimulating therapy becomes important. Recently a substance has been synthesized, called Thymopentin. It is identical with the active pentapeptid-sequence of the native thymic hormon thymopoetin. Thymopentin controls the early T-cell differentiation and is essential for the maturation, function and maintainance of cell mediated immunity. It restores immune deficiencies by increasing lymphocyte functions. It is known that patients with T-cell disorders are predisposed to chronic mycoses like chronic mucocutaneous candidosis or chronic recurrent vaginal candidosis. In some few cases, where a sole antifungal therapy failed, thymopentin induced a rapid improvement. Therefore, a combined therapy with antimycotics and thymopentin seems to be a therapeutic principle of the future, accelerating the healing process in chronic mycoses.

REFERENCES

1. Y. Banno, T. Yamada and Y. Nozawa, Secreted phospholipases of the dimorphic fungus, Candida albicans; separation of three enzymes and some biological properties, Sabouraudia, 23 : 47 (1985).
2. K. Barett-Bee, Y. Hayes, R.G.Wilson and J.F. Ryley, A comparison of phospholipase activity, cellular adherence and pathogenicity of yeasts, J.Gen.Microbiol., 131 : 1217 (1985).
3. H. Bussey and N. Skipper, Membrane-mediated killing of Saccharomyces cerevisiae by glycoproteins from Torulopsis glabrata, J.Bacteriol., 124 : 476 (1975).
4. I.A. Critchley and L.J. Douglas, Role of glycosides as epithelial cell receptors for Candida albicans, J.Gen.Microbiol., 133 : 637 (1987).
5. I.A. Critchley and L.J. Douglas, Differential adhesion of pathogenic Candida species to epithelial and inert surfaces, FEMS Microbiol.Lett. 28 : 199 (1987).
6. L.J. Douglas, Surface composition and adhesion of Candida albicans, Biochem.Soc.Trans., 13 : 1982 (1985).
7. L.J. Douglas, Adhesion of pathogenic Candida species to host surfaces, Microbiol Sci., 2 : 243 (1985).
8. A.Ismail and D.M. Lupan, Utilisation of siderophores by Candida albicans, Mycopathol., 96 : 109 (1986).
9. J.S. Kandel and T.A. Stern, Killer phenomenon in pathogenic yeast, Antimicrob. Agents Chemother., 15 : 568 (1979)
10. M.H. Kearns, P. Davies and C.H. Smith, Variability of the adherence of Candida albicans strains to human buccal epithelial cells, Sabouraudia, 21 : 93 (1983).
11. R.D. King, J.C. Lee and A.L. Morris, Adherence of Candida albicans and other Candida species to mucosal epithelial cells, Infect.Immun., 27 : 667 (1980).
12. K.J. Kwon-Chung, D. Lehman, C.Good and P.T. Magee, Genetic evidence for role of extracellular proteinase in virulence of Candida albicans, Infect.Immun., 49 : 571 (1985).
13. H.C. Makrides and T.W. MacFarlane, An investigation of the factors involved in increased adherence of Candida albicans to epithelial cells mediated by E.coli, Microbios, 38 : 177 (1983).
14. F. Mc Donald and F.C. Odds, Virulence for mice of a proteinase-secreting strain of Candida albicans and a proteinase-deficient mutant, J.Gen.Microbiol., 129 : 431 (1983).
15. F. Odds, Candida albicans proteinase as a virulence factor in the pathogenesis of Candida infection, Zbl.Bakteriol. A, 260 : 539 (1985).
16. M.F. Price and R.A. Cawson, Phospholipase activity in Candida albicans, Sabouraudia, 15 : 179 (1977).
17. M.F. Price, I.D. Wilkinson and L.O. Gentry, Plate method for detection of phospholipase activity in Candida albicans, Sabouraudia, 20 : 7 (1982).
18. D. Pugh and R.A. Cawson, The cytochemical localisation of phospholipase A and lysophospholipase in Candida albicans, Sabouraudia, 13 : 110 (1975).

19. H. Remold, H. Fasold and F. Staib, Purification and characteri-
    sation of a proteolytic enzyme from <u>Candida albicans</u>, <u>Biochim.
    Biophys.Acta</u>, 167 : 399 (1968).
20. R. Rüchel, R. Tegeler and M. Trost, A comparison of secretory
    proteinases from different strains of <u>Candida albicans</u>,
    <u>Sabouraudia</u>, 20 : 233 (1982).
21. L.P. Samaranayake and T.W. MacFarlane, The effect of dietary
    carbohydrates on the in-vitro-adhesion of <u>Candida albicans</u> to
    epithelial cells, <u>J.Med.Microbiol</u>., 15 : 511 (1982).
22. B. Schreiber, C.A. Lyman, J. Gurevich and C.A. Needham, Proteo-
    lytic activity of <u>Candida albicans</u> and other yeasts,
    <u>Diagn.Microbiol.Infect.Dis</u>., 3 : 1 (1985).
23. N. Sharon and H. Lis, Lectins : cell-agglutinating and sugar-
    specific proteins, <u>Science</u>, 177 : 949 (1972).
24. J.D.Sobel, P.G. Myers, D.Kaye and M.C. Levison, Adherence of
    <u>Candida albicans</u> to human vaginal and buccal epithelial cells.
    <u>J.Infect.Dis</u>., 143 : 76 (1981).
25. F.Staib, Proteolysis and pathogenicity of <u>Candida albicans</u>
    strains, <u>Mycopath.Mycol.Appl</u>., 37 : 345 (1969).
26. E. Sundal, ed., Thymopentin in experimental and clinical
    medicine, in : "Survey of Immunologic Research", Vol. 4, Suppl.
    1, S. Karger A.G., Basel (1985).
27. R.B. Wickner, Killer of <u>Saccharomyces cerevisiae</u> : a double-
    stranded ribonucleic acid plasmid, <u>Bacteriol.Rev</u>., 40 : 757
    (1976).
28. D.R. Woods and E.A. Revan, Studies on the nature of the killer
    factor produced by <u>Saccharomyces cerevisiae</u>, <u>J.Gen.Microbiol</u>.,
    51 : 115 (1968).

HOST-PARASITE INTERACTION-MECHANISMS OF PATHOGENESIS

H.P.R. Seeliger   C. Patzelt

Institute of Hygiene and Microbiology, University of
Würzburg, Würzburg, The Federal Rebuplic of Germany

INTRODUCTION

The asymptomatic presence on mucous membranes of <u>Candida</u> cells, especially of <u>Candida albicans</u>, is widespread in healthy persons. A sensible disturbance of such host-parasite relationship has to occur to make this opportunist a microbial aggressor. While transitory superficial infections by <u>Candida</u> species represent the bulk of such disorders, serious clinical forms as well as generalized visceral invasion with sometimes fatal outcome may result. The bulk of clinical observations indicates that changes on the side of the affected carrier are primarily responsible in destabilizing this equilibrium: such as disturbances of cellular defence mechanisms and the immune system, the alteration of the normal microbial flora and the hormonal state. These factors seem to be primarily involved in the steadily increasing frequency of <u>Candida</u> infections. The growing incidence and the therapeutic constraints in severe <u>Candida</u> invasion represent an alarming problem in medical care. In order to develop prophylactic strategies, a profound understanding of the molecular events that accompany these disorders will be necessary.

The following is a brief survey dealing with known pathogenic factors of <u>Candida</u>, the human defence mechanisms directed against it, and the interactions between both of them in stable and destabilized host-parasite relationships.

PATHOGENIC FACTORS IN <u>CANDIDA</u>

The following factors are preferentially associated with the ability of <u>Candida</u> species to provoke human infections:

1) Adherence to mucous membranes (adhesion),

2) Invasion of the tissue by breaching the epithelial layers (penetration),

3) Transition from the yeast or blastospore <u>via</u> a pseudomycelial form to the hyphal or true mycelial form, the latter being considered as the state of higher virulence (dimorphism).

Like many pathogenic bacteria, <u>Candida</u> species are able to adhere to mucous membranes. The involvement of some component of the yeast cell wall in this process is self-evident. However, the true nature of this ligand is still unknown. In various experimental systems, monosaccharides did not show consistently an effect in competitively interfering with this binding process nor did obligosaccharides or lectins. Since enzymatic interference with Candida adhesion in vitro is rather demonstrable for proteases than for glycosidases, Candida adhesin may be a protein in nature. Therefore, the mannoprotein component of the Candida wall is now thought to represent the adhesive ligand with the protein component playing some crucial role.

The nature of the receptor on mucosal surfaces is even less clear. Among several proteins that show some affinity to Candida cell walls, fibrinogen has most often been discussed to function as an adhesion receptor. Whereas this may hold true for intravasal thrombi or plastic implants coated with a fibrin layer, the omnipresence of fibrinogen or fibrin on mucous epithelia cannot be anticipated. Alternatively, the widely distributed cell anchoring protein "fibronectin" equally shows a high affinity to Candida. However, its function as the sole adhesin receptor in vivo will need more experimental confirmation.

The ability of Candida to penetrate epithelial layers has been widely demonstrated by in vivo and in vitro systems. An involvement of cytopathogenic toxins in this process is not evident. Therefore, hydrolytic enzymes may preferentially be operative in tissue invasion. Whether the primary step for intrusion is a receptor-mediated uptake of the yeast cell, possibly initiated by adhesion, is not clear so far. From experimental observations, tissue breakdown cannot be attributed to one single Candida enzyme. There are several enzymes released from enzymes released from the yeast that might act together such as an acidic protease, a phospholipase A, a lysophospholipase as well as keratinolytic and collagenlytic enzymes. However, the contribution of these various activities as well as the involvement of still other enzymes in tissue invasion remains to be astablished.

An even more elusive matter is the role of Candida dimorphism in pathogenicity. Increased virulence of the hyphal form has been originally deducted from the prevalence of germ tube outgrowth and the formation of pseudomycelia and mycelia in infected tissue. Whether this morphologic transition is the cause of tissue invasion or simply the consequence of increased Candida proliferation in this particular environment is still controversial. It has been shown three decades ago that surface active agents, Tween 80 and reduced oxygen tension encourage in vitro the formation of pseudomycelium which seems to be the first important step of penetration. Changes in cell adhesion and in various metabolic activities have been reported to accompany the transition from blastospores to true hyphae. However, also the yeast form clearly seems to be able to adhere to and eventually to invade epithelia.

Taken together, all three mechanisms, i.e. adhesion, penetration and morphologic transition, contribute to Candida pathogenicity. Experimental data on these phenomena are in accordance with the known pathogenicity of various Candida species. Candida albicans, as the most frequent pathogenic species due to clinical observations, also is the most active one with respect to the discussed phenomena. Nevertheless, additional experimental work will be necessary to unravel the underlying molecular mechanisms in more detail.

HOST DEFENCE MECHANISMS

The human body protects itself from Candida invasion, like from invasion by other infectious agents, through the concerted action of nonspecific defence mechamisms. In addition, also a specific immune response directed against Candida can be observed.

The nonspecific mechanisms include inflammation and phagocytosis. The latter process has amply been studied in vitro. Primarily polymorphnuclear leukocytes (PMNs) and monocytes are involved in Candida uptake and destruction. Moreover, there is good evidence that this process is initiated by the alternative complement pathway activated by the yeast mannan. Via attachment of the C3-component Candida cells are tagged (opsonized) as target for phagocytosis. Highly reactive ions and radicals that are generated by myeloperoxidase in phagocytic vacuoles, have a fungicidal effect on Candida. The final hydrolytic break-down will occur through the action of lysosomal enzymes.

Other components of the complement system, i.e. C3a, C4a and C5a activate mast cells that initiate the inflammatory response including increase in capillary permeability, tissue edema and cell infiltration.

With respect to the specific immune response, cellular defence mechanisms apparently dominate over humoral ones in Candida invasion. Like for other fungal infections, this response consists in a delayed-type hypersensitivity (DIH). Specific activation of T-cells requires processing of fungal antigens and their presentation by MHC-positive accessory cells. Effector cells that initiate DIH are among the CD4-population of lymphocytes. Through the action of lymphokine (e.g. interleukin 2) programmed T- lymphocytes are stimulated to proliferate. Their released soluble factors (e.g. macrophage activation factor) initiate efficient phagocytosis and activation of natural killer cells. Whether this scheme applies in every respect to Candida infections remains to be established. Experimental activation of the monocyte-macrophage lineage has resulted in non-phagocytosing monocytes with strong Candida-killing propensity.

The often reported immunosuppresive action of Candida appa-rently is also related to its mannan component. The cellular immune response outlined above is thus modulated by activation of a specific suppressor cell subset.

In candidomycotic patients as well as in experimental animals exposed to Candida, titers of specific circulating antibodies are often low. Though a humoral response can clearly be demonstrated, immunoglobulins may only have some supportive role in defence. Their function can be imagined in neutralizing released Candida enzymes or in blocking adhesins. Whether such effects can be attributed to surface IgA remains unknown. Moreover, specific antibodies may be able to tag Candida cells for antibody-directed cellular cytotoxicity of phagocytosis.

Still another protective system of apparently high efficiency is represented by the normal microbial flora. There is no evidence so far that bacteria normally colonizing mucous membranes do compete with Candida in adhesion on epithelial surfaces. Therefore, their protective effect may simply be due to competition for nutrients in the respective microenvironment keeping thus Candida proliferation under control.

HOST-PARASITE INTERACTIONS IN THE INCIDENCE OF CANDIDAMYCOSIS

A delayed-type hypersensitivity response (type IV hypersensitivity reaction) to Candida antigen preparation is very common in healthy persons. It indicates that already the carrier state is associated with an immune response in contrast to the silent tolerance of other commensal microorganisms.

The eventual persorption of Candida cells from the intestinal tract into the blood stream may trigger this reaction. Nevertheless, an encounter between Candida and immune cells on the surfaces of mucous membranes is equally conceivable.

The wealth of clinical observations in immunocompromised patients gives evidence that the equilibrium between the parasite and the immune surveillance is of delicate nature. Apparently, all alterations of the immune system seem to favour the incidence of Candida infection. Its most severe form, namely generalized visceral infection, appears to be primarily associated with dysfunction or paucity of PMNs, as can be seen in patients with agranulocytosis.

There is also ample evidence that iatrogenically induced changes of the normal microbial flora can promote Candida infection. Since immunocompromised patients often are treated with broad spectrum antibiotics for therapeutic or prophylactic purposes, both, i.e. immunodeficiency and changes of the microbial terrain, may frequently act together as aggravating factors. The increasing invasive application of plastics which do represent a preferred target for Candida adhesion, may be a third factor making these patients particularly prone to Candida infection.

Although these disturbances of the host-parasite relationship are not yet understood in every detail, their causative role in candidamycosis is evident and generally accepted. In contrast, changes in the host's metabolic or hormonal state are far less well understandable as pathogenic co-factors. Vaginal infestation with Candida is widespread in women during their reproductive age. The incidence is particularly high in pregnant women during the last trimester and in those on oral contraceptives. Changes of the vaginal pH, the epithelial glycogen content, the progesterone level, the normal microbial flora or still unkonwn mechanisms may cause increased proliferation and subsequent adhesion of Candida. Similarly, the high incidence of Candida infection in diabetic patients is still without satisfying explanation. Disorders of the diabetic immune system, though often cited, need more specification to substantiate their possible pathogenic role. An abundance of glucose on mucous membranes in poorly controlled diabetics may be considered an additional factor in favouring Candida cell proliferation. Several other endocrinologic disorders, e.g. those concerning th thyroid or the adrenal glands, seem to promote candidamycoses. Taken together, there is a striking connection between hormones and Candida pathogenicity. Therefore, some direct action of the host's hormonal system on the biology of Candida should be envisaged.

Whereas all changes discussed so far concern the host, alterations in the parasite's state have been noticed only infrequently in context with Candida infection. It is generally assumed that a given strain, which already may be acquired during delivery, may persist for the whole life and eventually causes overt infection. A few reports, however, give evidence for strain variability during prolonged Candida infection. Criteria for this variability are patterns of sugar assimilation, phenotype switching and DNA analyses. Therefore, the possible existence of strains of different virulence deserves attention. These strains may superseed a preexisting Candida population of low virulence causing thus a destabilization of a balanced host-parasite relationship.

Modern methods in molecular genetics will hopefully open new routes for research on Candida pathogenicity and virulence. Experimental systems employed so far suffer from some severe methodological constraints. Thus, parasite variables to be introduced in these systems are limited to different Candida strains and rarely observed mutants. Genetically manipuled organisms, e.g. yeast of low virulence transfected with genes for possible pathogenic factors (adhesins, enzymes), can be imagined to be soon available as research tools. Approaches of this type are likely to add a new dimension in experimental versatility and in our understanting of Candida pathogenicity.

REFERENCES

1. R.D.Diamond, Mechanisms of host resistance to Candida albicans, in: "Microbiology", D.Schlessinger, ed., American Society for Microbiology, Washington, D.C. (1981).
2. J.E.Edwards et al., (1978): Severe Candida infections: Clinical perspective, immune defense mechamisms, and current concepts of therapy, Ann.Intern.Med., 89: 91 (1978).
3. J.B.Epstein et al: Oral candidiasis: Pathogenesis and host defense, Rev.Infect.Dis., 6: 96 (1984).
4. Y.Fukazawa and K.Kayaga, Host defence mechanisms against fungal infections, Microbiol.Sci. 5: 124 (1988).
5. S.A.Klotz al., Adherence and penetration of vascular endothelium by Candida yeasts, Infect.Immun., 42: 373 (1983).
6. J.C.Lee and R.D.King, Adherence mechanisms of Candida albicans, in: "Microbiology", D.Schlessinger ed., P.269, American Society for Microbiology, Washington, D.C. (1983).
7. P.Lehrer et al., Phagocytosis in: "Microbiology", D.Schlessinger, ed., p.273, American Society for Microbiology, Washington, D.C. (1983).
8. F.C.Odds, (1988): "Candida and Candidosis", Baillière Tindall, London (1988).
9. T.J.Rogers and E.Balish, (1980): Immunity to Candida albicans, Microbiol.Rev. 44: 660 (1980).

CANDIDOSIS IN HEROIN ADDICTS AND AIDS:

NEW IMMUNOLOGIC DATA ON CHRONIC MUCOCUTANEOUS CANDIDOSIS

E. DROUHET   B. DUPONT

Institut Pasteur, Unité de Mycologie, Paris, France

INTRODUCTION

Recent reports illustrate that the problem of opportunistic fungal infections is ever-increasing. Invasive fungal infections have been reported in more than 25 % of patients who are chronically and intensively immunosuppressed by reason of underlying diseases and modern drug therapy. However, since 1980 the incidence and severity of opportunistic mycoses, particularly Candida infections has been even higher because of new pathological conditions produced by AIDS[1-4] and heroin addiction[5-7].

There have been many developments in our understanding of immune systems and of factors leading to their disturbance, in recent years. The nonspecific immune mechanisms, including cellular (i.e. phagocytic leukocytes) and humoral (i.e. immunoglobulin production, complement system) and the specific cell mediated and humoral immunity all play important, but unequal roles in host defences against Candida albicans, which remains a growing clinical problem, subject of a rich literature[8].

New pathological aspects of candidosis have appeared since 1980 in heroin addicts, AIDS patients and patients with other immuno-suppressive conditions.

The new septicaemic syndrome of C. albicans infection in heroin addicts is characterized by fever, followed immediately by cutaneous disseminated lesions (folliculitis with hair invasion by candidal hyphae not previously described in classical systemic candidosis, pustulosis, deep seated scalp nodules), associated with ocular and osteoarticular lesions. We first described an outbreak of this form of infection in detail in the early 1980's[5-7] and there have now been 250 reported cases in France. In Spain there have been 280 published cases[9-15] and it has been reported in Italy[16], Belgium[17], Great Britain[18-20] and even in Australia[9,21]. It has not been reported in the U.S.A.. The outbreak coincided with the introduction of a new brown heroin, of Asian origin. C. albicans was not isolated from the drug, but rather, was introduced by a contaminated syringe from the common spoon in which the crude heroin was dissolved (all these patients are C. albicans carriers). The role of lemon juice used by some to dissolve the heroin in questionable [8,18,20] but an immunosuppressor contaminant of the brown heroin may be at the origin of this new pathology [1,7]. Alterations of T and null lymphocyte frequencies in the peripheral blood of human opiate have been observed with evidence _in vivo_ for opiate receptor sites on T lymphocytes[22]. Elevated titres of antibodies to _C. albicans_ are in contrast to the low titres usually found in immunosuppressed patients with disseminated candidosis. The syndrome is, in fact, now decreasing because of the use of individual sterile syringes and to the use of purified heroin.

The oropharyngeal and oesophageal candidosis observed since long ago during antibiotic therapy, chemotherapy-induced agranulocytosis of cancers and leukaemia, is the principal marker of AIDS or ARC patients related to the cellular immunodeficiency[2,4,23,24]. No systemic candidosis is observed in AIDS, as in chronic mucocutaneous candidosis (CMCC), where thrush and oesophagitis are severe. In both syndromes (AIDS, CMCC) protective andibodies to a 47 KD antigen _C. albicans_ might explain the absence of invasive candidosis[25].

Immunological disorders observed in CMCC patients include an impairment of cell-mediated immunity to _C. albicans_ antigens, as demonstrated by an absence of delayed type hypersensitivity, and an abnormal _in vitro_ proliferative response to _C. albicans_ antigens contrasting with normal or increased levels of anti-_Candida_ antibodies. Mannan, the main polysaccharidic component of the cell wall of _C. albicans_, has been shown to exert an immunosupressive effect _in vivo_ and _in vitro_ on T cells in most patients with CMCC[26,27]. _In vivo_, the antigen-mediated T cell suppression disappeared in patients cured by prolonged effective therapy with ketoconazole or amphotericin B. Normal cellular responses to Candida antigen have now been restored for more than 10 years in most of these patients.

NEW SEPTICAEMIC CUTANEOUS, OCULAR AND OSTEOARTICULAR _CANDIDA ALBICANS_ SYNDROME IN HEROIN ADDICTS

Since 1940, when Joachim and Polayes reported the first case of _Candida_ endocarditis as a complication of heroin addiction, numerous cases of fungal endocarditis and endophthalmitis after intravenous drug abuse have been reported[28].

Out of a total of 319 cases of fungal endocarditis reviewed, 25 % are among heroin abusers. The majority are due to _Candida_ species, fewer to _Aspergillus_ and in rare instances to other fungi, but in most of these latter cases the source of fungi was contaminated by heroin and syringes. Among 55 cases of _Candida_ endocarditis in drug addicts, _C. albicans_ was isolated only in 5 cases, _Candida_ species other than _C. albicans_ in 46 cases, and _Candida sp._ (unidentified) in 4 cases. By contrast, in endophthalmitis, exclusively _C. albicans_ is observed among heroin users.

The epidemic new syndrome. The septicaemic cutaneous, ocular and osteoarticular syndrome of candidosis that we have observed since 1980[5-7] is produced by _C. albicans_ exclusively, but the samples of heroin seized from the drug addicts and the powder seized from the drug dealers were negative for _C. albicans_.

Between June 1980 and December 1982 we first observed 11 cases among heroin addicts in the Paris region (reported in 1981)[5], then 34 cases reported in 1983[6] and in 1985[7] we gave complete details of 38 cases. This febrile septicaemic syndrome begins with elevated temperature, chills, severe headache, and profuse sweating, beginning 2 to 24 hours after drug injection, in contrast with the self-healing syndrome known as "dust" by heroin addicts; this episode of 1-3 days is followed by metastatic cutaneous lesions (disseminated folliculitis or pustulosis in hairy zones, deep seated scalp nodules) and by ocular localizations (mainly chorioretinitis) appearing generally between 3-5 days. The cutaneous nodules, 0.5 - 1 cm in diameter, appear suddenly, in great number (up to 100) and the scalp feels sometimes like a sack of marbles. It is very painful to the patient. Numerous pustules, 2-3 mm in diameter, with frank pus (containing yeasts and filaments of C. albicans) and disseminated folliculitis on the scalp, face and other hairy zones (chest, axillae, pubic regions, thighs) suggesting a streptococcal or staphylococcal aetiology accompany the scalp nodules. A transitory alopecia is observed.

In the biopsy of follicular lesions a marked inflammatory involvement is observed with septated, bifurcated filaments of C. albicans. The disseminated cutaneous lesions associated with ocular and osteoarticular lesions have not previously been described in classical systemic candidosis.We also observed hair invasion by candidal hyphae, the aggressive form of C. albicans (the only species found in this syndrome).

The ocular involvement (chorioretinitis, uveitis, hyalitis, even panophthalmia) is the most serious component of this syndrome, because it can result in loss of vision in the affected eye.

Subsequently (between 15 days and 5 months), osteoarticular lesions (vertebrae, costal cartilages, knees, sacroiliac articulations), particularly spondylodiscitis appeared.

Since our first description[5-7] other cases have been reported in France, with cutaneous[29,30] and ocular and osteoarticular manifestations[7,8]. Other cases have been observed in Italy[16], but the most important epidemics have occurred in Spain, where in 2 years between 1982-1984 no fewer than 236 cases were reported and published[9-15] (the actual number might be up to 300 by now). Most cases presented cutaneous disseminated lesions (80-100 %) : nodular (100 %), follicular(52 %), abscesses (22 %), alopecia (20 %) localized at the the scalp (100 %), beard (57 %), axillae (23 %), pubis (15 %), other hairy regions (5 %). The ocular lesions occurred in 40-100 % of Spanish epidemic series and the osteoarticular in 9 to 40 % of cases. Other European cases have been published in Switzerland[31], Belgium[17], England[8,18] but overseas only in Australia,[21,32] where Collignon et al.[21] reported initially 7 well-documented cases (7 with cutaneous and ocular lesions, 3 with osteoarticular localizations) and 2 additional cases. No cases have been reported in the U.S.A., where usually the white, well-purified heroin is used.

C. albicans is the only species responsible of this syndrome[7]. The serotype A, biotype 15 3/7 was found exclusively among 21 intravenous heroin abusers from Madrid or Valencia with cutaneous and ocular candidosis[33], while in a control group the prevalence of this serotype and biotype was only 11.1 % (5 out 45)[30].

Candida endophthalmitis on its own is not a novel condition in intravenous drug abusers and C. albicans arthritis and osteoarthritis of haematogenous origin were also known in at least two heroin addicts in 1981[8]. However, the febrile septicaemic syndrome, with disseminated cutaneous follicular and nodular lesions associated in most cases simultaneously with chorioretinal ocular lesions and later with osteoarticular localizations (mostly spondylodiscitis) was a new epidemic event and was most often related to a brown (occasionally white) heroin of Asian origin. Recently, anecdotal reports[34,35] have mentioned cutaneous, ocular and osteoarticular candidosis, in patients who are not heroin addicts: Aguado et al.[34] reported a case in which a young man of 25 years, with parenteral alimentation for acute pancreatitis had septicaemia, chorioretinitis and chondrocostal abscesses and folliculitis with C. albicans, but the case of Feuillade Chauvin[35] showed only folliculitis of the beard and scalp after treatment with corticotherapy and parenteral alimentation for chronic disease.

In our first cases[5-7] we mentioned the important humoral immunological response to C. albicans antigen using counterimmunoelectrophoresis. Bisbae et al.[36], in a recent well-documented study, showed high indirect haemagglutination and immunofluorescence titres of antibodies but no C. albicans-mannan circulating antigen in sera of heroin addict patients with candidosis.

The lemon juice used by some drug abusers as the heroin solvent has been considered by various authors[8,18,20] as the likely source of infection. Those reports speculate or give anecdotal evidence that lemon juice is the likely source. The fact that C.albicans is able to grow in lemon juice is not proof that the juice is inevitably the source of infection; indeed some strains are not able to grow in lemon juice or vinegar[7] and dissolution of heroin in lemon juice contaminated with C. albicans has a sterilizing effect.

Because C. albicans is a natural resident of the gastrointestinal tract, there is a possibility that the organisms spread haematogenously from a gastro-intestinal source to produce the severe mouth infections we have found in most addicts. But the heroin factor is the new determinant for concomittant, nodular and follicular disseminated lesions in the other localizations. An immunosuppressor contaminant of brown heroin (quite exclusively implicated in this syndrome) may also be the origin of this syndrome, another characteristics of which is the sudden transformation of the yeast of C. albicans into the mycelial aggressive form observed in the hair of the pilous follicles. A few cases of localized folliculitis were observed in the past[7] after topical steroid therapy. This new pathological syndrome may be related to a depressive action of Candida on cellular immunity: alterations of T and null lymphocyte frequencies in the peripheral blood of human opiate receptor sites on T lymphocytes have been reported[22] as well as immunological dysfunction[37].

CANDIDOSIS IN AIDS PATIENTS

Among the opportunistic fungi implicated in AIDS, Candida albicans infections, with oral thrush and oesophagitis as clinical manifestations are the most frequent. The high incidence varies between 40 to 90 % according to the risk groups. The first clinical definition of AIDS (1981) as well as the most recent (revised in 1988) by CDC/WHO recognized candidosis of the oesophagus, trachea, bronchi and lungs and meningeal cryptococcosis as "major" opportunistic infections (category Cl) and important "indicator" diseases.

In 1986 the Walter Reed Army Institute of Research[24] adopted a staging classification for HIV infection, applicable to adults only (Table 1), based on HIV antibodies and/or virus isolation, chronic lymphadenopathy, T helper cells/mm^3, delayed hypersensitivity, appearance of thrush and finally other opportunistic infections. The quantitative depletion to <400 T helper cells/μl (CD4+), target of HIV virus, with cutaneous anergy to C. albicans antigen and development of thrush, indicate the stage WR 5 (corresponding to ARC IV C2), preceding the final stage WR6 (AIDS). A reduction in CD4 + lymphocyte count, in combination with oral candidosis, indicates a poor prognosis with HIV infection. A CD4+ less than 400/ml indicates severe immunodeficiency, a hallmark of advanced HIV infection, which according to Moss represents a 90 % risk of progression in 3 years.

Table 1. Walter Reed staging classification for HIV infection.

	HIV antibody or virus +	CLA	CD4	DHS	Thrush	Opport. infection
WR 0	−	−	>400	N1	−	−
WR 1	+	−	>400	N1	−	−
WR 2	+	+	>400	N1	−	−
WR 3	+	+	<400	N1	−	−
WR 4	+	+	<400	p	−	−
WR 5	+	+	<400	C or/and	+	−
WR 6	+	+	<400	Pc	+	+

CLA: Chronic lymphadenopathy. CD4: T helper cells/mm^3. DHS: Delayed hypersensitivity. NL: Normal P: Partial cutaneous anergy, defined as an intact cutaneous response to only one of the four test antigens: Candida, Trichophyton, tetanus, mumps. C: Complete cutaneous anergy to the four antigens.

The thrush, defined as clinical oral candidosis (creamy-white, curd-like patches on the tongue or other oral mucosal surfaces) may extend to the oesophageal mucosae to produce oesophagitis, diagnosed by chest pain, dyspagia, endoscopy, radiology and mucosal biopsy. Whereas, the prevalence of oral thrush in AIDS is between 40 90 %, the prevalence of oesophageal candidosis is only between 4 to 14 %. In patients with manifest AIDS, oral thrush is a marker of oesophageal candidosis[4]. However, the absence of oral thrush does not exclude oesophageal candidosis, which can be clinically inapparent, but proven by biopsy.

It is notable that the impaired cellular immunity in chronic mucocutaneous candidosis (CMCC) and the defective T cell cellular immunity in AIDS are accompanied by mucosal C. albicans involvement only, without further invasion or disseminated candidosis. Such invasive, disseminated candidosis occurs only when a profound granulocytopenia occurs or when other underlying risk factors are present (intravenous catheters, corticotherapy, etc.). In AIDS and in CMCC the polymorphonuclear leukocyte counts and phagocytic functions are generally unaltered, but recently Matthews et al.[25] found high levels of antibodies against a 47 KD antigen of C. albicans, which they considered as protective for AIDS and CMCC patients. A polyclonal activation of B cells is observed in AIDS as well as in CMCC (an argument will develop further).

Fungal infections such as candidosis and cryptococcosis may play a cofactor role in the development of AIDS. It is well known that the polysaccharide antigen of C. albicans, the mannan of the cell wall, has an immunosuppressor effect on T lymphocytes in chronic mucocutaneous candidosis[26,27,38]. The candidosis may itself aggravate a pre-existing T lympocyte deficiency. Even in non-CMCC, but in severe chronic oral candidosis in the absence of HIV, a reduced CD4 + T cells (<400/ml), indicating a severe immunodeficiency, may be observed[39].

CHRONIC MUCOCUTANEOUS CANDIDOSIS (CMCC)

Clinical and histopathological data

Chronic mucocutaneous candidosis (CMCC) (name proposed by Children, 1967) was described as monilial (Candida) granuloma by Rothman and Hauser in 1950, granulomatous candidosis. It is a rare disorder of childhood characterized clinically by recurrent chronic C. albicans infections of the mucous membranes and skin. The lesions progress from erythema to hyperkeratosis and crusts. Granulomatous lesions predominate on the scap and face; pranoychia and onychia may occur.

Histologically, CMCC is characterized by hyphal invasion of mucosal epithelium and by dermal inflammatory granulomas around the yeasts and filaments of C. albicans. Cellular immunity disorder are often associated with genetic and endocrinologic alterations[8]. However, we have observed in recent years numerous children and young adolescents with CMCC but with no endocrinological or genetic alterations. The CMCC developed from oral thrush following antibacterial antibiotic therapy that was resistant to classical antifungal treatment[6,40]. The chronic thrush became accompanied by paronychia with nail involvement of hands and feet and by granulomatous parakeratotic and hyperkeratotic crusting lesions affecting the face, scalp, ears, neck, shoulders, etc..

Alterations in cell-mediated immunity

While invasive C. albicans infections are observed in neutropenic subjects, probably because phagocytic cells are necessary to kill C. albicans yeasts, mucocutaneous C. albicans infection are found in patients with an impairment of cell-mediated immunity to C. albicans antigens in particular. The impairment takes the form of

- negative delayed-type hypersensitivity in skin tests with C. albicans antigens (Joulia and Le Couland, 1939, see ref. 8), an abnormal finding.
- deficiency in lymphocyte transformation to C. albicans antigens in vitro, although most of CMCC tested patients, responded normally to mitogens.
- increased antibody production to C. albicans antigens, especially mannan[26,27,38], which is now considered as a polyclonal activation of B cells, related to a deficient T-lymphocyte regulation[26,38].
- detection in a large group of patients of a serum inhibitory activity specifically blocking T-cell responses to Candida. The serum inhibitor has been identified as mannan[27] and the free circulating mannan could block the mannan antigen

66

Table 2. Cellular responsiveness to <u>Candida</u> antigens and other antigens of 6 patients with CMC befora and after ketoconazole (Ktz) treatment.

| Patient | Circulating inhibitory factor (mannan) | | Before Ktz intradermal reaction to candidin | | After Ktz intradermal reaction to candidin | | Transformation lymphocyte test | | | |
| | | | | | | | C. albicans | | PHA | Con A |
	Before Ktz	After Ktz	30'	48h	30'	48h	Before Ktz	After Ktz	Before Ktz	After Ktz
1 G.H.	+	−	+	−	+	+	−	+	+	+
2 S.G.	−	−	−	−	+	+	−	−	+	+
3 A.V.	+	−	+	−	+	+	−	+	+	+
4 B.C.	+	−	++	−	+	+	−	+	+	+
5 G.F.	+	−	+	−	+	+	−	+	+	+
6 L.F.	−	−	+	−	+	+	−	+	+	+

After Drouhet et al.[40]

presentation by monocytes to T lymphocytes - a suppressor T-cell activity was induced by antigens of <u>C. albicans</u> <u>in vitro</u>[26,38] and is comparable to the suppressor T cell activity observed <u>in vivo</u> in patients affected with CMCC. Successful treatment by a strong prolonged anifungal therapy such as ketoconazole[6] resulted in a complete correction of the immune abnormalities, primarily induced by mannan antigen accumulation (Table 2).

Mannan, as immunosuppressor of cell-mediated immunity in vivo and in vitro

In CMCC patients, the serum factor inhibitory of normal sensitivity to <u>C. albicans</u> antigens in the T lymphocyte transformation test <u>in vitro</u>, was identified as circulating <u>C. albicans</u> mannan by Fischer et al.[27]. This free mannan which accumulates in patients with active <u>C. albicans</u> infections, exerts a specific inhibitory effect in the cellular response to <u>C. albicans</u> antigen.

In addition to the serum inhibitory activity due to circulating mannan, Durandy et al.[26,38] also observed a cellular suppressor acitivity mediated by T lymphocytes. In CMCC patiens in the acute phase of the disease, they found an unresponsiveness of T lymphocytes to mannan antigen, that led to an absence of <u>in vitro</u> anti-mannan antibody production (Fig.1).

The lack of T cell proliferation was secondary to the presence of T suppressor (TS) lymphocytes (Fig. 2). A paradox remains that the suppressor T cells induced <u>in vitro</u> in normal individuals and found <u>in vivo</u> in CMCC patients, are able to block strongly the <u>in vitro</u> antibody response to mannan while <u>in vivo</u> antimannan antibody product is high in CMCC patients. Some experimental data[38] suggest that B cell activation by mannan requires only T cells to be amplified by the production of lymphocytes.

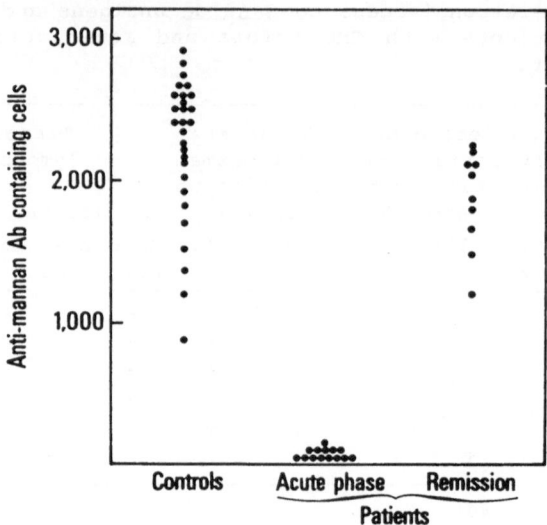

Fig. 1. Absence of anti-mannan antibody production in patients with CMCC in acute phase. After Durandy et al.[26]

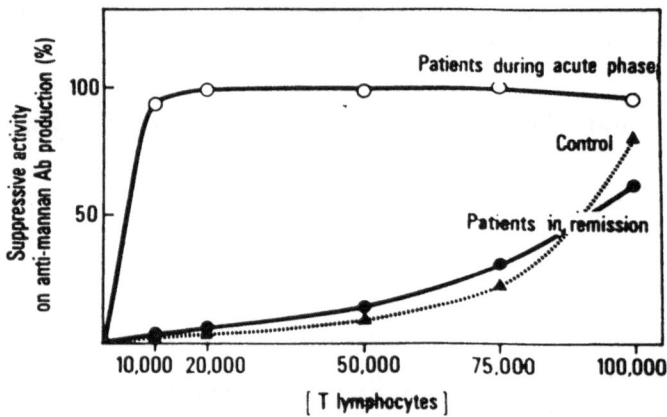

Fig. 2. Evidence for a T suppressor lymphocyte activity in patients in acute phase of CMCC. After Durandy et al.[26]

Polyclonal activation of B cells in CMCC

The hypergammaglobulinaemia observed in most patients with CMCC showing anti-C. albicans serum antibodies of the IgM, IgG or IgA classes, and demonstrated by immunoelectrophoresis, immunofluorescence or enzymoimmunological (ELISA) techniques, indicates a process of B lymphocyte polyclonal activation similar to that observed in AIDS patients. The various B lymphocyte abnormalities seem to be a consequence of a deficient T-lymphocyte regulation.

Recently, Matthews et al.[21] showed, by immunoblot technique, a major IgM and IgA response to the 47 KD antigen of C. albicans in sera of patients with CMCC or with oral and oesophageal candidosis with AIDS. This is a contrast with sera from fatal cases of systemic candidosis or other non-AIDS patients. In systemic C. albicans infections only those patients maintaining a good antibody response to the immunodominant 47 KD antigen survive. Matthews et al.[25] conclude that the reason is that the antibodies against 47 KD can be protective.

The absence of dissemination in CMCC and in mucosal candidosis in AIDS may be because of the elevated level of these antibodies, in addition to other factors such as the normal number and functions of polymorphonuclear cells.

Restoration of alterated cell – immunity by strong prolonged antifungal chemotherapy

Chronic mucocutaneous candidosis, a useful model in the study of cellular immunology[41], can be considered also as a model for understanding and evaluating antifungal therapy[6]. In the past, intravenous amphotericin B was the primary drug treatment used to prevent the evolution of this chronic disease, but relapses occurred frequently within a few months or years. Reversal of anergy was reported in some cases.

The advent of systemically active oral azole antifungal therapy has revolutionized the management of CMCC and permitted a complete and permanent restoration of the cellular abnormalities suggesting that they are not due to an intrinsic lymphocyte dysfunction. Thus, in 6 documented cases of children and adults with CMCC of varying ages (6 to 22 years), a remarkable cure has been obtained without relapse for up to 10 years for most patients. The cure is manifested by rapid disappearance of circulating mannan inhibitor, reappearance of lymphocyte transformation responses in vitro, reappearance of cutaneous allergy to C. albicans antigens and reduction of precipitating antibodies against C. albicans to undetectable levels. The restoration of the immunity and the absence of relapses are due to prolongation of treatment after the clinical and mycological cure. However, various authors[41] have reported some relapses, in instances of to strains less sensitive to azoles or of particular host conditions. Various types of immune reconstitution therapy have been attempted, but currently strong prolonged antifungal therapy seems to be effective.

Proposed mechanisms of the specific immune deciency observed in chronic mucocutaneous or mucosal candidosis

As shown in Fig. 3, mannan seems to be the major cause of the specific immune deficiency observed in CMCC. When for any reason, such as defective handling of mannan by monocytes[26,38], mannan accumulates in infected patients, it would induce a transient but strong specific suppressor effect.

Since mannan or cell wall glycoprotein of C. albicans[26,38] can suppress the cellular immune response, so candidosis in AIDS patients

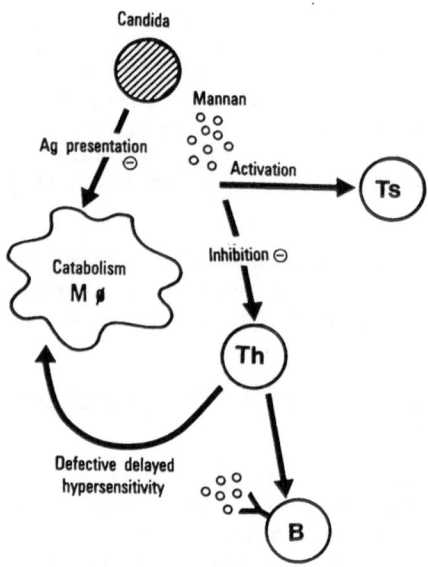

Fig.3. Model for immune abnormalities in patients with CMCC.

may itself aggravate a pre-existing T lymphocyte deficiency. If candidosis is a cofactor in the development of a manifest acquired immunodeficiency syndrome, antifungal therapy might be of particular benefit to HIV-infected patients with mucosal candidosis.

REFERENCES

1. E.Drouhet, B.Dupont, Mycotic infections complicating heroin adaddicts, AIDS and other immunocompromised host conditions, Ann. Ist. Super. Sanita, 23:735 (1987).
2. R.S.Klein, C.A.Harris, C.B.Small et al., Oral candidiasis in high-risk patients as the initial manifestation of the acquired immunodeficiency syndrome, New Engl. J. Med., 311:354 (1984).
3. C.Pedersen, J.Gerstoft, B.O.Lindhardt, J.Sindrup, Candida esophagitis associated with acute human immunodeficiency virus infection, J. Infect. Dis., 156:529 (1987).
4. A.Tavitian, J.P.Raufman, L.E.Rosenthal, Oral candidiasis as a marker for esophageal candidiasis in the acquired immunodefeciency syndrome, Ann. Intern. Med., 104:54 (1986).
5. E.Drouhet, B.Dupont, C.Lapresle, P.Ravisse, Nouvelle pathologie: candidose folliculaire et nodulaire avec des localisations ostéoarticulaires et oculaires au cours des septicémies à Candida albicans chez les héroinomanes. Mono et polythérapie antifongigue, Bull. Soc. Fr. Mycol. Med., 10:179 (1981).

6. E.Drouhet, B.Dupont, Laboratory and clinical assessment of keto-conazole in deep-seated mycoses, Am. J. Med., 74 (suppl. 1B):30 (1983).

7. B.Dupont and E. Drouhet, Cutaneous, ocular, and osteoarticular candidiasis in heroin addicts: new clinical and therapeutic aspects in 38 patients, J. Infect. Dis., 152:577 (1985).

8. F.C.Odds, "Candida and Candidosis. A Review and Bibliography, Second edition, Baillière Tindall Publ., London (1988).

9. Y.Cruces Prado and C. de la Torre Fraga, Folliculitis in a couple of heroin addicts, Revista Iberica di Micologia, 3 (Suppl. 1):579 (1986).

10. J. de la Cuadra Oyanguren, A. Aliaga Boniche, J.L. Sanchez Carazo, Skin disorders in systemic candidiasis of parenteral drug addicts, Revista Iberica de Micologia, 3 (Suppl. 1):S67 (1986).

11. J.M.Mascaro, C.Herrero, I.Bielsa, M.Cedra, Candidiasis en drog-adictos por via parenteral, Revista Iberica de Micologia, 3 (Suppl. 1): 575 (1986).

12. R.Perez Vidal, L.Forcé, A.Verdaguer, J.M.Torres Rodriquez, Giménez Camarasa, Candidiasis sistemica en heroinomanos. Resentacion de 13 cases, Revista Iberica de Micologia, 3 (Suppl. 1): S85 (1986).

13. D.Podzamczer and F.Gudiol, Systemic candidiasis in heroin abusers, J. Infect. Dis., 153:1182 (1986).

14. L.Zubiri, C.Cocojuela, J.Pinol, F.J.Carapeto, Candidiasis disemi-nada en heroinomanos. Estudio de ocho casos, Acta Dermo-sif., 78:345 (1987).

15. F.Vanaclocha Sebastiaan, R.Gil Martin, R.Diaz Diaz, P.Jaen Olasolo, L.Iglesias Diez, Systemic candidiasis in heroin ad-dicts: Clinical and pathological study of 22 patients, Revista Iberica de Micologia, 3 (Suppl. 1):S57 (1986).

16. A.Lopoz, N.Vescia, A.De Carolis, G.Gualdi, Le candidosis sistemiche dell heroinomane possono essere imputabili ad in quinaments delle miscelle stupefaceili? Nuovi Annali di Igiene e Microbiol., 34:109 (1983).

17. M.Etienne, A.Nemery, J.M.Darcis, G.E.Pierard, J.Demonty, Disseminated candidiasis in heroin addicts, Acta Clinica Belgica, 41:18 (1986).

18. R.J.Hay, Systemic candidiasis in heroin addicts, Br. Med. J., 292:1096 (1986).

19. I.F.Rowe, E.D.Wright, C.S.Higgens, J.P.Burnie, Intervertebral infection due to Candida albicans in an intravenous heroin abuser, Ann. Rheum. Dis., 47S6:522 (1988).

20. G.S.Shankland and M.D.Richardson, Possible role of preserved lemon juice in the epidemiology of Candida endophtalmitis in heroin addicts, Eur. J. Clin. Microbiol. Infect. Dis., 8:87 (1989).

21. P.J.Collignon and T.C.Sorrell, Candidiasis in heroin abusers, J.Infect. Dis., 155:595 (1987).

22. R.J.McDonough, J.J.Madden, A.Falek, D.A.Shafer, M.Pline, D.Gordon, P.Bokos, J.C.Kuehnie, J.Mendelson, Alteration of T and null lymphocyte frequencies in the peripheral blood of human opiate addicts: in vivo evidence for opiate receptor sites on T lymphocytes, J.Immunol., 125:2539 (1980).

23. B.Dupont and E.Drouhet, Fluconazole in the management of oro-pharyngeal candidosis in a predominantly HIV antibody-positive group of patients, J. Met. Vet. Mycol., 26:67 (1988).

24. R.R.Redfield, D.C.Wright, F.C.Tramont, Walter Reed staging classification for HTLV-III/LAV infection, New Engl. J. Med., 314:131 (1986).

25. R.Matthews, J.Burnie, D.Smith, J.Midgley, et al., *Candida* and AIDS. Evidence for protective antibody, *Lancet* (July 30): 263 (1988).

26. A.Durandy, A.Fischer, E.Drouhet, C.Griscelli, Mannan antigen of *Candida albicans* and cellular immune responses *in vitro* and *in vivo*, *in*: "Fungal Antigens", E. Drouhet, G.T. Cole, L. De Repentigny, J.P. Latgé, B. Dupont, eds., Plenum Press, New York(1988).

27. A.Fischer, J.J.Ballet, C.Griscelli, Specific inhibition of *in vitro Candida*-induced lymphocyte proliferation by polysaccharidic antigens present in the serum of patients with chronic muco-cutaneous candidiasis, *J. Clin. Invest.*, 62:1005 (1978).

28. R.McLeod and J.S.Remington, Fungal endocarditis, *in*: "Infective Endocarditis", S.H. Rahimtoola, ed., pp. 211-290, Grune and Stratton, New York (1978).

29. G.Badillet, P.Pietrini, A.Puissant, Pustuloses chez des héroïnomanes, *Ann. Dermatol. Vénéréol.*, 110: 691 (1983).

30. M.Feuillade de Chauvin and R.Touraine, Folliculite a *Candida albicans* chez trois heroïnomanes, *Bull. Soc. Franc. Mycol. Med.*, 10:179 (1983).

31. T.Calandra, P.Francioli, M.P.Glauser, F.Baudraz-Rosselet, C.Ruffieux, D.Grigoriu, Disseminated candidiasis with extensive folliculitis in abusers of brown Iranian heroin, *Eur. J. Clin. Microbiol.*, 4:340 (1985).

32. P.J.Collignon and T.C.Sorrell, Disseminated candidiasis: evidence of a distinctive syndrome in heroin abusers, *Br. Med. J.*, 287:861 (1983).

33. F.C.Odds, A.Palacio-Hernanz, J.Cuadia, J.Sanchez, Disseminated *Candida* infection syndrome in heroin addicts - dominance of a single *Candida albicans* biotype, *J. Med. Microbiol.*, 23:275 (1987).

34. J.M.Aguado, C.Barros, M.Fernandez-Guerrero, Cutaneous, ocular and osteoarticular candidosis in patients who are not heroin addicts, *J. Infect. Dis.*, 155:1082 (1987).

35. M.Feuillade de Chauvin, A.Cosnes, F.Benkhraba, R.Touraine, Folliculite du cuir chevelu et de la barbe à *Candida albicans*. Localisation septicémique chez un patient non toxicomane, *Ann. Dermatol. Vénéréol.*, 115:1162 (1988).

36. J.Bisbe, J.M.Mino, J.M.Torres, X.Latorre, C.Alia, M.Amarol, D.Sitivill, J.M.Mallolas, A.Trilla, Diagnostic value of serum antibody and antigen detection in heroin addicts with systemic candidiasis, *Rev. Infect. Dis.*, 11:310 (1989).

37. S.M.Brown, B.Stimmel, R.N.Taub, S.Kochwa, R.E.Rosenfield, Immunologic dysfunction in heroin addicts, *Arch. Intern. Med.*, 134:1001 (1974).

38. A.Durandy, A.Fisher, F.Le Deist, E.Drouhet, C.Griscelli, Mannan-specific and mannan-induced T-cell suppressive activity in patients with chronic mucocutaneous candidiasis, *J. Clin. Immunol.*, 7:400 (1987).

39. C.Pankhurst and M.Peakman, Reduced CD_4 + T cells and severe oral candidiasis in absence of HIV infection, *Lancet* (March 25) : 672 (1989).

40. E.Drouhet, B.Dupont, T.Dikeacou, Antifungal Agents and Immunity; *Z. Gl. Bakt. Suppl.* 13, 1, *in*: "Chemotherapy and Immunity", G. Pulverer and J. Jeliaszewicz, eds., Gustav Fischer Verlag, Stuttgart, New York (1985).

41. C.H.Kirkpatrick, Chronic mucocutaneous candidiasis: antibiotic and immunologic therapy, *in*: "Antifungal Drugs", V.S. Georgiev, ed., *Ann. New York Acad. Sci.*, 544:471 (1988).

CANDIDAMYCOSIS IN GYNAECOLOGY

H. Van der Pas

Department of Obstetrics and Gynecology, University of
Ghent and St. Elisabeth Hospital, Turnhout, Belgium

INTRODUCTION

Despite Wilkinson's description of Candida as a causative agent
of fungal vaginitis already in 1849 in The Lancet, it is still a
current topic in medical literature.

Indeed, 75% of all women are confronted with Candida, at least
once in their lifetime.

The procreator of vaginal candidosis is very well-known: Candida
albicans in 80% of the cases. In-vitro studies suggest that this is
because Candida albicans has great ability to adhere to vaginal
epithelial cells. In addition, several studies using scanning and
transmission electron microscopy have demonstrated the invasive
capacity of germinated yeasts. In other cases Torulopsis glabrata is
mostly responsible for the disease [1,2] (Table 1).

The causative mechanism, however, is no that obvious at all. One
out of ten women is an asymptomatic carrier of Candida. Among healthy
asymptomatic pregnant women up to 30% carriers are found. This incidence increases also with women who frequently visit venereal disease
clinics. The cervical mucus is probably a very effective barrier
against the invasion as the yeast has never been found either
in the non-pregnant uterus or in the tubae. On the whole, very
little is known about the natural defence mechanism of the vaginal

Table 1. Distribution of yeast species in patients with
vaginal candidosis.

Yeast species	Distribution
Candida albicans	78.9%
Candida guilliermondii	3.7%
Candida krusei	1.8%
Candida tropicalis	0.9%
Candida parapsilosis	0.4%
Other Candida species	1.8%
Torulopsis glabrata	8.8%

yeast cells against _Candida_ infection. The biology of the vagina is characterized by the enzymatic breakdown of glycogen into sugar and lactic acid, an acid environment of pH 4 being a favourable condition for the presence of the Döderlein's bacteria. Unter these normal circumstances, _Candida albicans_ has virtually no chance of becoming invasive. The vaginal bacterial flora plays a very protective role in preventing the conversion of the saprophyte form to the pathogenic mycelial form of the yeast.

TRANSFORMATION FROM SAPROPHYTE INTO PATHOGEN

Colonization can develop into symptomatic vaginitis. Suggested triggering factors may be divided into "host" factors and yeast factors (Table 2).

Host Factors

Several "host" factors have been put forward. Most of these factors are questionable [3,4].

a) _Pregnancy_. Vaginal yeast prevalence is rather frequent and increasing during pregnancy. Consequently, the newborn child can be affected during delivery and be contaminated in the first three to four weeks by oral or anogenital candidosis. Special attention must be paid to intensively treated prematures in the incubator. Therefore, it is recommendable to systematically make fungal cultures from any pregnant woman during the last weeks of pregnancy. It allows treatment of the infected woman. Thus, the risk of infection for the newborn, and especially for prematures, can be reduced.

b) _Oral contraceptives_. Although there are several reports of the increased occurrence of both asymptomatic vaginal carriage of yeasts and symptomatic vaginal candidosis in women taking oral contraceptives, the precise role of these drugs in predisposing to symptomatic vaginal infection remains controversial. Most epidemiological studies have documented at least an increased colonization rate of 20-45% which is thought to result from the effects of female hormones on epithelial cell adherence or receptivity, on glycogen and substrates available to the micro-organisms, as well as from the direct effect of oral contraceptives on yeast virulence.

c) _Antibiotics_. The rapidity which vaginal candidosis can make its appearance after oral intake of antibiotics suggests that there

Table 2. Predisposing factors implicated in the conversion of _C. albicans_ from a saprophyte to the pathological mycelial form (after Sobel, 1985) [3,4].

"Host" Factors	Yeast Factors (Virulence)
- pregnancy	- capacity to adhere to vaginal epithelial cells
- oral contraceptives	- amount of protease production
- antibiotics	- germ tube formation
- diabetes mellitus	
- tight clothing	

is a stimulating effect on the yeasts. However, some investigations [5,6] prove the contrary. The only hypothesis that has stood up to the facts and remains to be refuted refers to the effect of antibiotics on the bacterial flora of the vagina. The postulated mechanism is that a reduction of the number of lactobacilli causes a flare-up of an already present but not yet symptomatic Candida infection.

d) Diabetes mellitus. Uncontrolled diabetes mellitus with accompanying glycosuria and increased glucose concentrations in vaginal secretions may precipitate symptomatic vaginitis. Clinicians invariably consider subclinical or unrecognized diabetes in patients with recurrent vaginitis, and give these women 3- and 5-hour glucose tolerance tests. The rate of diagnosis of diabetes in such women is extremely low and does not justify the perpetuation of this practice.

e) Tight clothing. Many gynaecologists have identified tight, insulating clothing, particularly nylon underwear and tights, as factors that precipitate symptomatic bouts of vaginal candidosis. Poor ventilation and increased temperature and moisture of the perineum encourage yeast proliferation. A wet bathing suit is commonly thought to elicit symptoms, but it may be the chlorine in swimming pools acting as a vaginal mucosal irritant that precipitates the symptoms. Many gynaecologists feel that deodorant sprays, perfumed toilet paper and commercial douches similarly act to exacerbate symptoms. The role of vaginal tampons in causing minor friction of the mucosa and precipitating attacks has not been clarified. All the above factors may sensitize the mucosa to the pathogenic mechanisms of resident yeasts in the vagina and induce symptoms.

Although the various factors listed above have been implicated in the transformation from colonization to symptomatic vaginitis, in over 50% of cases no such predisposing factors can be identified. This raises the question of whether different strains of Candida albicans vary in their vaginopathic tendencies.

Yeast Factors

It has been suggested that prolonged asymptomatic colonization might merely be the result of colonization with a relatively less virulent strain of Candida[7]. Proposed virulance factors include capacity to adhere to vaginal cells, amount of protease production and germ tube formation of C. albicans.

a) Adherence. The capacity of the infective organism to adhere to the vaginal epithelial cells may affect its virulence. C. albicans adheres to these cells in larger numbers than the other fungi causing vaginitis. It is possible that this adherent quality accounts, at least in part, for the predominance of C. albicans vaginitis. Virtually all strains of C. albicans are adherent in-vitro, but there appeared to be minor variations in adherence between the various strains. Experimental studies show that a mutant strain of C. albicans that lacks the ability to adhere also fails to cause vaginitis in guinea-pigs or rats. Non-adherent strains are non-virulent and it is likely that, the less adherent the strain, the less virulent it may be.

b) Protease production. Protease production appears to be important not so much in colonization, but in increasing the invasiveness of the organism. Mutant strains that do not produce protease appear to be less adherent and less virulent in the animal model, while organisms isolated from cases of active vaginitis secrete higher levels of protease [7]. The protease produced by the organisms may be important in creating "holes" that allow the germinating organisms to invade the superficial layers of the vaginal mucosa.

c) <u>Germ tube formation of C. albicans</u>. Germ tube formation is associated with both greater adherence of the organism to the vaginal epithelial cells and greater invasiveness [6]. In colonization, the organism is in the non-germinating form suggesting that, in healthy women, germ tube formation is actively inhibited. Whether this is a function of the vaginal bacteria flora, other factors in the local vaginal enviroment, or of cell-mediated immunity has not yet been analysed.

SYMPTOMS

Patients with vaginal candidosis predominantly complain of vaginal or vulval itching. They also present discharge with a characteristic clumpy, white, cottage-cheese appearance, mostly adhering to the vaginal wall. In addition, erythema develops due to the infection in the vaginal mucosa or on the skin of the perineum. Thus, vulvar pruritus is the main symptom of candidosis in 90% of the patients, whereas leukorrhoea is not alays a classic symptom. Inspection of the vagina through a speculum reveals a white granular discharge adhering to the vaginal wall. The vagina is reddened and, after removal of the patches, superficial ulcerations with oozing of blood are sometimes apparent.

DIAGNOSIS AND DIFFERENTIAL DIAGNOSIS

Effective treatment of acute vaginitis requires accurate diagnosis and the identification of the aetiologic microorganimsms. In general, the differential diagnosis of acute vaginitis does not rely on elaborate technology but rather requires inexpensive and readily available office equipment and supplies a detailed history and an adequate examination of the external genitals, vagina and cervix. Only after the aetiology of vaginitis has been identified, can appropriate therapeutic action be taken.

The three major causative agents of vulvovaginitis are : <u>Trichomonas</u>, <u>Gardnerella</u> and <u>Candida</u>.

<u>Trichomonas</u>

The main complaint of trichomoniasis, in most cases, is a florid discharge sometimes, however, dyspareunia or chronic cystitis form the primary complaint. At examination one notices a thin yellow-green-white discharge with small air bubbles. The foul odour is sour-sweet. The vagina is velvet-reddish with clear upward red specks (strawberry vaginitis).

A fresh preparation usually allows rapid diagnosis. The protozoon can be recognized in the fresh preparation by the rapid, somewhat sharp movements; in general, flagellae and ondulating membranes are visible. They are greater than leucocytes, smaller than epithelial cells. They also look oval-pointed extended to a spot at which very long flagellae are present. Trichomoniasis is a sexually transmittable disease.

Gardnerella

In a large of cases with foetid vaginal discharge one finds numerous bacteria -in general in clue cells- without typical infection characteristics, such as redness or proliferation of leucocytes. Probably an anaerobic digestion of secreted epithelial cells take place, whereby the pH of the vagina increases, followed by a colonisation with Gardnerella. The released amines initiate a specific stench resembling rotten fish. The grey-white cream thin adherent discharge with an appalling smell usually shows no bubbles. In the fresh smear one finds squamous cells with numerous bacterial colonies, apparent as specks, sometimes also as loose clots besides the cells. The cells are not identifiable in stained sections. Remarkable is the absence of leucocytes and Döderlein's bacteria.

Candida

Besides a diagnosis on the basis of the above-mentioned and well known symptoms, vaginal candidosis is diagnosed microscopically and with the help of cultures.

a) Microscopical diagnosis. The KOH preparation is the simplest way of demonstrating fungal elements in vaginal secretion. A smear is suspended in a drop of a KOH 10% solution and covered with a covered a cover-glass. The material is then examined on the slide at a magnification of 400. A phase-contrast microscope may be helpful.

Microscopic diagnosis is for orientating purposes only. Demonstration of the pathogenic organism by culture is essential for a definitive identification of the agent of vaginal mycosis.

b) Culture diagnosis. Identification of the pathogen is only possible after culturing. For culturing, a Sabouraud glucose medium or a Nickerson medium are used. For Torulopsis glabrata, a Sabouraud medium is preferred. False-positive reactions are caused in about 5% of the cases, recognizable as brown black colonies on the Nickerson medium. If the culture on Sabouraud medium seems to be posivite, one of these colonies is suspended in a drop of saline and microscopic examination is performed.

If yeast cells are found, the colony is transferred to rice cream agar and covered with a cover-slide to create a micro-aerobic condition. After 24 to 48 hours at room temperature, the Petri plate is observed under the light microscope to recognize the pseudomycelium and chlamydospores in the case of Candida albicans. As said before, in 80% it is the causative factor by this first examination. If, however, chlamydospores are not observed, identification is performed by fermentation tests, by auxanogram and by biochemical tests. Identification of the pathogen is only possible by culturing, a procedure which is important above all for therapeutic reasons, since an infection with Torulopsis glabrata is particularly difficult to eliminate. However, patients with a high rate of recurrence must also be checked to establish whether the same yeast is always involved.

Table 3. A selection of drugs most commonly used for the treatment of vaginal candidosis.

TOPICAL THERAPY	ORAL THERAPY

- <u>Unspecified color agents</u>
  - Gentian violet
  - Brilliant green
	- <u>Imidazole</u>
	  - Ketoconazole

- <u>Polyenes</u>
  - Nystatin
  - Amphotericin B
  - Pimaricin
	- <u>Triazoles</u>
	  - Itraconazole
	  - Fluconazole

- <u>Topical azoles</u>
  - Miconazole
  - Clotrimazole
  - Econazole
  - Isoconazole
  - Terconazole
  - Butoconazole	

TREATMENT

There are probably more pharmaceutical agents currently available for the treatment of vaginal candidosis than for any other type of <u>Candida</u> infection (Table 3). Currently, the most commonly used are topical antimycotic agents. Formulations available for local treatment include pessaries, foaming pessaries, vaginal tablets, creams, lotions, coated tampons or gelatine capsules.

Dyes and polyenes

Between 1932 and 1952 dyes were commonly used as disinfectants of the vulvovaginal area. Gentian violet 2%, brilliant green 2%, and also boric acid were common antifungal area. Gentian violet 2%, brilliant green 2%, and also boric acid were common atifungal medications. From 1953 on, specific topical antimycotics of the polyene type have been used such as nystatin, amphotericin B and pimaricin.

Azoles

With the advent of the topical imidazoles, therapy length could be reduced remarkably. Patient compliance increased considerably as it was no longer necessary to apply drugs once or twice daily for several weeks. Miconazole, clotrimazole, econazole ad isoconazole were introduced as short and even single dose topical (i.e.vaginal) forms. Compared with the topical preparations, there are very few oral preparations available. With the advent of the orally active antimycotic, ketoconazole, there are two possible modes of treatment of vaginal candidosis, i.e. topical and oral administration. Two orally active agents are now available in several countries-fulconazole and itraconazole.

Oral therapy versus topical therapy

Other gynaecological infections such as trichmoniasis and bacterial infections have been treated orally for many years; this is in contrast with other vaginal conditions, which until recently were exclusively treated topically. Indeed, only in recent years has the orally active imidazole, ketoconazole become available for use in vaginal conditions [5].

However, oral therapy has obvious advantages over topical therapy. Firstly, Candida albicans in the deeper mucosal layers (Figure 1) might be partly protected from topical antifungal agents and re-emerges into the vaginal lumen some weeks or months later when epithelial cells are shed under the normal maturation process that occurs each month. With oral therapy all infecting organisms may be reached. Secondly, most patients prefer oral therapy because topical therapy is cumbersome to apply and cosmetically unacceptable [8] (Figure 2). Oral therapy is likely to lead to higher patient cmopliance than topical therapy. Obviously, a short oral treatment course would further enhance patient compliance and, more importantly, minimize the risk of unwanted side-effects sometimes associated with prolonged oral therapy.

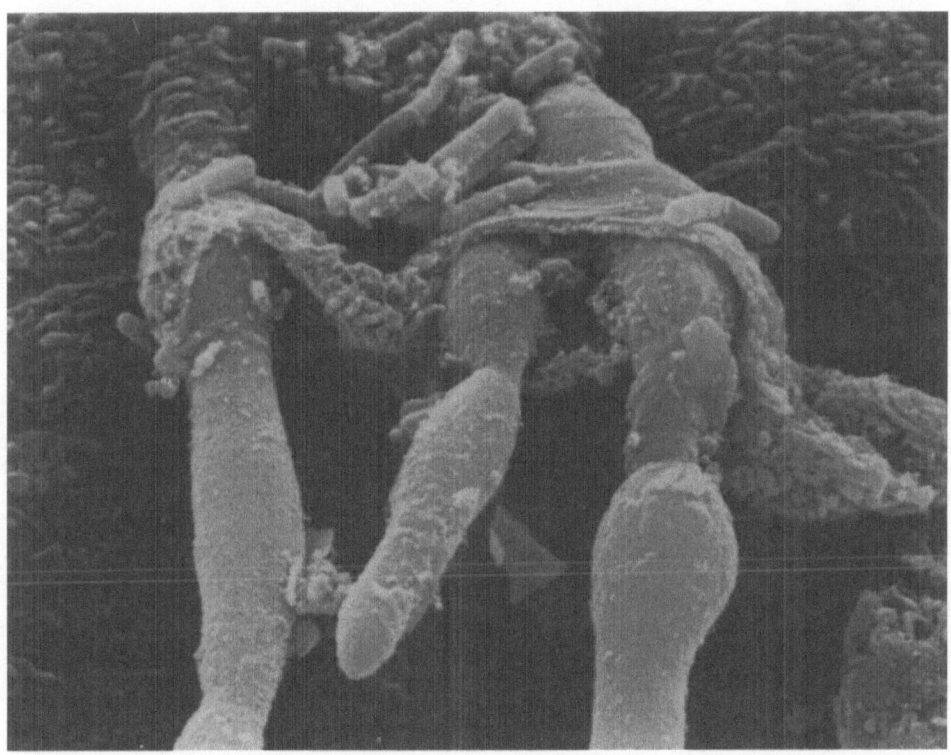

Fig. 1. Scanning electron microscopy of C. albicans showing the invasive capacity of the yeast into the vaginal epithelium (x 6,600).

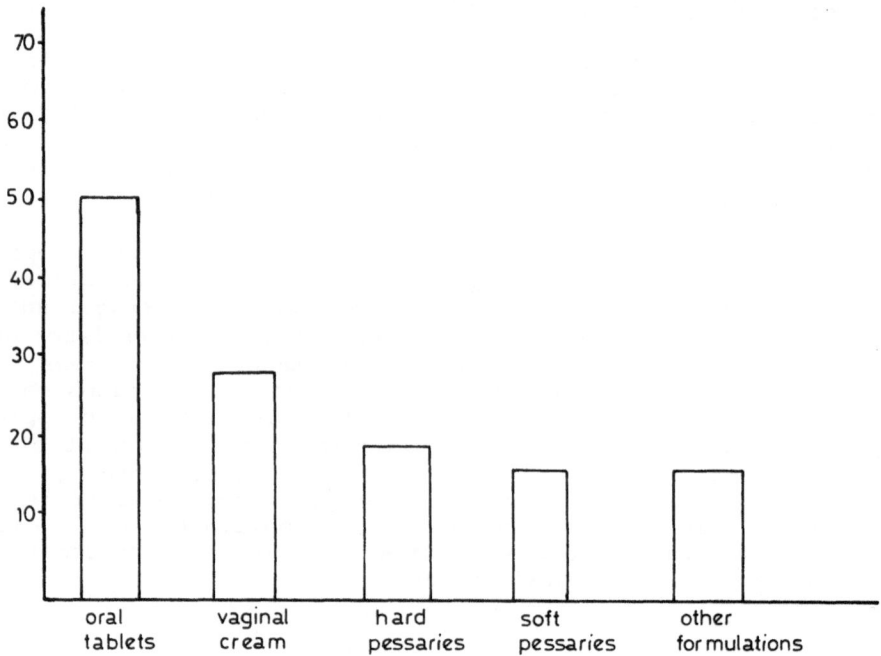

Fig. 2. Patient's preference for type of treatment for
vaginal candidosis[8].

Itraconazole

Itraconazole, a new orally active antimycotic [9] seems to be almost
the ideal agent, at least from a pharmacokinetic point of view[10].

Intraconazole is an oral broad-spectrum antifungal agent hightly
active *in-vitro* and *in-vivo*, and with a propensity to leave the blood
stream and achieve very high and persistent levels in tissues where
fungi reside . Indeed, the highly lipophilic character of itraconazole results in a favourable tissue/blood ration; most tissue
concentrations exceeding by far concentrations in the blood.

The epidermis and the vaginal epithelium prove to be separate
compartments in the body from the point of view of distribution of
the drug. In these tissues, itraconazole has a depot effect without
leaking into the plasma.

The high lipophilicity and the specific molecular shape result
in the ability to easily penetrate Candida cell membranes and in a
high selectivity for Candida cytochrome P-450.

At therapeutic doses there is no known interference with human
enzyme systems.

This promising pharmaokinetic profile has been challenged in
extensive clinical trials [11].

In acute vaginal candidosis, 3 different dosage schedules have
been compared: a single day tretament with 2x200 mg itraconazole, a
two day treatment with 200 mg daily and a three-day treatment with

200 mg daily. Total dose was either 400 or 600 mg. The 3 dosage schedules analyzed show overall clinical and mycological cure rates of more than 80% in a total of over 500 patients.

From this, we can also conclude that in vulvaginal yeast infections a minimum total dose of 400 mg itraconazole seems to be needed to reach about 80% mycological cure rates, and to maintain that level 1 month after end of therapy.

Adverse reactions reported in over 120 patients treated for vaginal candidosis are rare and mild and in line with what has been seen in other indications such as pityriasis versicolor and dermatomycoses: i.e. 6-8%. Only 0.5% of the patients in the vaginasl candidosis group discontinued therapy for side-effects. The most frequently reported adverse reactions including treatment schedules up to 1 month are: 1.3% nausea, 1.2% abdominal pain and 1% headache.

In view of the rare hepatotoxity problems with orally active azoles, the possibility of side-effects occurring with itraconazole was thoroughly investigated. No itraconazole induced hepatitis has been reported in over 150.000 patients treated today.

In over 15.000 well documented patients with various fungal infections no asymptomatic liver enzyme increase above background incidence (1 to 2%) has been reported. This includes about 1000 patients with treatment lengths of up to 2 years.

In contrast to some other azoles available or under development no induction or inhibiton of other xenobiotics has been reported in doses up to 160 mg/kg in rats [12,13]. This is 30x times the therapeutic dose in vaginal candidosis.

Investigations went even further: cirrhotic patients and patients who had developed liver problems on ketoconazole had no liver abnormalities with itraconazole.

CONCLUSIONS

Fungal vaginitis is usually due to C. albicans. Many healthy women are asymptomatic carriers, and it is thought that this colonization can persist for months or years without causing symptoms. However, it can also transform to symptomatic vaginitis. Triggering factors can be divided into host factors and yeast factors.

In contrast with other gynaecological infecions such as trichomoniasis and bacterial infections, vaginal candidosis was exclusively treated topically until recently. Ketoconazole was the first orally active agent to be used in the treatment of vaginal candidosis.

If given the choice, many women prefer the comfort and convenience of oral therapy. This natural preference improves patients compliance, which should indirectly reduce the chance of recurrence.

Obviously, an ultra-short oral treatment course, which is now available with the single day itraconazole therapy, further enhances the patient's compliance and, more importantly, minimizes the risk of unwanted side-effects.

REFERENCES

1. F.C. Odds, "Candida and Candida Vaginitis", Baltimore University Park Press, Baltimore (1979).
2. F.C. Odds, Candidosis of the genitalia, in: "Candida and Candidosis", 2nd. ed., pp.124-135, Baillière Tindall Publishers, London (1988).
3. J.D. Sobel, Pathogenesis of vaginal conditions, in: "Oral therapy in vaginal condiditons", pp.1-11, The Medicine Publishing Foundation, Oxford (1985).
4. J.D. Sobel, Epidemiology and pathogenesis of recurrent vulvovaginal conditions, Am. J. Obstet. Gynaecol., 152:924 (1985).
5. M.P.J.M. Bisschop, J.M.W.M. Merkus, H. Scheijgrond, J. Van Cutsem and A. Van de Kuy, Treatment of vaginal candidosis with ketoconazole, a new orally active antimycotic, Eur. J. Obstet. Gynaecol. Reprod. Biol., 9:253 (1979).
6. M.P.J.M. Bisschop, J.M.W.M. Merkus, and J. Van Cutsem, The influence of antibiotics on the growth of Candida albicans in the vagina, Eur. J. Obstet. Gynaecol. Reprod. Biol., 20:113 (1985).
7. J.D. Sobel, New insights into the pathogenesis of vaginal conditions, in: "Vaginal Conditions and Oral Antifungal Therapy", Proceedings of a Satellite Symposium to te IVth International Workshop for Infections in Gynaecology and Obstetrics, Munich (in press).
8. P.J.H. Tooley, Patient and doctors preferences in the treatment of vaginal conditions, Practitioner, 229:655 (1985).
9. J. Van Cutsem, F. Van Gerven and P.A.J. Janssen, Activity of orally, topically and parenterally administered itraconazole in the treatment of superficial and deep mycoses: Animal models, Rev. Infect. Dis., 9:15 (1987).
10. J. Heykants, M. Michiels and W. Meuldermans, The pharmacokinetics of itraconazole in animals and man: an overview, in:"Recent Trends in the Discovery, Development, and Evaluation of Antifungal Agents", R.A. Fromtling, ed., pp.251-263, J.R. Prous Science Publishers, Barcelona (1987).
11. G. Cauwenbergh, Itraconazole: the first active antifungal for single-day treatment of vaginal candidosis, Cur. Ther. Res., 41:210 (1987).
12. K. Lavrijssen, J. Van Houdt, D. Thijs, W. Meuldermans, and J. Heykants, Induction potential of antifungals containing an imidazole or triazole moiety. Miconazole and ketoconazole, but not itraconazole are able to induce hepatic drug metobilizing enzymes of male rats at high doses, Biochem. Pharmacol., 35:1867 (1986).
13. K. Lavrijssen, J. Van Houdt, D. Van Dijk, W. Meuldermans, and J. Heykants, Study on the induction and/or inhibition potential of itraconazole and fluconazole towards drug-metabolizing enzymes in the liver of the mouse. Preclin. Res. Report R 51 211/58, Jansen Pharmaceutica, Beerse, Belgium (1988).

ANTI-CANDIDA DRUGS - MECHANISMS OF ACTION

H. Vanden Bossche

Department of Comparative Biochemistry, Jannsen
Research Foundation, B-2340 Beerse, Belgium

INTRODUCTION

During recent years considerable advances have been made in the
identification of potential targets for antifungal agents (for a review
see Kerridge and Vanden Bossche, in press). A number of them are in
the cell wall, others are in the plasma membrane, endoplasmic
reticulum, nucleus, mitochondria, cytoskeleton or cytosol (Fig. 1).

The plasma membrane of yeast cells such as Candida albicans
retains its structure and properties by virtue of a complex cell wall
that is composed of glucan, mannan, chitin, proteins and lipids. The
cell wall is a highly dynamic structure, its size and shape continu-
ally change according to the needs of reproduction. These changes
require the localized and controlled action of both biosynthetic and
hydrolytic enzymes [1].

Any impairment of the balance between the synthesis of new wall
subunits and the controlled and localized hydrolysis of pre-existing
polysaccharides, to create the addition sites, will lead to an imbal-
ance in the growth, bud formation and/or septation. Considering the
essential functions and the particular nature of the fungal cell
wall, this organelle is, in principle, the ideal target for
antimycotic action.

The fundamental function of the plasma membrane is to act as a
barrier between the extra- and intracellular components. It is involved
in cell wall synthesis and is responsible for the selective transport
of low molecular weight molecules [2].

The physiological role of the plasma membrane provides important
targets for a number of antifungals. The most significant difference
between the mammalian and Candida cell membrane resides in the
sterols present. Ergosterol, instead of cholesterol, is the major in
C. albicans and a number of other pathogenic fungi.

A considerable part of sterol synthesis is localized at the
smooth endoplasmic reticulum, a target for important antifungal
agents such as allylamines, morpholines and the pyridine, pyrimidine,
imidazole and triazole derivatives.

Other important targets are DNA ve RNA synthesis, oxidative

phosphorylation and the ATPases in mitochondrial and vacuolar membranes.

It is the aim of this paper to discuss the interaction of anti-fungal agents with some of the targets listed in Fig.1.

Targets for "antifungal agents"

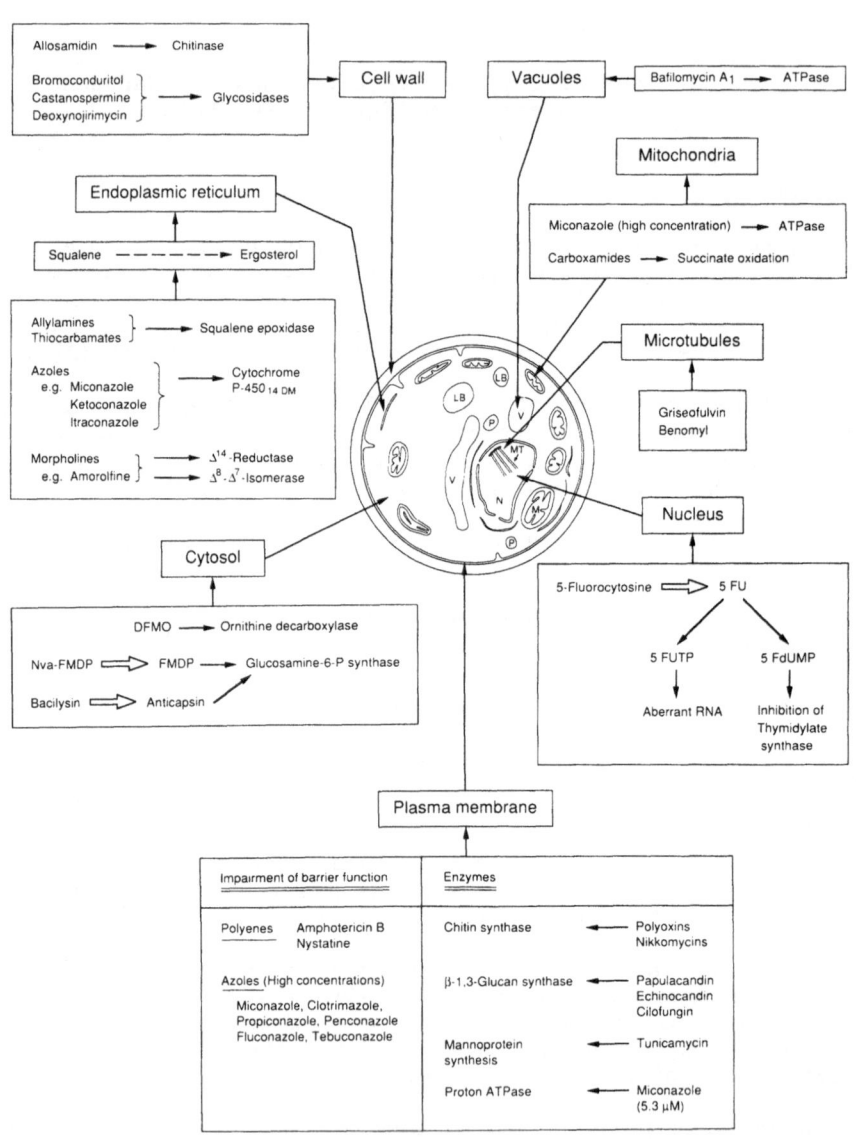

Fig. 1. LB= lipid body, P= peroxisome, 5 FU= 5-fluorouracil,
5 FUTP= 5-fluorouridine triphosphate,
5 FdUMP= 5-fluorodeoxyuridine monophosphate,
DFMO= DL-α-difluo-romethylornithine, Nva=FMDP= a peptide
of N^3-4 methoxyfumaroyl-L-2,3-diaminopropanoic acid (FMDP)

COMPOUNDS THAT INTERFERE WITH THE SYNTHESIS OR HYDROLYSIS OF CELL WALL SUBUNITS

Chitin is an important component of the septum, separating mother and daughter cells and in C. albicans, it is important for yeast-mycelium transition. It has been shown that the proportion of chitin in the mycelial form is about 3 times that of the budding form [3]. The crucial role of chitin and the absence of chitin from mammalian and plants cells makes chitin synthesis and hydrolysis interesting targets for antifungal therapy.

The chitin synthase of human pathogenic fungi, associated with the plasma membrane, is highly sensitive to polyoxins (micromolar concentrations) in cell-free systems [4]. However, growth is inhibited by these antibiotics only at millimolar concentrations [5]. The poly-oxins possess a peptide backbone and are therefore susceptible to be actively taken up through the peptide transport systems present in C. albicans. However, this also means that in media containing peptides the uptake of the polyoxine will be inhibited. Indeed, growth inhibition of C. albicans, highly depends on the medium composition; these polyoxins are almost inactive in rich medium containing peptides, whereas in liquid minimal medium, devoid of peptones, MICs of 5-10 mg/ml of polyoxin D are obtained.

Studies by Yadan et al [6]. indicate that the activity of a closely related substance, nikkomycin Z, was much less sensitive to compe-tition with peptides for the peptide transport systems. Their studies revealed that C. albicans is using two distinct systems for peptide transport and that polyoxin D and nikkomycin Z are not transported by the same system. They suggest that nikkomycins and tripeptidyl deri-vatives should prove useful for inhibition of both wild-type and dipeptide transport-deficient mutants of C. albicans. However, the same authors prove that microorganisms are able to change their peptide transport ability. For example, they isolated a nikkomycin-resistant mutant of C. albicans that showed a decreased dipeptide uptake but the trimethionine uptake was markedly increased. Thus, this mutant seems to lack the general peptide transport system of the wild type but still satisfies its nutritional requirements by developing a transport system more specific for oligopeptides. The latter system is less efficient in transporting nikkomycin Z. This already presents some doubts on the future of this kind of compounds in the treatment of candidosis.

In C. albicans chitinase, an enzyme associated with chitin degradation, has a molecular weight of about 70 kDa and an optimum pH of 6.5 [7]. In the more carefully examined Saccharomyces cerevisiae, part of the mannan-associated chitinase is present in the periplasmic space. It might have a function in cell division, perhaps it plays a role in the final fission of septa which leads to cell separation [8,9].

Chitinase inhibitors may have an advantage over chitin synthase inhibitors as a target. Indeed, chitinase may be accessible to drugs without requiring transpot across the plasmalemma. Allosamidin is a streptomycete antibiotic which in 1987 in 1987 was reported by Koga et al [10]. to inhibit insect chitinase with an IC_{50}-value of about 0.4 µM in a competitive and selective way. This C-3 epimer of N-acetyl-β-d-glusosamine is also a potent inhibitor of the chitinase of Neurospora crassa with an IC50-value of 0.5 µM (Prof.G.W.Gooday, personal communication). This compound has yet to be tested against C.albicans and other human pathogenic fungi.

Another target is the plasma membrane associated synthesis of glucans. A number of glucan synthase inhibitors are available. The best known are papulacandin B, echinocandin B, aculeacin A, and cilofungin. Papulacandin B is an amphophilic molecule in which the hydrophilic portion contains glucose and galactose residues and the hydrophobic portion two unsaturated partially hydroxylated fatty acids.

Elorza et al.[11] studied the effect of papulacandin B on the distribution of label from [^{14}C]-glucose into various polysaccharide fractions of regenerating C. albicans protoplasts. At 5 µg/ml and after 5 h of generation, they found more than 85% of inhibition of mannoprotein and glucan incorporation. Papulacandin B seems to have a narrow spectrum of activity and might be more important as a biochemical tool than as an antifungal agent. This is also true for the two related antifungal antibiotics, echinocandin B and aculeacin A. Both cyclic-peptides are active against some species of yeasts but inactive against filamentous fungi. Echinocandin B causes lysis of the budding yeast cells and of elongating mycelium at the apex. Both echinocandin B and aculeacin A are inhibitors of glucan synthesis. However, they are less potent than papulacandin B.

Of greater interest might be the semisynthetic antifungal antibiotic LY121019 or cilofungin [12-14].

As can be expected from its structure, which is derived from echinocandin, cilofungin competitively inhibits (1-3)-β-d-glucan synthase activity of C. albicans and Neurospora crassa. The apparent inhibitor constant is [K_i (app.)] 2.5 µM and 16 mM, respectively [15-17].

Cilofungin treatment leads to the following changes in the C. albicans cells:

- the glucan layer is dispersed and loses its normal texture
- the cytoplasmic membrane is dislocated and ruptured
- cytoplasmic material is released in large quantities into the extracellular space
- the destruction of the cytoplasmic membrane and the efflux of cytoplasmic constituents lead to the rapid death of the Candida cells.

This echinocandin derivative possesses in vitro and in vivo antifungal acivity against C. albicans isolates and is at least 20-fold less toxic than amphotericin B. Parenterally administered cilofungin at doses of 50 mg/kg significantly reduced the recovery of C. albicans from infected mouse kidneys [14]. Isolates of Candida (Torulopsis) glabrata are less susceptible to cilofungin than to amphotericin B and MIC90-values for isolates of Candida parapsilosis are mostly higher than 20 µg/ml[14,18]. Thus, the spectrum of antifungal activity of this compound might be too narrow to be of help in the treatment of mycoses other than candidosis due to Candida albicans.

COMPOUNDS WHICH IMPAIR THE BARRIER FUNCTION OF PLASMA MEMBRANES

The polyene macrolide antibiotics such as amphotericin B (Ampho B) and nystatin interact with sterol containing plasma membranes of sensitive organisms changing the physical state of the membrane, causing an impairment of membrane functions, resulting in an enhanced permeability to protons and leakage of internal constituents such as K^+, Ca^{2+} and PO_4^{3-} [19,21].

Differences in the relative affinities of Ampho B for ergosterol-containing fungal membranes and cholesterol cantaining mammalian membranes are thought to be at the origin of Ampho B's selectivity.

Studies on intact cells, in which the membrane composition has been modified, and on artificial membrane systems, as reviewed by Kerridge [22], indicate that the original hypothesis, that the fungistatic and fungicidal effects originate from the association of Ampho B with membrane, sterols, is no longer satisfactory. Both the nature of the sterol and the physical state of the phospholipids play an important role in the disruptive interaction of polyenes with membranes. This calls into question that there is a single molecular target for Ampho B and that its interaction with this target is responsible for the biological effects, inclusive cell death [22].

Recent studies by Medoff [23] suggest that the involvement of amphotericin-induced oxidative damage to the membrane is an important possibility to consider as a cause of cell death. Indeed, it has been shown that protoplasts of C. albicans were protected against polyenes by exogenous catalase [23]. Recently, the investigators isolated a laboratory mutant resistant to the lethal effects of Ampho B, which has 20-fold higher catalase levels than the wild strain [23].

Thus, almost 4 decades after the first description of nystatin [24] and more than 3 decades after the isolation of Ampho B by Gold et al [25]. in 1956, it is still unclear what mechanisms are at the origin of both the fungistatic and fungicidal activity of the polyene antifungals.

Azole, allylamine and thiocarbamate antifungals also impair the barrier function of fungal plasma membranes. The most important interaction of the azole, pyrimidine and pyridine antifungals with the fungal plasma membrane is an indirect one resulting from an interaction with ergosterol synthesis (see later).

Some azole derivatives also have a direct effect on artificial and cellular membranes causing an impairment of the barrier function (for a review see reference 26). Examples are clotrimazole [27], miconazole [28-30], econazole [30], propiconazole and penconazole [31], fluconazole and terbuconazole [32]. Imazalil has a slight effect at 50 mg/ml only [31]. However, ketoconazole, itraconazole and flutriafol do not affect membranes directly [29,31]. The direct effects of miconazole might results from its conformation in the lipid bilayer. The mean molecular area calculated by projection of miconazole on the lipid water interface is 90 A^2 i.e. much higher than that occupied per dipalmitoylphosphatidylcholine (DPPC) molecule (60A^2). Such a conformation should results in a destabilizing effect of high miconazole concentrations. It should be noted that even when a disruptive interaction occurs with these derivatives, this is at concentrations in excess of those required to inhibit growth.

COMPOUNDS AFFECTING ENZYMES OF THE ENDOPLASMIC RETICULUM

An important group of modern antifungals interferes with the biosynthesis of ergosterol. The ergosterol biosynthesis pathway can be divided into three segments. The first part, the synthesis of squalene from acetyl-CoA, is common to the biosynthesis of sterol in fungal, plant and mammalian cells. This part of the pathway leads via 3-hydroxy-3-methylglutaryl-CoA (HMG) and mevalonate to squalene. A key

Acetyl-CoA
HMG-CoA
HMGR
Mevalonate
Isopentenyl pyrophosphate
Farnesyl pyrophosphate
Ubiquinone
Dolichol
Squalene
Electron transport
Glycoproteins
Glycolipids
Sterols

Fig. 2. Biosynthesis of isoprenoid lipids. HMG= 3-hydroxy-3-methylglutaryl-CoA, R= reductase.

Squalene epoxidase
Squalene
Naftifine
Terbinafine
Thiocarbamates
Squalene-2,3-Oxide
HO
14
Lanosterol

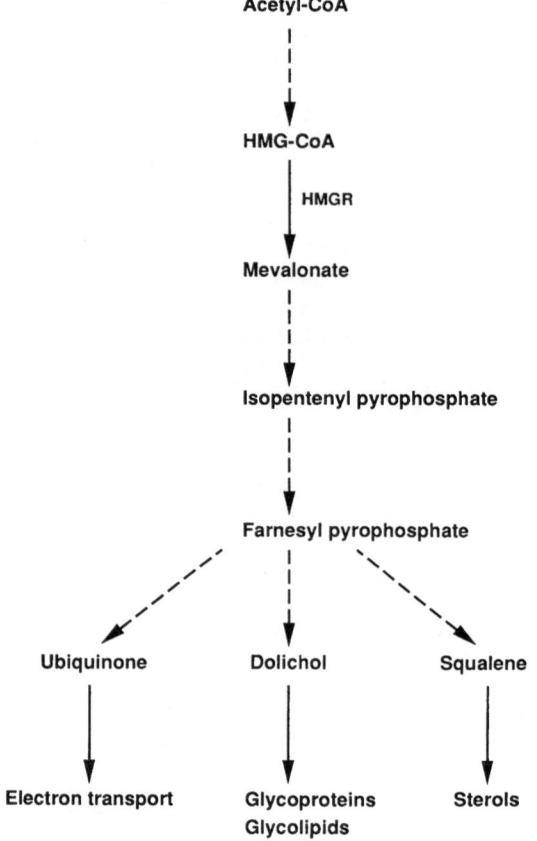

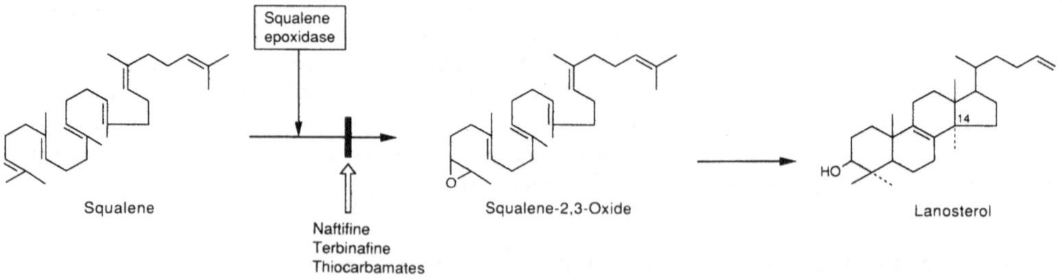

Fig. 3. The synthesis of lanosterol from squalene.

enzyme is the MHMG-reductase (HMGR). It catalyses a reaction which is common to the synthetic pathways of three isoprenoid lipids: sterols, dolichols and the side chain of ubiquinone (Fig.2). Dolichols, poly-isoprenoid alcohols containing 12-22 isoprene units, are involved in the synthesis of glycoproteins such as the mannoproteins of the cell wall. Ubiquinone is a component of the mitochondrial electron-transport chain. Compounds that inhibit HMGR will not only decrease the supply of sterol but may also deplete the cell of key non-sterol intermediates [34]. Berg et al. [35] claim that bifonazole at concent-rations above 3.2×10^{-6} M inhibits the microsomal HMGR of Trichophyton mentagrophytes by about 80%. However, at concentrations up to 10-5 M, our experiments did not reveal an effect of bifonazole on the HMGR of T. mentagrophytes nor S. cerevisiae microsomes (unpublished results). The HMGR from S. cerevisiae is similar in several ways to its mamma-lian counterpart [36]. Therefore, it is not surprising that this yeast enzyme is sensitive to compactin [37] and mevinolin [36], two inhibitors of the mammalian HMGR.

The second segment of the pathway consists of the enzymes involved in the conversion of squalene into lanosterol (Fig. 3). The first step, catalysed by the squalene epoxidase, is the target for the allylamines and thiocarbamates. Both naftifine and terbinafine inhibit this endoplasmic enzyme in C. albicans. Fifty % inhibition is achieved at 1.2×10^{-6} M and 2.7×10^{-8} M, respectively; 95% inhibition is obtained at 2.6×10^{-5} M and 6.86×10^{-7} M. These allylamines are weak inhibitors of the rat liver microsomal squalene epoxidase. The latter enzyme differs in many aspects from the C. albicans enzyme. For example, the liver enzyme has a strong requirement for a factor of the cytosol, it has a preference for NADPH instead of NADH, the preferred cofactor of the Candida enzyme. Furthermore, allylamines are noncompetitive inhibitors of the Candida enzyme and competitive inhibitors of the liver enzyme and the squalene analogue, 2-aza-2,3-dihydro-squalene, is a much better inhibitor of the liver enzyme [38]. Major differences not only exist between the epoxidase of Candida and the rat enzyme. Important differences in sensitivity to terbinafine are also found between the enzyme present in rat and guinea-pig liver, the latter being 23 times more sensitive [38]. Tolnaftate and the closely related tolciclate, both active against dermatophytes, also inhibit the Candida squalene epoxidase. The lack of activity of these thiocarbamates against Candida might originate from difficulties to cross the weaker effects of allylamines on C. albicans as compared with their effects on Trichophyton spp [38].

The third segment of the ergosterol synthesis pathway is the target for a long list of important antifungal compounds useful in plant protection as well as against animal and human pathogenic fungi. By a multiple process, catalysed by membrane-bound enzymes, lanosterol is converted into ergosterol (Fig.4). In S. cerevisiae lanosterol is the substrate for a cytochrome P-450-dependent 14α-demethylation (P450$_{14DM}$). In most fungal cells and in a number of C. albicans isolates, lanosterol is first converted into 24-methylenedihydrolanosterol. To achieve this 24-alkylation, lanosterol has to be transported from the endoplasmic reticulum to the mito-chondria. The 24-alkylsterol thus formed is then transferred back to the endoplasmic reticulum where the P450$_{14DM}$ is localized. In both yeast and fungi this P450$_{14DM}$ is the primary target for a number of pyrimidine, pyridine, imidazole and triazole antifungals (for reviews see: References 26,40,41 and 42).

The activity and selectivity of these antifungal agents depend on their affinity for the heme iron ion and for the apoprotein moiety

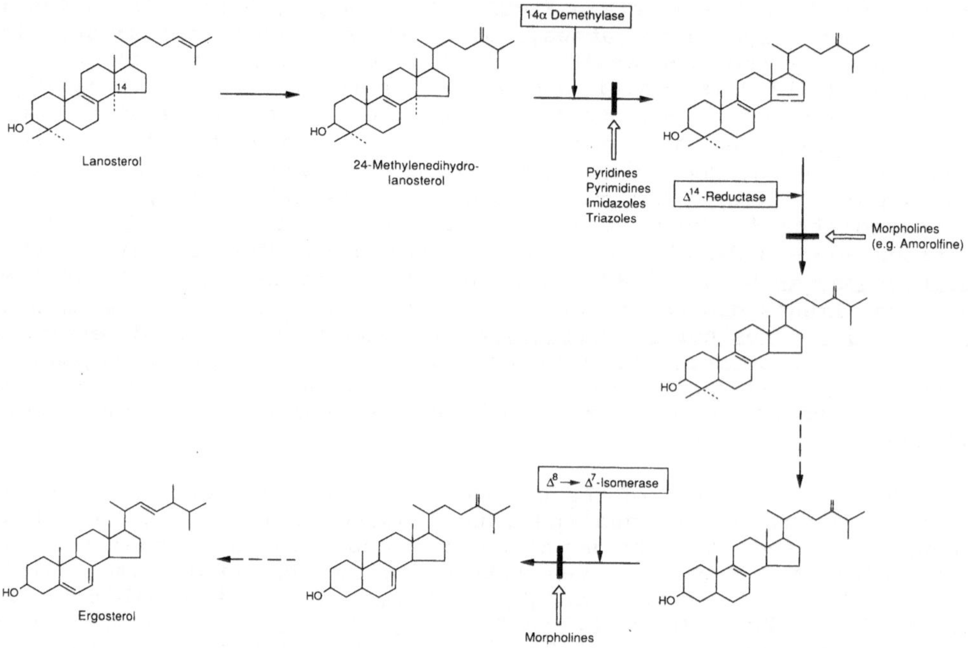

Fig. 4. Simplified pathway of ergosterol biosynthesis. Bars represent sites of inhibition of the 14-demethylase inhibitors, the DMIs (pyridine, pyrimidine, imidazole and triazole antifungals) and the Δ^{14} reductase and/or $\Delta^8 \rightarrow \Delta^7$ isomerase inhibitors (the morpholines).

of the P450 isozymes. The imidazole and triazole derivatives bind via the N^3 of the imidazole ring or the N^4 of the triazole ring to the sixth coordination position of the heme iron of P450 interfering in this way with oxygen binding and activation. This results in an inhibition of the monooxygenase reaction catalysed by P450 according the following (simplified) reaction scheme:

$$RH + NADPH_2 + O_2 \rightarrow ROH + NADP^+ + H_2O$$

(RH = a lipophylic substrate e.g. lanosterol). The oxygen needed in this reaction is bound by the iron ion and then activated to generate an iron-oxene species that is able to transfer its oxygen to the substrate to yield for example an alcohol. $P450_{14DM}$, purified to homogenity [43-45], catalyses three oxidative steps [42]:

1. the hydroxylation of the C-32-methyl (14 α-methyl) group of lanosterol
2. the oxidation of the C-32 alcohol to the C-32 aldehyde
3. the oxidative elimination of the aldehyde as formic acid

Thus, compounds that bind to the heme iron ion of $P450_{14DM}$ will inhibit one of the 3 steps. Studies of Yoshida and Aoyama [42] seem to indicate that ketoconazole interacts with the 1st step (hydroxylation). It is of interest to note that using hepatic microsomes both miconazole and ketoconazole promote accumulation of the C-32 aldehyde intermediate [46]. This means that in liver microsomes these azole antifungals inhibit the lyase instead of the hydroxylase activity suggesting differences in the $P450_{14DM}$ of yeast and mamallian microsomes.

Reduced P450 has a marked affinity not only for oxygen but also for carbon monoxide (CO). The $CO-Fe^{2+}$ complex displays a split absorption at about 360 nm and 450 nm (for yeast P450:448 nm). Compounds that compete with CO for binding to the iron ion will reduce absorption at 450 nm. This property has been used to compare the effects of different azole antifungals on the microsomal P450 from C. albicans, piglet testes and rabbit liver and on mitochondrial P450 from bovine adrenal cortex (Table 1).

A striking difference in potency and selectivity of the different imidazole and triazole antifungals has been found. Bifonazole and fluconazole show the lowest affinity for the C. albicans P450. The topically active, bifonazole, has even a greater effect on P450 in testes and liver microsomes than on its target P450 in Candida microsomes. Itraconazole and its fluor derivative, saperconazole, both orally and topically active antifungals, combine high activity (IC_{50} value$<5x10^{-8}$M) with good selectivity (IC_{50} values$>10^{-5}$M).

The interaction of azoles with the P45014DM results in an inhibition of ergosteral synthesis. For example, in C. albicans, complete ergosterol depletion is achieved, after 4 h of incubation, at $5x10^{-9}$M or $7.5x10^{-9}$M of itraconazole and ketoconazole, respectively. Since the presence of ergosterol in fungi is a prerequisite for cell proliferation, the observed azole-induced depletion might be the prime cause for the effect on fungal growth. Inhibition of ergosterol synthesis coincides with the accumulation of 14-methylated sterols. Conclusive evidence has been collected showing that these sterols induce permeability changes, membrane leakiness, changes in membrane-bound enzymes, inhibition of growth and cell death (for reviews see References 26 and 40).

Table 1. Effects of azole antifungals on cytochrome-P-450 isozymes (IC_{50}-values x 10^{-8}M[a]).

Azole derivatives	C.albicans[b]	Testis[c]	Adrenals[d]	Liver[e]
Bifonazole	117.0	6.9	212.0	97.2
Fluconazole	24.8	>1000	>1000	>1000
Nor-ketoconazole[f]	9.6	87.5	785.0	>1000
Miconazole	7.6	48.7	125.0	>1000
Azaconazole	6.4	>1000	>1000	>1000
Propiconazole	6.1	>1000	>1000	>1000
Econazole	5.3	16.6	58.3	141.0
Imazalil	5.0	393.0	492.0	540.0
Terconazole	4.1	571.0	>1000	>1000
Clotrimazole	3.6	43.0	33.4	587.0
Parconazole	3.6	39.5	438.0	>1000
Saperconazole	3.5	>1000	>1000	>1000
Itraconazole	3.1	>1000	>1000	>1000
Ketoconazole	3.0	38.7	>1000	>1000
Penconazole	1.9	316.0	>1000	>1000

[a] IC_{50}-values = concentration needed to inhibit CO binding to the heme iron for 50%. Cytochrome P-450 content = 0.1 nmole/ml
[b] C. albicans (ATCC 28516) microsomes
[c] Piglet testis microsomes
[d] Bovine adrenal cortex mitochondria
[e] Rabbit liver microsomes
[f] Nor-ketoconazole = deacylated ketoconazole

COMPOUNDS INTERFERING WITH MACROMOLECULAR SYNTHESIS

Of all the compounds that inhibit macromolecular synthesis in fungi only 5-fluorocytosine has been exploited in the treatment of patients suffering from fungal infections. 5-fluorocytosine is illicitly transported across the plasma membrane of sensitive fungi by a cytosine permease, deaminated to 5-fluorouracil and this latter compounds is metabolised by the pyrimidine salvage pathway to 5-fluorouridine triphosphate (5-FUTP) and 5-fluorodeoxyuridylate (5-FdUMP). 5-FUTP is incorporated into fungal RNA in place of uridylic acid. This alters amino acylation of tRNA and disturbs the amino pool and protein synthesis. 5-FdUMP is a potent inhibitor of thymidylate synthase, a key enzyme in the synthesis of DNA [22,47]. Selectivity results from the fact that man lacks a cytosine deaminase responsible for converting cytosine to uracil; 5-fluorocytosine is itself non-toxic. Yet this well-tolerated drug suffers from an important drawback i.e. the high frequency with which resistant variants appear during therapy [48]. A common enzyme deficiency associated with resistance of C. albicans isolates is the deficiency of the UMP: pyrophosphate phosphoribosyltransferase, the enzyme responsible for the synthesis of uridylate in the pyrimidine salvage pathway [22].

OTHER TARGETS FOR ANTIFUNGALS

Polyamines (e.g. putrescine, spermidine, spermine) are essential metabolites in microorganisms. They play a role in cell growth and multiplication. The first inhibitor of polyamine synthesis which was developed was DL-α-difluoromethylornithine (DFMO), a potent inhibitor of ornithine decarboxylase (ODC) and hence putrescine synthesis [49]. DFMO is being used therapeuticalliy for Pneumocystis carinii pneumonia [50] (studies of Edman et al. [51] showed Pneumocystis carinii to be a member of the fungi).Polyamine synthesis in a number of Candida, Trichophyton and Microsporum species is inhibited by DFMO in vitro [52,53]. DFMO is considered to be well tolerated by man. The side effects that do occur are reversible and inhibitors of polyamine synthesis might prove suitable for future development as antimycotic drugs.

CONCLUSION

Between the impressive number of targets for antifungal compounds, those on the smooth endoplasmic reticulum, provided the clinicians with an amazing number of potent, broad spectrum antifungal agents. An evaluation of the interaction of the triazole derivatives, itraconazole and saperconazole, with the fungal cytochrome P-450-dependent ergosterol synthesis and with cytochrome P-450 isozymes of the host has revealed that these agents might combine potent antifungal activity with selectivity.

REFERENCES

1. G.H.Fleet, The occurence and function of endogenous wall-degrading enzymes in yeasts, in: "Microbial Cell Wall Synthesis and Autolysis", J.R.Sabine, ed., Elsevier Science Publishers, Amsterdam, (1984).
2. M.G.Shepherd, Cell envelope of Candida albicans, CRC Crit.Rev.Microb., 15: 7 (1987).

3. D.R.Soll, _Candida albicans,_ _in_: "Fungal Dimorphism. With Emphasis on Fungi Pathogenic for Humans", P.J.Szaniszlo, ed., Plenum Press, New York (1985).

4. Y.Y.Chiew, M.G.Shepherd, P.A.Sullivan, Regulation of chitin synthesis during germ-tube formation in _Candida albicans,_ _Arch. Microbiol.,_ 125: 97 (1980).

5. J.M.Becker, N.L.Covert, P.Shenbagamurthi, A.S.Steinfeld, F.Naider, Polyoxin D inhibits growth of zoopathogenic fungi, _Antimicrob. Agents Chemother.,_ 23: 926 (1983).

6. J.-C.Yadan, M.Gonneau, P.Sarthou, F.Le Goffic, Sensitivity to nikkomycin Z in _Candida albicans_: role of peptide permeases. _J.Bacteriol.,_ 160: 884 (1984).

7. K.Barret-Bee and M.Hamilton, The detection and analysis of chitinase activity from the yeast form of _Candida albicans,_ _J.Gen.Microbiol.,_ 130: 1857 (1987).

8. J.U.Correa, N.Elango, I.Polacheck, E.Cabib, Endochitinase, a mannan-associated enzyme from _Saccharomyces cerevisiae,_ _J.Biol. Chem.,_ 257: 1392 (1982).

9. N.Elango, J.U.Correa, E.Cabib, Secretory character of yeast chitinase, _J.Biol.Chem.,_ 257: 1398 (1982).

10. D.Koga, A.Isogai, S.Sakuda, S.Matsumoato, A.Suzuki, S.Kimura, A.Ide, Specific inhibition of _Bombyx mori_ chitinase by allosa-midin, _Agric.Biol.Chem.,_ 51: 471 (1987).

11. M.V.Elorza, A.Murgui, H.Rico, F.Mirogall, R.Sentandreu, Formation of new cell wall by protoplasts of _Candida albicans_: Effect of papulacandin B, tunicamycin and nikkomycin, _J.Gen.Microbiol.,_ 133: 2315 (1987).

12. M.Debono, B.J.Abbott, J.R.Turner, L.C.Howard, R.S.Gordee, A.S. Hunt, M.Barnhart, R.M.Molloy, K.E.Willard, D.Fukuda, T.F.Butler, D.J.Zeckner, Synthesis and evaluation of LY121019, a member of a series of semisynthetic analogues of the autifungal lipopeptide echinocandin B, _Ann.N.Y.Acad. Sci.,_ 544: 152 (1988).

13. R.S.Gordee, D.J.Zeckner, L.F.Ellis, A.L.Thakkar, L.C.Howard, _In vitro_ and _in vivo_ anti-_Candida_ activity of LY1201019, _J.Antibiot.,_ 37: 1054 (1984).

14. R.S.Gordee, D.J.Zeckner, L.C.Howard, W.E.Alborn Jr., M.Debono, Anti-_Candida_ activity and toxicology of LY121019. A novel semi-synthetic polypeptide antifungal antibiotic, _Ann.N.Y.Acad.Sci.,_ 544: 294 (1988).

15. C.S.Taft, C.P.Selitrennikoff, LY121019 inhibits _Neorospora crassa_ growth and (1-3)-β-D-glucan synthase, _J.Antibiot.,_ 51: 697 (1988).

16. C.S.Taft, T.Stark, C.P.Selitrennikoff, Cilofungin (LY121019) in-hibits _Candida albicans_ (1-3)-β-D-glucan synthase activity, _Antimicrob.Agents.Chemother.,_ 32: 1901 (1988).

17. J.Müller and I.Scheidecker, Immunoelectronmicroscopic studies on the influence of an echinocandin B analog on the cell wall antigenicity of _Candida albicans,_ _in_: "Proceedings of the Xth Congress of the International Society for Human and Animal Mycology", p.152 (1988).

18. H.G.Hall, C.Myles, K.J.Pratt, J.A.Washington, Cilofungin (LY120019), an antifungal agent with specific activity against _Candida albicans_ and _Candida tropicalis,_ _Antimicrob.Agents Chemother.,_ 32: 1331 (1988).

19. D.Kerridge, M.Fasoli, F.J.Wayman, Drug resistance in _Candida albicans_ and _Candida glabrata,Ann.N.Y.Acad.Sci.,_ 544: 245 (1988).

20. D.Kerridge, H.Vanden Bossche, Drug discovery: a biochemist's approach, _in_ "Handbook of Experimental Pharmacology. Chemotherapy of Fungal Diseases", J.F.Ryley, ed., Springer Verlag, Berlin (in press).

21. H.Vanden Bossche, G.Willemsens, P.Marichal, Anti-_Candida_ drugs-
    the biochemical basis for their action, _CRC Crit.Rev.Microbiol._,
    15: 57 (1987).
22. D.Kerridge, Polyene macrolide antibiotics, _in_: "Aspergillus and
    Aspergillosis", H.Vanden Bossche, D.W.R.Mackenzie, G.Cauwenbergh,
    eds., Plenum Press, New York & London (1988).
23. G.Medoff, The mechamism of action of amphotericin, _in_:
    "Aspergillus and Aspergillosis", Vanden Bossche, D.W.R.Mackenzie,
    G.Cauwenbergh, eds., Plenum Press, New York & London (1988).
24. E.L.Hazen and R.H.Brown, Two antibiotics produced by a soil
    actinomycete, _Science_, 112: 423 (1950).
25. W.Gold, H.A.Stout, J.F.Pagano, R.Donovick, Amphotericins A and
    B, antifungal antibiotics produced by a streptomycete. I. _In vitro_
    studies. _Antibiotics Annual._, 567 (1956).
26. H.Vanden Bossche, Mode of action of pyridine, pyrimidine and
    azole antifungals, _in_ "Sterol Biosynthesis Inhibitors.
    Pharmaceutical and Agrochemical Aspects", D.Berg, M.Plempel,
    eds., Ellis Horwood Ltd., Chichester, England (1988).
27. K.Iwata, H.Yamaguchi, T.Hiratani, Mode of action of clotri-
    mazole, _Sabouraudia_, 11: 158 (1973).
28. J.Cope, Mode of action of miconazole on _Candida albicans_: effects
    on growth, viability and K$^+$ release, _J.Gen.Microbiol._, 119: 245
    (1980).
29. H.Vanden Bossche, J.M. Ruysschaert, F.Defrise-Quertain,
    G.Willemsens, F.Cornelissen, P.Marichal, W.Cools, J.Van Cutsem,
    The interaction of miconazole and ketoconazole with lipids.
    Biochem. Pharmacol., 32: 2175 (1982).
30. N.H.Georgopapadakou, B.A.Dix, S.A.Smith, J.Freudenberger,
    P.T. Funke, Effect of antifungal agents on lipid biosynthesis and
    membrane integrity in _Candida albicans_, _Antimicrob.Agents
    Chemother._, 31: 46 (1987).
31. H.Dahmen, H.C.Hoch, T.Staub, Differential effects of sterol
    inhibitors on growth, cell membrane permeability and ultrastructure
    of two target fungi, _Cytol.Histol._, 78: 1933 (1988).
32. D.Berg, K.-H.Büchel, W.Kramer, W.Plempel, H.Scheinpflug,
    Mechanistic studies as a tool for the development of new
    compounds, _in_: "Sterol Biosynthesis Inhibitors. Pharmaceutical
    and Agrochemical Aspects", D.Berg and M.Plempel, eds., Ellis
    Horwood Ltd., Chichester, England (1988).
33. R.Brasseur, C.Vandenbosch, H.Vanden Bossche, J.M.Ruyschaert,
    Mode of insertion of miconazole, ketoconazole and deacylated
    ketoconazole in lipid bilayers. A conformational analysis,
    Biochem.Pharmacol., 32: 2175 (1983).
34. R.Fears, Pharmacological control of 3-hydroxy-3-methylglutaryl-
    coenzyme A reductase activity, in: "3-Hydroxy-3-methylglutaryl-
    coenzyme A reductase", J.R.Sabine, ed., CRC Press, Inc., Boca
    Raton, Florida (1983).
35. D.Berg, E.Regel, H.E.Harenberg, M.Plempel, Bifonazole and
    clotrimazole. Their mode of action and possible reason for the
    fungicidal behaviour of bifonazole, _Arzneimittelforsch._, 34: 139
    (1984).
36. M.Bard, N.D.Lees, A.S.Burnett, R.A.Packer, Isolation and charac-
    terisation of mevinolin resistant mutants of _Saccharomyces
    cerevisiae_, _J.Gen.Microbiol._, 134: 1071 (1988).
37. C.E.Nakamura and R.H.Abeles, Mode of interaction of 3-hydroxy-3-
    methylglutaryl coenzyme A reductase with strong binding inhi-
    bitors: compactin and related compounds, _Biochemistry_, 24: 1364
    (1985).
38. N.S.Ryder, Mechanism of action and biochemical selectivity of
    allylamine agents, _Ann.N.Y.Acad.Sci._, 544: 208 (1988).

39. N.S.Ryder, M.C.Dupont, I.Frank, Ergosterol biosynthesis inhibition by the thiocarbamate antifungal agents tolnaftate and tolciclate, Antimicrob. Agents Chemother., 29: 858 (1986).

40. H.Vanden Bossche, Biochemical targets for antifungal azole derivatives: hypothesis on the mode of action, in: "Current Topics in Medical Mycology", Vol.1, Springer Verlag, New York (1985).

41. H.Vanden Bossche, P.Marichal, J.Gorrens, H.Geerts, P.A.J.Janssen, Mode of action studies - Basis for the search for new antifungals, Ann.N.Y.Acad.Sci. 544: 191 (1988).

42. Y.Yoshida and Y.Aoyama, Interaction of azole fungicides with yeast cytochrome P. 450 which catalyzes lanosterol 14 α-demethylation, in: "In Vitro and In Vivo Evaluation of Antifungal Agents", K.Iwata and H.Vanden Bossche, eds., Elsevier Publishers BV (Biomedical Division), Amsterdam (1986).

43. Y.Aoyama and Y.Yoshida, The 14 α-demethylation of lanosterol by a reconstituted cytochrome P-450 system from yeast microsomes, Biochem.Biophys.Res.Commun., 85: 28 (1978).

44. Y.Aoyama, Y.Yoshida, R.Sato R, Yeast cytochrome P-450 catalysing lanosterol 14 α-demethylation, J.Biol.Chem., 259: 1661 (1984).

45. Y.Aoyama, Y.Yoshida, Y.Sonoda, Y.Sato, Metabolism of 32-hydroxy-24,25-dihydrolanosterol by purified cytochrome $P-450_{14DM}$ from yeast. Evidence for contribution of the cytochrome to the whole process of lanosterol 14 α-demethylation, J.Biol.Chem., 262: 1239 (1987).

46. J.M.Trzaskos, R.T.Fischer, M.F.Favata, Mechanistic studies of lanosterol C-32 demethylation. Conditions which promote oxysterol intermediate accumulation during the demethylation process, J.Biol.Chem., 261: 16937 (1986).

47. A.Polak, Mode of action of 5-fluorocytosine in Aspergillus famigatus. in: "Aspergillus and Aspergillosis", H.Vanden Bossche, D.W.R.Mackenzie, G.Cauwenbergh, eds., Plenum Press, New York & London (1988).

48. H.J.Scholer, Flucytosine, in: "Antifungal Chemotherapy,", D.C.E. Speller, ed., J.Wiley & Sons Ltd., Chichester (1980).

49. P.S.Mamont, M.-C.Duhesne, J.Grove, P.Bey, Anti-proliferative properties of DL-α-difluoromethylornithine in cultured cells. A consequence of the irrversible inhibition of ornithine decarboxylase, Biochem.Biophys.Res.Commun., 81: 58 (1978).

50. P.P.McCann, C.J.Bacchi, A.B.Clarkson, P.Bey, A.Sjoerdsma, P.J.Schester, P.D.Walzer, J.L.R.Barlow, Inhibition of polyamine synthesis by α difluoromethylornithine in African trypanosomes and Pneumocystis carinii as a basis of chemotherapy: biochemical and clinical aspects, Am.J.Trop.Med.Hyg., 35: 1153 (1986).

51. J.C.Edman, J.A.Kovacs, H.Masur, D.V.Santi, H.J.Elwood, M.L.Sogin, Ribosomal RNA sequence shows Pneumocystis carinii to be a member of the fungi, Nature 334: 519 (1988).

52. M.A.Pfaller, T.Gerarden, J.Riley, Growth inhibition of pathogenic yeast isolates by a-difluoromethylornithine: an inhibitor of ornithine decarboxylase, Mycopathol. 98: 3 (1987).

53. S.M.Boyle, N.Spiranganathan, D.Cordes, Susceptibility of Microsporum and Trichosporum species to suicide inhibitors of polyamine biosynthesis, J.Med.Vet.Mycol., 26: 227 (1988).

INFLUENCE OF AZOLE COMPOUNDS ON ADHESION, GERM TUBE FORMATION AND VIRULENCE OF C. ALBICANS IN CELL CULTURES AND INFECTED ANIMALS

K.D.Bremm   J.Hawkins   M.Plempel   D.Berg

Institute for Chemotherapy, Bayer Research Centre, Aprather Weg, 5650 Wuppertal, The Federal Republic of Germany

Three stages of fungal infection have been described; attachment of the fungal cells, production of germ tubes and penetration of the epithelial surface by hyphae. It was the purpose of our investigations to study the influence of azole compounds on these three stages of infection.

Adherence of microorganisms to host tissues is regarded as a prerequisite for tissue invasion. For example, the ability of bacterial pathogens to adhere to mammalian tissue has been shown to equate with pathogenicity in the case of Neisseria gonorrhoeae and enterotoxigenic E.coli infections. Although investigations of fungal adherence has been more limited, Candida species have been shown to adhere to human buccal epithelial cells, human vaginal epithelial cells, the gastrointestinal epithelium and materials used in dentures. In our laboratory we employed in addition HeLa or Vero cells for adhesion studies. With this model we were able to show that there is a hierarchy among Candida species with regard to their ability to adhere to cell cultures. Candida albicans is clearly the most adherent, which corresponds well to their virulence in comparison to the other species. This result is in accordance with results from Klotz et al.who used vascular endothelium as target, and these results demonstrate that there is a correlation between adherence and virulence.

In order to study adherence of C. albicans to cell cultures we used 2.5×10^5 HeLa-cells as monolayer or in suspension and incubated them with 2.5×10^6 C. albicans which were grown in liquid media. The influence of incubation time on adhesion was investigated using Vero-cells as targets. During the first hour there is a clear increase in adhesion which turns into a plateau during the second hour. For this reason, we used an incubation time of one hour for our tests. It was also obvious from our studies that the adherence of untreated yeasts to human culture cells was always round 50% with MEM as medium. There was no difference in this value whether we used monolayers or cell suspensions. There was only a sharp increase up to 70% adherence when germ tube producing cells were used compared to 50% with budding cells. With this test system we investigated the influence of azoles on adherence of C. albicans to cell cultures when C.albicans (2.5×10^6) were pretreated for 30 minutes with the three Bayer-azoles

-clotrimazole, bifonazole, R3783- within a concentration range from 0.01 to 100 mcg/ml. We observed that beginning with 0.1 mcg/ml three is a significant decrease of adherence by pretreated yeasts. It was not possible with our test system to see a decreased adhesion of C.albicans to cell cultures when the azoles were added simultaneously with the yeasts. Furthermore, there was no diminished adhesion when germ tubes were pretreated with azoles and subsequently incubated with cell cultures. With other substances like chlorhexidine it was also possible to inhibit adhesion when they were added simultaneously with the yeasts to the cell cultures. A comparison of chlorhexidin pretreated and simultaneously treated yeast with regard to their ability to adhere to HeLa-cells showed that there is nearly no difference between preincubation and simultaneous addition. Furthermore, chlorhexidin induced also a diminished adherence when germ tubes were incubated with cell cultures in the presence of that compound. Taken together the results of adherence of yeasts to cell cultures demonstrate that there is an influence of azoles on fungal adhesion. It seems that it is mainly a problem of sensitivity to show further effects. There remain some questions concerning size of inoculum, physiological substrate of adhesion and role of germ tube formation which have to be clarified in order to establish the influence of azoles on fungal adhesion.

In order to examine these problems tissue cultures of ectocervical epithelial cells in monolayer culture were used as a model. The process of invasion of the cells by Candida was observed by light and scanning electron microscopy and the effects of the antifungal drugs was assessed. Tissue cultures of ectocervical epithelial cells were set up by taking small biopsies of tissue from the uterine cervix of women having abdominal hysterectomy at Hammersmith Hospital. The cultures were grown in Eagles minimal essential medium, 15 % fetal calf serum, 1% amino asids, 1% L-glutamine and antibiotics. After four days incubation of these fragments a bud of epithelial cells can be seen growing out from the explant, and after 8 days in culture the cells have divided into a monolayer of about 5000 cells which completely surround the explant. Under scanning electron microscopy the cells have the typical pavement like appeerence of squamous epithelial cells in culture. It is at the eight-day stage that the cultures were used in experiments. Clinical isolates of Candida albicans were maintained in Sabouraud's dextrose agar. Subcultures from a single strain were made 24 hours prior to using in experiments.The Sabouraud's medium was washed off by centrifugation at 1000xg for 5 minutes and the fungal cells were resuspended in either Hank's balanced salt solution or antibiotic free epithelial cell medium depending on the experiment. The process of attachment of Candida cells to the epithelial cells, production of germ tubes and invasion of the cells was observed by fixation at 2, 4, 6 and 24 hours after inoculation. Within the first two hours, yeasts settle singly or in groups. After 4 hours the the yeasts cells start to produce germ tubes which either invade immediately or grow out across the surface of the epithelial cell before invading. Six hours after inoculation hyphae weaving in and out of the cells are usually seen. After 24 hours almost complete destruction of the monolayer has occured and large cracks in the upper surface of the epithelial cells reveal interlacing hyphae within the cells. The action of azoles was assessed on each of these stages of infection. Quantitative studies of attachment of C. albicans yeast cells to epithelial cells were made after placing the dishes containing inoculated cultures on a rotating table set inside a dry incubator for 2 hours, decanting the salt solution containing unattached yeast cells, washing twice with phosphate buffer saline, fixing and staining with Giemsa. The number of yeast cells attached to epithelial cells per unit area of monolayer culture was counted using the light microscope.

In experiments where the action of the drugs on attachment was studied the drugs were incorporated in Hank's balanced salt solution at the time of inoculation of the Candida onto the cultures. Clotrimazole in concentrations between 0.01 and 0.9 mcg/ml reduced the number of yeast cells adhering to epithelial cells by 80% compared to the controls. Bifonazole was slightly more potent, producing the same results between 0.01 and 0.3 mcg/ml.

A study of the next stage of infection, that of germ tube production, was carried out in Eagle's medium with serum. In these experiments, the drugs were added 2 hours after addition of the yeasts to the epithelial cells and the cultures were incubated for a further 2 hours.

Under the light microscope the number of attached yeast cells which had produced germ tubes was the same in the controls and treated cultures. There was a difference between controls and treated cultures in the length of germ tube produced. In the controls this was 0.05-0.07 mcm. With the azoles (clotrimazol, bifonazole) tested concentration range of 1ng/ml-1mcg/ml the length of the germ tubes was shorter than in the controls, being 0.01-0.02 mcm.

In conclusion, what we have now demonstrated is that very low concentrations of antifungal drugs, like azoles, between 0.001 and 0.8 mcg/ml can interfere with adhesion of the fungal cells and inhibit production of normal germ tubes and hyphae by yeast cells that have attached.

The results of this in vitro studies correlate well with in vivo studies which were performed with cells of C. albicans exposed to subinhibitory concentrations of clotrimazole or bifonazole. They were found to have a reduced capacity to divide normally or produce germ tubes and mycelia under appropriate conditions, although treated cultures produced a cell mass equivalent to that of untreated controls. Animals infected with the non-pretreated control cells exhibited high levels of mortality while the animals receiving cells pretreated with an azole for 16-24 h had survival rates of 90-100% by the 8th day after infection. After pretreatment with 2 and 4 mcg/ml of azole for 16-24 h, the cellular morphology was examined microscopically and compared to nonpretreated control cultures. Pretreatment periods of 1-6 hours resulted in no differences in mortality in comparison to control cells, whole pretreatment for 10 h resulted in an intermediate loss of virulence. In a further experiment we compared C. albicans cell counts in the kidney homogenates of mice 24, 48 and 72 hours after the injection of pretreated and nonpretreated cells. A striking feature was the extraordinary reduction of the CFU in the kidney of the animals infected with pretreated cells as compared to the counts in animals infected with nonpretreated cells. It appears that previously damaged cell of C. albicans can be eliminated by host defence mechanisms to a much greater extent and more rapidly than normal C. albicans cells.

The pretreated C. albicans cells exhibited pronounced conglomeration. The aggregation consisted of 10-30 cells. This effect was more pronounced with bifonazole than with clotrimazole. We consider conglomerate formation and separation during cell division. Nonpretreated C. albicans cells from shake cultures and cells that had been pretreated for 16-24 h were centrifuged and washed and then transferred to bovine serum, incubated at 37 °C, and examined under the microscope at 2-hours intervals to establish the frequency and the extent of germ tube formation. Compared te control cells, the pretreated microorganisms exhibited a marked delay and reduced frequency of germ tube formation. Germ tubes from pretreated cells after incubation for 24 h were shorter and appeared collapsed in comparison to the controls. This effect was more pronounced after treatment with clotrimazole than with bifonazole. Furthermore, reduction of germ-tube formation by bifonazole seemed to be also a secondary phenomenon to conglomerate formation, in that only cells at the edge of a conglomerate were able to develop germ tubes. When azole pretreated and nonpretreated C. albicans cells were transferred into and incubated in liquid Eagle 's medium containing fetal calf serum at 37 °C the mycelium formation was reduced compared with the controls after incubation for 10 and 24 hours. The mycelia that did develop were small and exhibited little branching. Clotrimazole and bifonazole proved more or less equally effective under these conditions. These results are in agreement with the findings of other authors that concentrations in the ng-range of azoles effected a complete inhibition of the transformation of budding cells of C. albicans into mycelia.

We started further investigations to demonstrate the reduced virulence of C. albicans in cell cultures. For this purpose, mono-layers of 2.5×10^5 HeLa-cells were grown overnight and subsequently stained with a vital dye which did not affect their properties. Incu-bation of these monolayers with C. albicans resulted in a loss of viability of the cell cultures what could be demonstrated by the appearence of stained cells in the supernatant. The amount of desquamated cells could be quantitated by photometric means. When chlorhexidin was co-incubated with the infected cell cultures, there was a dose dependent reduction of desquamated cells. This result gave us further evidence that adhesion of C. albicans is an important part of their virulence. In order to study the effect of azole in this test system the monolayer was then incubated over 4-24 h with 2.5×10^5 C.albicans in the presence and absence of antifungals or other drugs. Our tests revealed evidence that all tested azoles showed a protective effect in the concentration range between 0.1 mcg/ml and 10 mcg/ml. Furthermore, it was important to see that this model revealed differences between the different azoles and showed significant dose-response correlations. The most potent drugs were fluconazole and Bay R3783, which show already at 0.1 mcg/ml 95-100% protection. This can not be due to reduced Candida counts but to changed expression of virulence factors like adhesion, enzyme release, etc. Further investigations with this test system might give us new informations about the effect of azoles on fungal factors of pathogenicity.

In summary : We investigated the three early stages of fungal infection adhesion, germ tube formation and invasion with primary cell cultures, permanent cell cultures and in vivo models. The influence of azoles was demonstrated at all three stages in the different test systems. This means, as a consequence of the known inhibition of ergosterol biosynthesis, the virulence of C.albicans is changed by the azoles, because the effects were also seen in subinhibitory concentrations.

REFERENCES

For references please contact K.D. Bremm, the first author of this paper.

COMBINATION THERAPY IN EXPERIMENTAL AND CLINICAL CANDIDOSIS

A. Polak

F. Hoffmann-La Roche & Co., Ltd., Basel, Switzerland

INTRODUCTION

Antimycotic chemotherapy for deep-seated candidosis in patients with reduced immune-defense is still a problem. Only a few antimycotics are available for systemic use and all have their limitations. Thus, interest in drug combinations is increasing in the hope of improving the efficacy and the spectrum of antimycotic therapy.

MATERIAL AND METHOD

The animal model used has been described in detail earlier [1]. Mice weighing 20g were infected with Candida yeast cells. The single and combined drugs were administered 5 times, the first dose given immediately after the infection, the second 6h later, thereafter once daily. Ten animals were used per dosage group. The survival time of treated and untreated animals was observed during 20 days. The effect of the dosage combination was compared to the better of the partners in monotherapy, the effect of the triple combination was compared to the better double combination.

The following antimycotics have been investigated: the polyene amphotericin B (AmphB), flucytosine (5FC) and the imidazole and triazole derivatives ketoconazole (Keto), itraconazole (Itra) and fluconazole (Flu).

RESULTS AND DISCUSSION

Animal studies

It has long been known that combination of 5FC with AmphB exhibits an improved chemotherapeutic activity in deep-seated candidosis, and an especially strong synergy is observed against strains partially resistant to 5FC [1,2]. The combination of 5FC with imidazole or triazole derivatives also shows a beneficial effect in most instances and antagonism has never been observed [1]. The degree of synergism is strongly dependent on the 5FC sensitivity of the Candida strain and on the azole derivative used. As with the combination of 5FC + AmphB, the synergism is more pronounced in 5FC resistant strains than in highly sensitive ones. Ketoconazole, which shows only a weak anti-Candida activity, in monotherapy, exerts only an additive effect in

Candida and Candidamycosis, Edited by E. Tümbay et al.
Plenum Press, New York, 1991

101

Table 1.Combination therapy in experimental candidosis.

Drug Combination	Strain H12 5FC Sensitive	Strain 140/1 5FC Moderately sensitive	Strain Sh13 5FC Partially resistant
5FC+AmphB	ADD/SYN	ADD/SYN	SYN
5FC+Keto	IN/ADD	ADD/SYN	SYN
5FC+Itra	ADD/SYN	ADD/SYN	SYN
5FC+Flu	ADD/SYN	SYN	ADD/SYN
AmphB+Keto	ANT	ANT	ADD/SYN
AmphB+Itra	IN/ANT	-	IN

combination with 5FC. Itraconazole, and the water soluble fluconazole, which show good to high efficacy in our <u>Candida</u> model when given alone, show a genuinely synergistic effect in combination with 5FC, of a degree as good as with 5FC + AmphB (Table 1).

The earlier observation [3] that the combination of 5FC + Flu did not show the same high beneficial effect as the combination of 5FC + AmphB, has proved to be strain-dependent. When different strains are tested, the synergistic effect is more pronounced and compares favorably to those of 5FC + AmphB or 5FC + Itra.

The combination of AmphB with azoles showed a strong antagonistic effect in candidoses as well as in aspergillosis [1,3,4]. This antagonism can be explaind in terms of differences in the mode of interference with the fungal membrane ergosterol: all azoles inhibit the synthesis of ergosterol, therefore, AmphB loses its binding site in the membrane. Despite the fact that all tested azole derivatives inhibit ergosterol synthesis, the degree of antagonism varied. Ketoconazole exerted a significant antagonism even after a single prophylactic dose, [5,6] whereas with AmphB + Itra only a weak antagonism was observed. This difference may be explained by the better chemotherapeutic activity of Itra compared to Keto in our <u>Candida</u> model. It may be that that <u>in vivo</u> Itra influences the candidosis not only by the inhibition of ergosterol but by another as yet unknown mechanism.

In conclusion, the following (double) combination therapy for the deep-seated candidosis may be recommended for clinical use: 5FC + AmphB, 5 FC + Flu. The combination AmphB + azoles should be avoided.

In recent years the treatment of cryptococcosis in AIDS has become a major problem. The standard therapy of 5FC + AmphB [7] was in many cases not sufficient to prevent a relapse, so new ways to treat these infections are being sought. The combination of 3 different antifungals may have some advantages in such cases. However, a triple combination would usually involve the combination of AmphB with an azole derivative with the disadvantage outlined above [6]. We, therefore, studied various triple combinations in our animal models of <u>Candida</u> infection. Using ketoconazole in combination with 5FC + AmphB brought no benficial effect; on the contrary, the number of <u>Candida</u>

cells in the kidneys of mice treated with the combination 5FC + AmphB + Keto were significantly higher than that in mice treated with 5FC + AmphB [5].

In more recent experiments we used as our parameter the survival time of mice treated with triple combinations. When 5FC + AmphB was combined with Itra or Flu, the antagonistic effect was not so and a 5FC resistant strain. Activity enhancement was observed in 3 instances, indifference in 9 and a slight antagonistic effect was observed in 4 instances (Table 2). The antagonistic effect was never statistically significant, there was only a trend to a reduced survival time. Furthermore, this antagonistic effect was only seen when both the dose of AmphB and the dose of Itra were subinhibitory, i.e. the serum level of Itra was too low to inhibit the cell growth, but not too low to inhibit the ergosterol biosynthesis. In addition, the low dose of AmphB was not sufficient to inhibit cell growth, especially not in ergosterol depleted Candida cells.

Table 2. Chemotherapeutic activity of triple combinations in comparison to that of the better double combination.

Dose (mg/kg)	5FC-sensitive C.albicans	Dose (mg/kg)	5FC-resistant C.albicans
5FC+AmphB+Itra		5FC+AmphB+Itra	
6.25+0.06+15	S-ANT	12.5+0.06+6.25	ANT
6.25+0.06+30	ADD	12.5+0.06+12.5	ADD
6.25+0.12+15	INDIFF	12.5+0.12+6.25	INDIFF
6.25+0.12+30	INDIFF	12.5+0.12+12.5	INDIFF
12.5+0.06+15	S-ANT	25.0+0.06+6.25	INDIFF
12.5+0.06+30	ADD	25.0+0.06+12.5	INDIFF
12.5+0.12+15	S-ANT	25.0+0.12+6.25	INDIFF
12.5+0.12+30	INDIFF	25.0+0.12+12.5	INDIFF
5FC+AmphB+Flu		5FC+AmphB+Flu	
6.25+0.06+12.5	INDIFF	12.5+0.06+2.5	S-ANT
6.25+0.06+25	INDIFF	12.5+0.06+5	INDIFF
		12.5+0.06+10	*
6.25+0.12+12.5	INDIFF	12.5+0.12+2.5	S-ANT
6.25+0.12+25	ADD	12.5+0.12+5	ANT
12.5+0.06+12.5	ADD	25.0+0.06+2.5	S-ANT
12.5+0.06+25	ADD	25.0+0.06+5	INDIFF
12.5+0.12+12.5	INDIFF	25.0+0.12+2.5	INDIFF
62.5+0.12+25	ADD	25.0+0.12+5	ADD

S-ANT, slightly antagonistic; ANT, antagonistic; INDIFF, no additive effect; ADD, better than better double combination
* All animals survived with single Flu or double or triple combination.

When Flu was used as third partner, the results depended on the strain used. In 50 % of the cases a beneficial effect was seen; in 50 % indifference (Table 2). With the 5FC sensitive strain antagonism was never observed. With the 5FC resistant strain, Flu alone showed an extremely good activity-with 10 mg/kg all animals survived. Thus the combination therapy could only be evaluated with lower doses. Under these conditions antagonism was observed in 4 instances, indifference in 3 and a beneficial effect only in one instance. When a sufficiently high dose of Flu (5 mg/kg) and AmphB (0.12 mg/kg) is combined with 25 mg/kg of 5FC, a truly additive effect is seen. Combinations with higher doses always lead to 100 % survival using this strain (Table 2). Flu may be a better partner than Itra in a triple combination due to its pharmacokinetic properties. Flu is the only triazole derivative which is eliminated unchanged in the active form by the kidney, the organ where Candida is manifested in our mouse model.

In summary, our experiments showed that only certain triple combinations are beneficial. The expected antagonism between AmphB and azole derivatives was only marked in the combination with ketoconazole.

From our data in the animal model it may be concluded that a triple combination could also be used in the clinic. However, it must be noted that only high doses i.e. those normally used of all combination partners should be employed in order to reach sufficiently high serum and tissue levels to inhibit the growth of Candida albicans and not only to inhibit the ergosterol biosynthesis. The latter effect is induced in vivo and in vitro at extremely low concentrations of imdazole or triazole derivatives which by themselves do not inhibit cell growth.

Combination in human chemotherapy

The clinical usefulness of all the studies mentioned remains to be demonstrated. Only one combination, namely 5FC + AmphB, has been tested in controlled clinical trials and showed synergy in the therapy of meningeal cryptococcosis[7]. However, one should remember that it is not only a true synergistic effect on the level of the fungal cell that is important, but also the complementary antimicrobial and pharmacological properties of the two or three partners.

This complementary effect is especially evident with the combination of 5FC plus AmphB [8] and explains the favourable results obtained in human chemotherapy. Particular properties of individual triazoles may bring additional benefits to the combination with 5FC, e.g. the accumulation of Itra in the skin or its high and long-lasting tissue levels.

There have been no controlled clinical trials of combination therapy in Candida infections, but there have been several examples in which a combination therapy showed clinically advantages.

Candida sepsis, disseminated candidosis and Candida meningitis can prove lethal to premature babies. Monotherapy with 5FC has given a success rate (80 % in sepsis), but 5FC + AmphB has to be used for Candida meningitis [9,10].

Antifungal therapy for Candida sepsis in patients with neoplastic disease is not effective as long as the underlying disease is not controlled. In leukaemic patients therapy only leads to a permanent

Table 3. Candida infections in neutropenic and neoplastic patients.

Infective fungus	Therapy	No. of patients	Cure rate (%)
C. albicans	none	47	26
	AmphB	132	32
	AmphB+5FC	18	55
C. tropicalis	none	16	31
	AmphB	72	25
	AmphB+5FC	13	69

cure if bone marrow recovery occurs [11,12]. In Europe, the standard therapy for Candida infection in the area of neoplastic diseases is the combination of 5FC + AmphB. This therapy is employed with good success especially when the therapy is started early. In the USA, Candida infections have been treated mainly by AmphB monotherapy with a relative poor outcome (30 %) [11]. The use of combination therapy leads to higher cure rate, especially when the infective agent is Candida tropicalis (Table 3).

Hepatic candidosis has been increasingly diagnosed in immunocompromised patients. Combination therapy may be more successful than AmphB monotherapy in these cases as well; in some instances a maintenance therapy with Keto is added after a successful therapy [13].

Dupont and Drouhet [14] have described a new Candida syndrome in heroin addicts. As long as it is restricted to the skin, monotherapy with Keto or another newer triazole derivative is sufficient for a permanent cure. However, when an additional localization is diagnosed, monotherapy is less efficient. For example, when endophthalmitis is also involved, cure is only achieved after the eyesight has been lost; under combination therapy the sight can in most cases be saved. When the bone is also infected (osteoarthritis), a combination therapy has to be used to achieve a permanent cure. Dupont and Drouhet [14] have used the combination 5FC plus Keto, but others used 5FC + AmphB with success. In our opinion 5FC + Flu or 5FC + Itra may be also highly beneficial for this indication. In some instances the infected bone has to be surgically treated despite aggressive therapy.

Thus, details of the published cases in treatment of Candida infection reveal that combination therapy with 5FC + AmphB or 5FC + Keto may be superior to monotherapy. For Candida infection no clinical experiences with newer combinations i.e. 5FC + Itra or 5FC + Flu have been reported although these combinations have been used for human chemotherapy with success for other indications. Recently Viviani and Tortorano [15] demonstrated that 5FC + Itra has a beneficial effect over Itra monotherapy in cryptococcosis of AIDS patients. Borelli [16] reported that the high efficacy of Itra in chromomycosis is only apparent when it is caused by Cladosporium carionii. When the infection is caused by Fonsecaea pedrosoi, Itra monotherapy was not sufficient to induce a cure, but 5FC plus Itra were highly efficacious and cured all infections caused by F. pedrosoi. From this data it is apparent that 5FC plus Itra is well tolerated even in AIDS patients, and that this combination is more powerful than the monotherapy. Therefore, 5FC + Flu may show the same beneficial effect, and the triple combination may also provide a beneficial effect in some indications which are difficult to treat with the known monotherapies.

RERERENCES

1. A.Polak, H.J. Scholer, M.Wall, Combination therapy of experimental candidiasis, cryptococcosis and aspergillosis in mice, Chemotherapy, 28: 461 (1982).
2. A.Polak: Synergism of polyene antibiotics with 5-fluorocytosine, Chemotherapy, 24: 2 (1978).
3. A.Polak,: Combination therapy of experimental candidiasis, cryptococcosis, aspergillosis and wangiellosis in mice, Chemotherapy, 33: 381 (1987).
4. A.Polak, 5- Fluorocytosine et associations, Ann.Biol.Clin., 45: 669 (1987).
5. A.Polak and A.Schaffner, Association des antifongiques dans les mycoses experimentales, Med.Mal.Inf., 14: 553 (1980).
6. A.Schaffner and P.G.Frick, The effect of ketoconazole on amphotericin B in a model of disseminated aspergillosis, J.Infect.Dis., 151: 902 (1985).
7. J.E. Bennett, W.E.Dismukes, R.J.Duma, G.Medoff, M.A.Sande et al., A comparison of amphotericin B alone and combined with flucytosine in the treatment of cryptococcal meningitis, New Engl. J.Med., 301: 127 (1979).
8. A.Polak, Combined therapy of systematic mycoses, in: Proc. 13th Intern. Congr. Chemother., Vienna 1983, K.H.Spitzy and K.Karrer, eds., Symposia 49/1 and 49/2, part 20, pp. 2-9, H.Egermann, Vienna (1983).
9. J.E.Bennett, R.M.Kliegman, A.A.Fanaroff, Disseminated fungal infections in very lowbirth-weight infants: therapeutic toxicity, Pediatrics, 73: 153 (1984).
10. R.G.Faix, Systemic Candida infections in infants in intensive care nurseries: high incidence of central nervous system involvement, J.Pediatr., 105: 616 (1984).
11. D.Armstrogn, Problems in the management of opportunistic fungal infections, 9th Ian Murray Memorial Lecture, Brit.Soc.Mycopathol. (1986).
12. J.W.M.Gold, Opportunistic fungal infections in patients with neoplastic disease, Am.J.Med., 76: 458 (1984).
13. E.Haron,R.Feld, P.Tuffnell, B.Patterson, R.Hasselback, A.Matlow, Hepatic candidiasis: an increasing problem in immunocompromised patients, Am.J.Med., 83: 17 (1987).
14. D.Dupont and E.E.Drouhet, Cutaneous, ocular, and osteoarticular candidiasis in heroin addicts, new clinical and therapeutic aspects in 38 patients. J.Infect.Dis., 152: 577 (1985).
15. M.A.Viviani and A.M.Tortorano, Management of cryptococcosis in AIDS patients, Xth Congr. ISHAM, Barcelona 1988, poster N.RT17-1.
16. D.Borelli, A clinical trial of itraconazole in the treatment of deep mycoses and leishmaniasis, Rev.Infect.Dis., 9 (Suppl. 1) 557 (1987).

EXPERIMENTAL CANDIDOSIS IN ANIMALS AND CHEMOTHERAPY

J.Van Cutsem   F.Van Gerven   J.Fransen   P.A.J.Janssen

Janssen Research Foundation, B-2340   Beerse, Belgium

INTRODUCTION

Members of the genus Candida are widely spread. They are present as endosaprophytes in the gastrointestinal tract of man and the majority of animals or as exosaprophytes in the environment. Various Candida spp. may be responsible for pathological manifestations, but Candida albicans is considered to be the species with the highest potency for morbidity and mortality, mainly in immunocompromised individuals or when predisposing, known or unknown, factors or regimens are present. Any organ may be infected, but superficial infections constitute by far the highest incidence.

In order to study the various aspects of candidosis and to evaluate drug activity, several animal models have been proposed [1-17]. In this review, a number of important standardized animal models of candidosis that present the various clinical aspects and that allow good therapeutic extrapolation to clinical disease in man and animal, are presented. Special attention was drawn to the following experimental models:

1. A model of skin candidosis in guinea-pigs, that can be treated orally or topically was described [8] and will be more largely extended.

2. Vaginal candidosis was easily induced in ovariectomized and hysterectomized rats with oestrogen-induced permanent pseudo-cestrus [6]. Additional injection of streptozotocin, inducing diabetes [18], contributed to prolonged presence of Torulopsis glabrata in the vagina.

3. Invasion of placenta and foetuses in rodents was caused by oral and intravenous infection with high doses of C. albicans.

4. Candidosis of the gastrointestinal tract of guinea-pigs, predisposed by corticoids and antibiotics was attained by gavage of C.albicans [7].

5. Systemic candidosis was obtained in various animal species with emphasis to the systemic infection in guinea-pigs [9-12].

Skin candidosis

Albino guinea-pigs weighing more than 500 g were inoculated on the intact electrically clipped non-occluded skin of the back with 8×10^6 colony forming units (CFU) of C. albicans [8]. The blastospores

Candida and Candidamycosis, Edited by E. Tümbay *et al.*
Plenum Press, New York, 1991

were suspended in saline. Diabetes was induced 24 hours before infection by administering 200 mg per kg body weight (BW) of alloxan intramuscularly (IM). The inoculation resulted within 3 to 4 days in minute papules.

After 6 to 8 days small vesicles, containing a clear exudate, were present. Erythema and oedema appeared, followed by hyperkeratosis and desquamation.

Spontaneous healing was reached in 5 to 6 weeks. Large numbers of yeasts and (pseudo-) hyphae were present during the first two weeks after infections, but C. albicans was isolated in only 50 % of the animals after three weeks. Histological preparations stained with periodic acid Schiff (PAS) or Gomori's methenamine silver nitrate (GMSN) revealed the fungal elements in the stratum corneum and in the necrotic crusts of the epidermis. In total, the pathogenicity of 166 strains of C. albicans and C. stellatoidea was evaluated: 97.6 % of the strains produced pronounced skin infection and 1.8 % caused moderate skin infection, while only 0.6 % was non-pathogenic (Table 1).

In 20.6 % of 34 strains of five other Candida spp. evaluated (tropicalis, pseudotropicalis, krusei, parapsilosis and guilliermondii) and in 44.4 % of 9 Torulopsis glabrata strains, moderate skin eruptions appeared. In general, these eruptions healed in 7 to 10 days (Table 1).

C. albicans is the species of choice for skin infections. These can be treated orally or topically. The guinea-pig remains the indicated animal for Candida skin infections: it is highly sensitive and can be treated with success topically or orally with ketoconazole or itraconazole [15]. It will not touch the infection site on the skin of the back, nor remove any topically applied medication. Rats and mice are not highly sensitive, moreover, they always lick and scratch their skin. The rabbit is as sensitive as the guinea-pig, but the rabbit may have the same inconveniences as other rodents. Moreover, clipping of rabbit hairs is more fastidious.

Table 1. Pathogenicity of Candida and Torulopsis on the intact skin of alloxan-diabetic guinea-pigs.

Species	No. of strains	% of pathogen strains	
		highly	moderately
C. albicans & C. stellatoidea	166	97.6	1.8
Other Candida spp.*	34	0	20.6
T. glabrata	9	0	44.4

* Candida: tropicalis (8), pseudotropicalis (8), krusei (6) parapsilosis (6), and guilliermondii (6)

Vaginal candidosis

Vaginal candidosis can be induced in several animal species, but persistence of infection without spontaneous cure is irregular in non-castrated female animals [3,6,10,12,13,15]. In spite of high doses of oestrogens weekly administered to non-castrated animals, spontaneous negativation occurred in 2-3 weeks in 100 % of the rabbits, 100 % of the guinea-pigs, 90 % of the mice and 70 % of the rats.

Ovariectomy and hysterectomy of rats, followed by weekly subcutaneous injection of oestrogens (0.1 mg of oestradiol undecylate in 1 ml of an oily solution) produced a permanent pseudo-oestrus. These animals were highly susceptible to a vaginal infection by C.albicans, when infected intravaginally with $8x10^5$ CFU in a 0.2 ml volume. More than 99 % of the rats were positive two weeks after infection. In a group of 60 female castrated rats treated every week with oestrogens and controlled weekly for 11 months, 98 % were positive one month after infection, 90 % after three months and 71 % after 11 months (Table 2).

A large number of strains of C. albicans and several strains of other Candida spp. were intravaginally infected with various inoculum sizes, but only with C. albicans a constant infection was possible. For all further experiments strain C. albicans ATCC 44858 was selected. Rats were treated orally or topically. Therapeutic oral once daily treatment started three days after infection for one day to three consecutive days. At least 12 animals were used in each group. Therapeutic topical treatment started three days after infection and was given twice daily for three days. In order to assess the infection, samples were collected from the vagina, 2, 6 and 9 days after the last treatment.

Results of oral treatment for one to three days are given in Table 3 and of topical treatment for three days in Table 4. High cure rates were obtained with the azoles used (ketoconazole, fluconazole and itraconazole), and after oral as well as after topical treatment, the highest percentages of cure at the same dosages were observed with itraconazole.

Table 2. Vaginal infection with C. albicans in castrated rats in permanent pseudo-oestrus (group of 60 rats). Observation for 11 months.

Control (months)	% of positive rats
0.25	100
0.5	100
1	98
3	90
6	75
9	73
11	71

Table 3. Oral therapeutic treatment of vaginal candidosis in
rats with ketoconazole, fluconazole or itraconazole.

Treatment		% of cured + % of improved rats		
mg.kg^{-1}	No. of days	Ketoconazole	Fluconazole	Itraconazole
Excipient	1	0-0	–	–
20	1	33+8	25+0	75+0
40	1	75+8	58+4	92+8
Excipient	2	0-0	–	–
2.5	2	ND*	ND	83+8
5	2	ND	ND	83+17
Excipient	3	0.2+0.2	–	–
2.5	3	19 +0	46+7	91+4
5	3	38 +4	75+8	94+3
10	3	95 +1	92+0	100+0

* ND: No data

Table 4. Topical therapeutic treatment of vaginal candidosis
in rats with ketoconazole, fluconazole or itraconazole for three days b.i.d.

Treatment concentration in %	% of cured + % of improved rats		
	Ketoconazole	Fluconazole	Itraconazole
excipient	0+2	–	–
0.125	6+3	58+25	87+2
0.5	92+3	92+0	96+4

Table 5. Vaginal torulopsidosis with T.glabrata in castrated
non-diabetic   and   diabetic rats in permanent
pseudo-oestrus (groups of 6).

T.glabrata strain	No. of infections	No.of positive animals on stated days after infections					
		Non-diabetic			Diabetic		
		7	11	14	7	11	14
B 51386	1	3	2	0	5	2	3
B 51386	3	2	1	1	6	2	2
B 51386	5	4	2	2	6	4	2
M13911-26.1	1	3	1	0	6	5	2
	3	4	2	2	6	5	2
	5	4	2	0	6	4	2

Streptozotocin at 40 mg.kg^{-1} was administered by the intraperitoneal (IP) route for three consecutive days: irreversible diabetes was obtained (serum glucose levels raised from 125-156 before, to 300-577 mg per dl 14 days after administration). When the diabetic rats in permanent pseudo-oestrus were infected intravaginally with T. glabrata, sampling revealed a prolonged presence of the yeasts in the vagina (Table 5). No higher yeast presence was obtained with repeated infections.

Infection in pregnancy

A. Oral infection

Administration by gavage of $2x10^8$ blastospores of C. albicans to pregnant guinea-pigs was without morbidity, while a dose of $8x10^8$ resulted in abortion in one out of six animals three days after infection and two out of three foetuses were highly positive. After gavage of $3.2x10^9$ blastospores of C. albicans, five out of six pregnant guinea-pigs aborted from day 2 to 5: the placentas that could be controlled were highly positive and 13 out of 16 foetuses were also positive. The sixth guinea-pig gave birth to two living young ones of which one was negative and one was clinically positive, presenting small eruptions on the skin, with a high content of fungal elements; two dead foetuses were also born and both were positive.

Three out of four pregnant rats receiving $8x10^8$ blastospores by gavage, aborted: only three foetuses could be examined and were positive. The fourth rat gave birth to three negative foetuses. After sacrificing the mother, seven implantations were noted and C. albicans was recovered from the uterus and the vagina.

Four pregnant mice were given $2x10^8$ C. albicans yeasts by gavage: two aborted, no foetuses could be examined. The two other mice were sacrificed and large amounts of hyphae were found on the epidermis of the foetuses.

B. Intravenous infection

A pregnant guinea-pig infected intravenously (IV) with 8.000 CFU of C. albicans per gram BW presented disseminated candidosis and invasion of almost all organs [11].

Foetal envelopes and foetuses of two pregnant rats infected IV with 10^7 CFU of C. albicans, remained free of infection. One rat had a fungal granuloma in the heart, its kidney was affected and one out of eight placentas was positive. In the second rat, no specific lesions were observed.

In two pregnant mice infected IV with 10^6 CFU of C. albicans, the kidneys were largely invaded with numerous granulomas. In the first mouse sacrificed 3 days after infection, 9 foetuses were found alive, 10 and 11 days old and without evidence of infection. In the second mouse sacrificed 6 days after infection, 7 living and 2 dead macerated foetuses of 15 days old were found. All foetuses showed infection with abundant formation of (pseudo-) mycelium generally restricted to the epidermis but with concomitant infection of envelopes and placenta.

Candidosis of the gastrointestinal tract

Guinea-pigs were immunocompromised by IM injection of 10 mg.kg^{-1} of prednisolone acetate, five doses weekly from day minus 8 on and by oral administration on alternating days of 50 mg.kg^{-1} of chloramphenicol and 40 mg.kg^{-1} of streptomycin, i.e. in total 6 days doses weekly, from day minus 2 on. C. albicans at a dosage of 2x10^7 CFU was given by gavage [9,10,12]. Faeces weekly collected and cultured were highly positive. At necropsy, tongue, oesophagus, stomach, small and large intestine were highly positive. The rectum contained in general very few yeasts. Ketoconazole and itraconazole given for two weeks, starting 5 days after infection were highly active in dosages from 2.5 to 10 mg.kg^{-1}.

The infection in mice was more irregular than in guinea-pigs. Crop candidosis was successfully established in young turkeys and treated with medicated feed at 16 to 31 ppm of ketoconazole, corresponding to a dose of 2.5 to 5 mg.kg^{-1} daily, respectively. Treatment by gavage was also performed and the results were comparable to those obtained with medicated feed.

Crop candidosis in chickens is always more discrete than in turkeys, due to lower sensitivity of the chicken for Candida.

Systemic candidosis

Systemic candidosis in various animal species has been described [2,7,9-12,17]. In these experiments blastospores of C. albicans were injected intravenously. The susceptibility was different from one animal species to another. Various animal species were infected in order to obtain a comparable intensity of infection (Table 6). After IV infection of the rabbit with 2.000 CFU of C. albicans per gram BW systemic candidosis was reached, comparable to that with 8.000 in the dog, the guinea-pig, the hamster, and the monkey, to that with 40.000 in the mouse, to that with 200.000 in the rat and to that with 40.000 in the chicken.

In the rabbit and especially in the guinea-pig, the infection was, compared to the other animal species, more regularly present in almost all organs. Moreover, only the guinea-pig among all animal species studied presented skin folliculitis and large erythematous and hyperkeratotic plaques on the skin. This phenomenon was not present in other species or if present, the incidence was largely reduced. In chickens however, the IV infection with C. albicans provoked lesions on the comb and on the wattles and intertrigo between the toes. The lesions on the comb and the wattles started as small eruptions. Rapidly these organs were covered with hyperkeratotic and powdery squamae, containing large amounts of yeasts and hyphae.

Intravenous inocula from 63 to 32.000 CFU of C. albicans ATCC 44858 per gram BW were used in guinea-pigs (Table 7). In order to evaluate the antifungal activity of compounds, the concentration of 8.000 CFU gram BW was selected [9,11,12]. The skin folliculitis in the guinea-pig infected by the IV route was in relation to organ invasion. Skin folliculitis occurred only after IV infection with C. albicans and C. stellatoidea. In total 159 strains were tested and skin folliculitis was not present in only one strain of C. albicans. This animal model of systemic candidosis in guinea-pigs covered the various aspects of human and animal candidosis and contributed to the evaluation of the activity of compounds by the oral and the parenteral

route. Itraconazole was dissolved for oral administration in poly-
ethylene glycol 200 and for parenteral treatment in hydroxypropyl-β-
cyclodextrin. The solvents were without effect on the infection and
were free of side-effects. Oral and parenteral treatment of systemic
candidosis with itraconazole was highly efficacious, safe and superior
campared to parenteral treatment with amphotericin B (Table 8).

Table 6. Systemic candidosis in various animal species after
IV infection with <u>C. albicans</u> ATCC 44858.

Animal		Inoculum $\times 10^3$ CFU per g BW	No. of animals				
Species	Breed or strain		Total	Skin folliculitis*		Positive cultures	
				Moderate	Pronounced	Skin	Kidneys
Rabbit	New Zealand	2	16	3	7	11	16
Dog	Canis vulgaris	8	6	2	0	2	5
Guinea-pig	Pirbright	8	920	1	919	919	917
Golden Hamster	Mesocricetus auratus	8	8	0	0	1	8
Monkey	Cercopithecus aethiops	8	6	2	1	3	5
Mouse	Swiss	40	80	2	0	5	80
Rat	Wistar	200	24	1	0	4	24
Chicken	Arbor acre	400	24	8	16	24	3

* <u>Candida</u> lesions on the comb, the wattles and intertrigo for
chickens, folliculitis for others.

Conclusion

In order to reproduce human and animal candidosis, experimentally
induced in laboratory animals, models that allow good extrapolation
from experimental to clinical disease, are needed. The models
described in this study met these requirements and contributed
largely to the selection of new efficacious drugs such as miconazole,
ketoconazole, itraconazole and others used not only in immuno-
competent patients, but also in patients that are highly immunocompromised, e.g. AIDS patients.

Table 7. Systemic candidosis in guinea-pigs after IV
infections with <u>C. albicans</u> ATCC 44858.

Inoculum No.of CFU per g BW	No.of Animals	Mortality %	Skin folliculitis		Positive cultures	
			Moderate	Pronounced	Skin	Kidneys
32.000	18	89	0	18	18	18
16.000	30	67	0	30	30	30
8.000	920	25	1	919	919	917
6.000	36	17	0	36	36	35
4.000	40	13	1	39	39	37
2.000	12	0	2	10	1	10
1.000	12	0	7	3	4	9
500	6	0	4	1	2	2
250	12	0	6	0	2	3
125	6	0	4	0	3	1
63	12	0	2	0	0	0

Table 8. Oral and parenteral treatment* of systemic candidosis
in guinea-pigs with itraconazole and parenteral
treatment with amphotericin B.

Drug	Treatment		No.of animals					
				Skin lesions			Positive cultures	
	Route	Exci-pient	No.of animals	Absent	Moderate	Pro-nounced	Skin	Kidneys
Untreated	-	-	12	0	0	12	12	12
Itraconazole	Oral	PEG 200	12	12	0	0	0	2
Itraconazole	Parenteral	HPβCD**	18	18	0	0	0	4
Amphoter. B	Parenteral	Saline	12	5	4	3	6	8

* Treatment: once daily at 2.5 mg.kg^{-1}, starting on the day of
infection for 14 consecutive days
** HPβCD: Hydroxypropyl-β-cyclodextrin

REFERENCES

1. G.Maestrone and R.Semar, Establishment and treatment of cutaneous
<u>Candida albicans</u> infection in the rabbit, <u>Naturwissenschaften</u>,
55: 87 (1968).
2. F.C.Odds, "<u>Candida and Candidosis</u>", 2nd ed.,pp.282-283, Baillière
Tindall, W.B. Saunders, London (1988).
3. A.Polak, Experimental <u>Candida</u> vaginitis (vaginal thrush), <u>in</u>:
"Experimental Models in Antimicrobial Chemotherapy", Vol.2, D.Zak
and M.A.Sande, eds., pp. 21-42, Academic Press, London (1986).

4. T.L.Ray, Animal models of experimental Candida infections of the skin, in: "Models in Dermatology", Vol.1, H.I.Maibach and N.J. Lowe, eds., pp. 41-50, Karger, Basel (1985).

5. M.G.Rinaldi, In-vivo models of the mycoses for the evaluation of antifungal agents, in: "Recent Trends in the Discovery, Development and Evaluation of Antifungal Agents", R.A.Fromtling, ed., pp.S1: 11-24, Prous J.R.Science Publ., Barcelona (1987).

6. H.J.Scholer, 1960, Experimental Vaginal-Candidiasis der Ratte, Pathol.Microbiol., 23: 62 (1960).

7. D.Thienpont, J.Van Cutsem, M.Borgers, Ketoconazole in experimental candidosis, Rev.Inf.Dis., 2: 570 (1980).

8. J.Van Cutsem, D.Thienpont, Experimental cutaneous Candida albicans infections is guinea-pigs, Sabouraudia, 9: 17 (1971).

9. J.Van Cutsem, The antifungal activity of ketoconazole, Am.J.Med., 74: 1B, 9 (1983).

10. J.Van Cutsem, F.Van Gerven, P.A.J. Janssen, Activité anticandidose des azoles, Ann.Biol.Clin., 45: 661 (1987).

11. J.Van Cutsem, J.Fransen, P.A.J.Janssen, Therapeutic efficacy of itraconazole in systemic candidosis in guinea-pigs, Chemotherapy, 33: 52 (1987).

12. J.Van Cutsem, F.Van Gerven, P.A.J.Janssen, Activity of orally, topically and parenterally administered itraconazole in the treatment of superficial and deep mycoses. Animal models, Rev.Infect.Dis., 9: SI, 15 (1987).

13. J.Van Cutsem, F.Van Gerven, P.A.J. Janssen, Oral therapeutic treatment of vaginal candidosis in rats with itraconazole, fluconazole and ketoconazole: comparison of one-, two- and three-day treatment schedules, in: "Proceedings of the 15th International Congress of Chemotherapy: Progress in Antimicrobial and Anticancer Chemotherapy, Istanbul, Turkey, 19-24 June 1987", B. Berkarda and H.P.Kuemmerle, eds., Vol.1, pp. 879-881 (1987).

14. J.Van Cutsem, Animal models for dermatomycotic infections, in: "Current Topics in Medical Mycology", Vol. 3, M.R.McGinnis and M.Borgers, eds., Springer-Verlag, Publ., New York, 1989 (in press).

15. Van Cutsem, J., 1989, In-vitro sensitivity of Candida spp. to antifungals. Treatment of experimental candidosis with itraconazole, in: "Proceedings of FEMS-Symposium on Candida and Candidamycosis, Alanya, Turkey, 24-28 April 1989", Plenum, London (1990).

16. A.Wildfeuer, Die Chemotherapie der vaginalen Trichomoniasis and Candidosis der Maus, Arzneimittelforschung (Drug Res.), 24: 937 (1974).

17. H.Yamaguchi, Opportunistic fungal infections, Candidiasis, in: "Animal Models in Medical Mycology", M.Miyaji, ed., pp. 111-112 and 144-151, C.R.C.Press, Inc., Boca Raton (1987).

18. P.M.Laduron and P.F.M.Janssen, 1986, Impaired axonal transport of opiate and muscarinic receptors in streptozotocin-diabetic rats, Brain Res., 380: 359 (1986).

ULTRASTRUCTURE AND GROWTH OF CANDIDA ALBICANS ASSOCIATED

WITH RAT VAGINAL EPITHELIUM

P.A. Hunter[1]   A. Everett[1]   J. Warrack[2]   M. Wilkinson[2]

[1]Biosciences Research Centre, Great Burgh, Yew Tree Bottom Road, Epsom, Survey, The United Kingdom
[2]Chemotherapeutic Research Centre, Brockham Park, Betchworth, Survey, The United Kingdom

INTRODUCTION

The rodent model of Candida vaginitis is widely used to examine the effects of antifungal agents, and several studies of the pathogenesis of the infection, particularly in rats, have been made[1-3]. Differences exist between the infection in rodents and humans, particularly the presence of an extensive squamous keratinised vaginal epithelium, higher vaginal pH and lack of an inflammatory response in rodents, and differences in the vaginal flora. However, a true vaginitis is produced in the rat and the model offers considerable advantages for testing drugs, since the infection is benign, causing little discomfort to the animal, is reasonably reproducible and persists for several months.

Early in our studies, we noted some discrepancies between our observations and some published work where some authors find that the infection peaks at 24 hours[4]. There are also conflicting reports in the literature as to the rate of clearance of the organism. However, most of these authors rely heavily on assessing the progress of the infection by sequential sampling from the vagina and culture of organism. Since the growth is predominantly hyphal, viable counts are not always reliable.

We describe here our findings using a variety of techniques to examine the progression of the infection.

MATERIAL AND METHOD

Rats. Ovarectomised female Sprague Dawley rats weighing 180-220 g were provided by Charles River (UK). 1mg Oestradiol (Benztrone-Paynes and Byrne Ltd) was administered subcutaneously at weekly intervals commencing approximately 1 week after ovarectomy. Swabs of the vagina were taken and smears examined to ensure that rats were in pseudo-oestrus.

Infection. Candida albicans B2630 (kindly provided by Dr.J.Ryley, ICI Pharmaceuticals) was grown for 48 hours on Sabouraud's dextrose

Candida and Candidamycosis, Edited by E. Tümbay *et al.*
Plenum Press, New York, 1991

agar slopes, washed and suspended in 10% gum arabic. 0.1ml of this suspension was inserted into the vaginal lumen of rats in pseudo-oestrus. Each rat received approximately 10^6 yeasts.

Vaginal Smears. Sterile swabs, moistened in water were inserted into the vaginal lumen and rotated gently several times. The swabs were removed and rubbed firmly onto the surface of microscope slides. Slides were fixed with 'Cytofix' (Paramount Reagents Ltd) and stained with either Giemsa, Papanicolaou or calcofluor white. The calcofluor white-stained smears were examined using fluorescent microscopy.

Histology. Small portions of vaginal tissue were fixed in formol saline prior to processing and embedded in paraffin wax using conventional means. Thick serial sections were cut (c. 5mm) and stained using either Haematoxylin and Eosin or Periodic Acid-Schiff's reagent (PAS).

Scanning Electron Microscopy. Approximately $10mm^2$ of vaginal tissue was mounted on a stub, quenched in sub-cooled liquid nitrogen and examined in a Philips 501 SEM equipped with a Hexland cryochamber and a cold stage operated at $-170°C$. Conventional SEM was carried out on glutareldehyde-fixed specimens which had been dehydrated in ethanol and critical point dried from liquid CO_2. Specimens were sputter-coated with gold.

A minimum of three animals were killed by an overdose of pentabarbital at various times after infection and the vaginal tissue removed immediately for histology and SEM. Vaginal swabs were taken from a total of approx. 40 rats from several experiments. Rats were sampled at 2 or 3 day intervals; those which were to be killed were not sampled by swabbing on that day.

RESULTS

Swabs. Examination of the swabs revealed that by 24 hrs hyphal growth was evident with few blastospores (Fig.1). Generally growth was relavitely sparse being absent on many areas of the smears. Some variation was seen in the degree of infection at 24 hrs with only a few animals being more heavily infected. Hyphae were frequently long with few branches. The areas of fungal growth were much easier to detect on the smears stained with calcoflour white than on those stained with either Geimsa or Papanicolaou.

By 3-6 days the fungal growth had increased in intensity all smears having dense areas of tangled masses of hyphae with few blastospores (Fig. 2). The infection remained at this level for several days with less variation between animals, but by 10 days more variation occurred, some animals producing sparser growth. By 20 days the level of infection became less but where areas of fungal growth were seen, they were relatively dense, unlike smears from the early days. Polymorphonuclear leucocytes were never numerous and there was no evidence even after several weeks, that polymorphs were clearing the infection.

Histology. In oestrous uninfected vaginal tissue the layers of squamous epithelial cells were densely packed (Fig. 3) whereas in infected animals the layers were disorganised with a loose matrix of desquamated cells. The intensity of the infection closely paralleled that seen in the vaginal swabs with predominantly hyphal growth

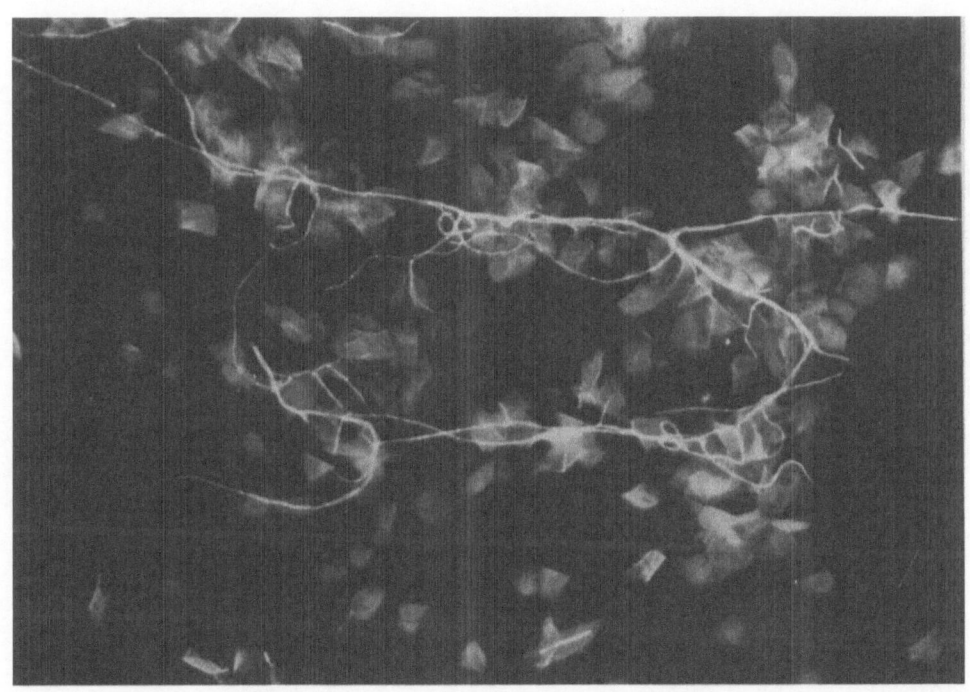

Fig. 1. Smear prepared from vaginal swabs, day 1. Stained with calcofluor
white.

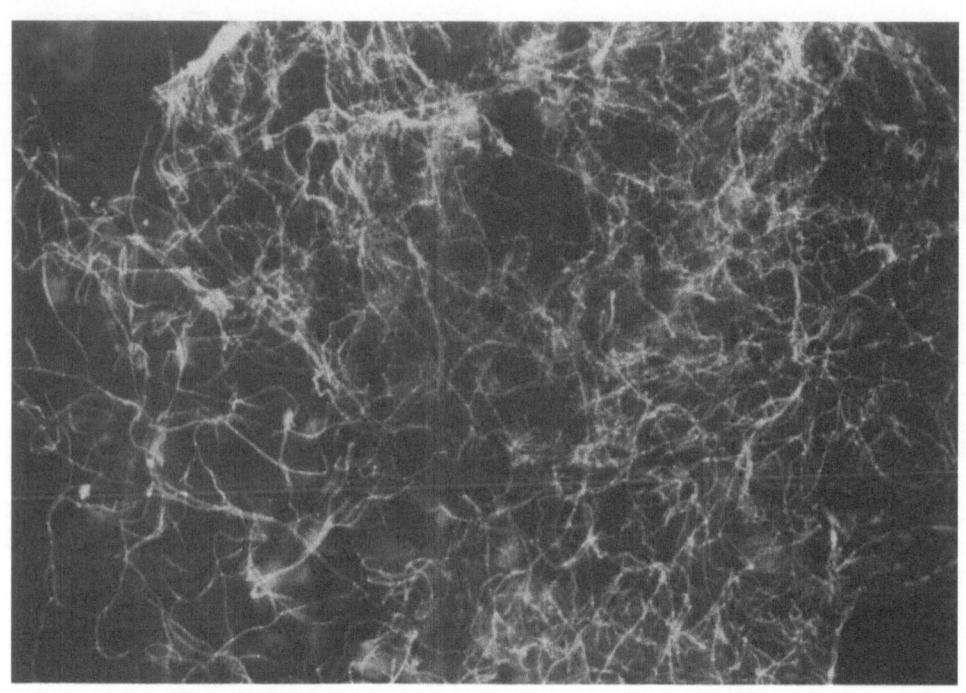

Fig. 2. Smear prepared from vaginal swab, day 5. Stained with calcofluor white.

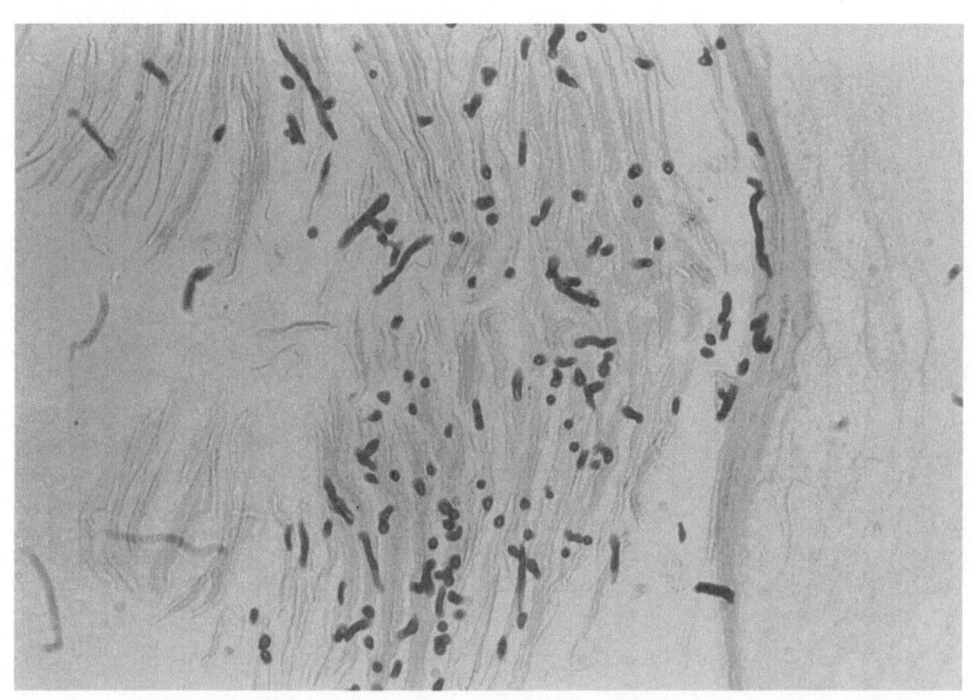

Fig. 3. Section of uninfected vaginal tissue. Stained H and E.

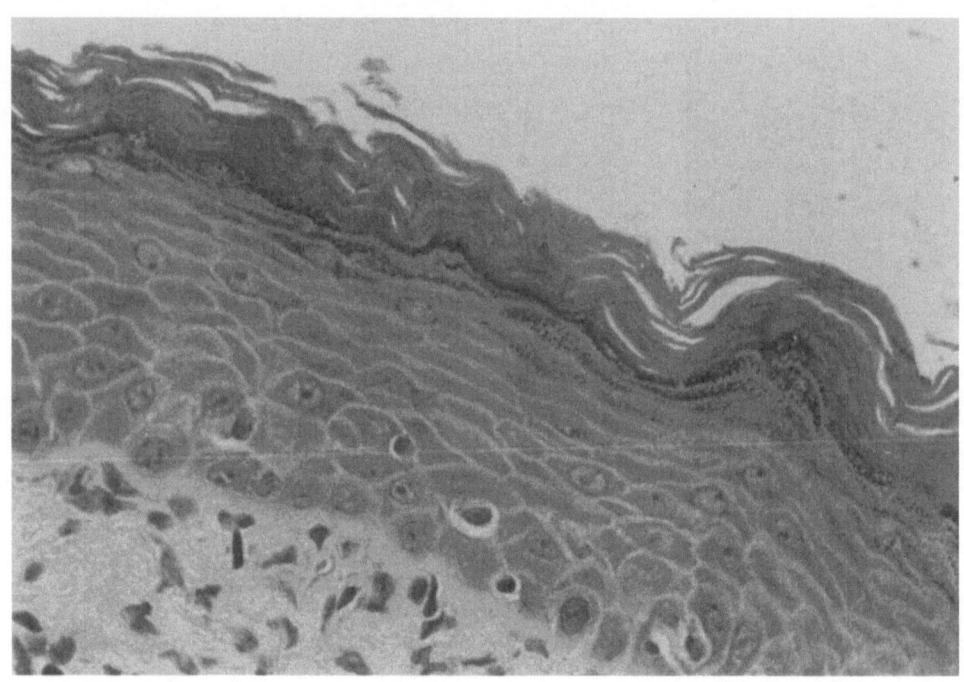

Fig. 4. Section of infected vaginal tissue, day 3. Stained PAS.

peaking at 3-6 days and remaining fairly constant for another 5-7 days. Hyphae were evident throughout the stratum corneum both in the deeper layers and in the looser desquamated areas (Fig. 4). There was little evidence of an inflammatory response. Some sections had lost much of the loose layers of desquamated cells and their associated hyphae.

SEM. Conventional SEM confirmed the differences noted above between infected and non-infected vaginal tissue. Again hyphal growth was evident at 24 hr but was more variable and more difficult to find than at later times where, particularly at 3-6 days, areas of dense hyphal masses were seen (Fig. 5). In some specimens, areas of the stratum corneum were missing and the smooth stratum malpigium was evident (Fig. 6).

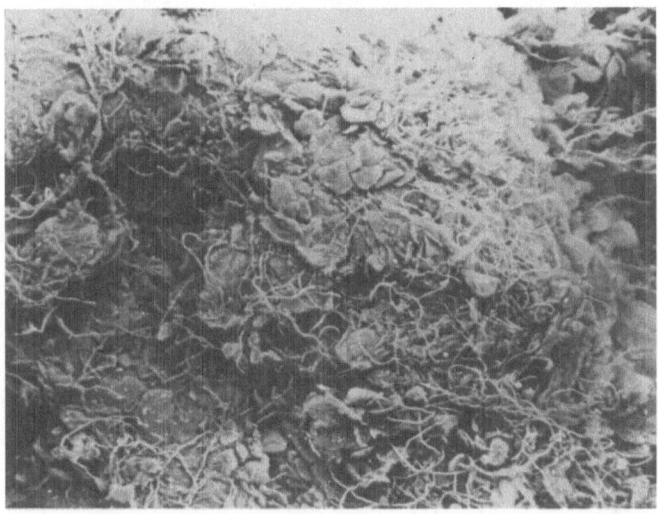

Fig. 5. Conventional SEM of infected vaginal tissue.

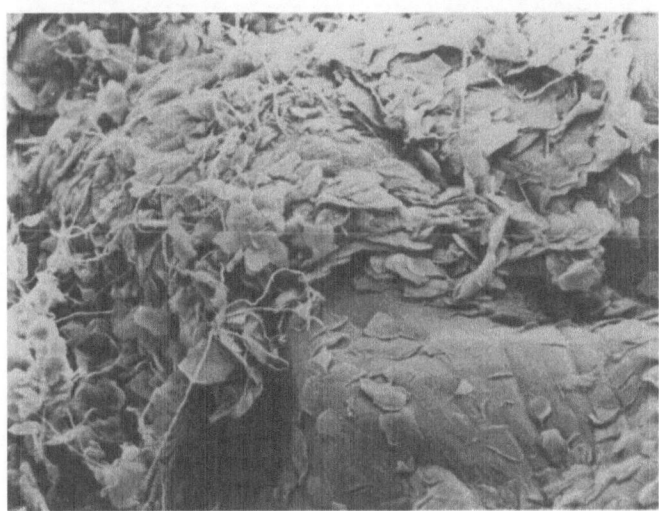

Fig. 6. Conventional SEM of infected vaginal tissue.

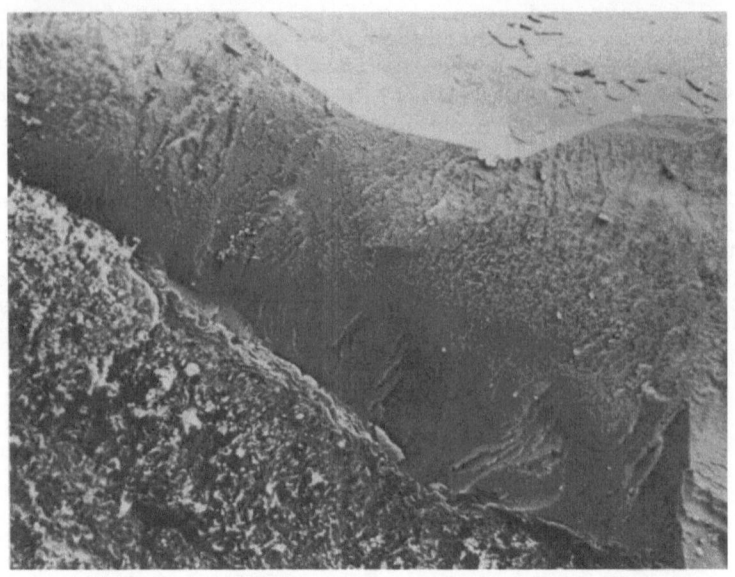

Fig. 7. Cryo SEM of uninfected vaginal tissue.

Cryo SEM revealed that a layer of mucus was present in both infected and uninfected rats. In the uninfected samples, very few squamous cells were seen in the mucus, and freeze-fracture showed that the squamous cells were densely packed (Fig. 7). In the infected rats, squamous cells and hyphae could be seen on the surface (Fig. 8) and throughout the mucus layer. Freeze-fracturing showed that the mucous layer was variable in thickness but in some specimens reached 150 m. Penetration of epithelial cells by hyphae was seen.

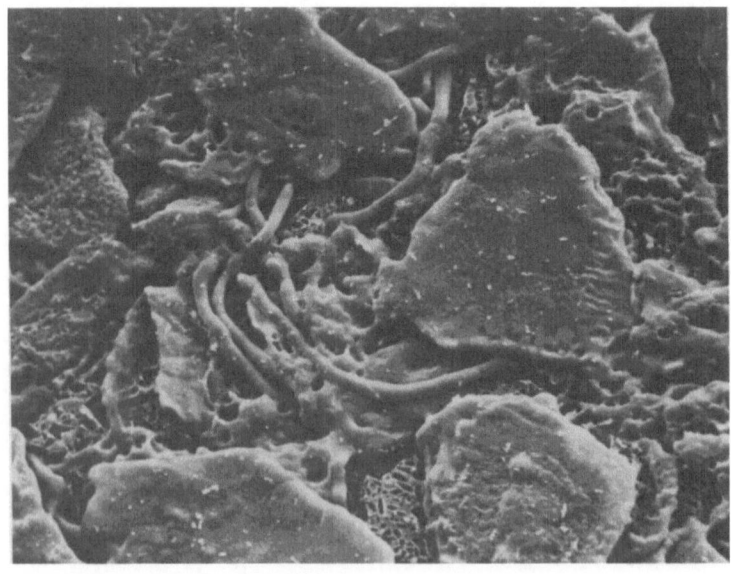

Fig. 8. Cryo SEM of infected vaginal tissue.

DISSCUSSION

Viable counts are frequently used to monitor the progress of infection in the rodent model; they are convenient and enable the same animal to be monitored sequentially, an important point in drug evaluation studies. Ryley and McGregor [5] have pointed out discrepancies between viable counting techniques and results obtained by removing the vagina and homogenising the tissue. An additional point not noted by most authors, is the problem of estimating colony forming units from hyphal growth. Both the SEM and the swabs revealed the presence of large clumps consisting of tangled masses of hyphae between 3 and 10 days post-infection. At 1 day post-infection, although hyphae were evident, the clumps were fewer and smaller. This difference is not always reflected in viable counts, presumably as a consequence of the larger clumps not being adequately broken up. A feature of the infection seen clearly in the smears made from swabs, is the lack of branching of the hyphae; portions of eight or nine cells with no branches were common.

Histological section confirm the findings of the swabs and SEM, namely that the infection does not reach maximum intensity until 3-6 days post-infection. Results at 1 day were more variable with all three methods, the infection being more patchy and generaly more difficult to detect.

The fluorescent stain calcofluor white proved to be an excellent way of finding organisms particularly when they were more sparsely distributed. Althoug PAS, Giemsa and Papanicolaou are all useful staining techniques, it can sometimes be difficult to distinguish Candida clearly from surrounding cells or tissues with these stains. The bright blue-white fluorescence of the fungus when stained with calcofluor white makes this distinction much easier.

Cryo-SEM maintained the integrity of the vaginal mucosa far better than either conventiona 1 SEM or histology and revealed a suprisingly thick layer of mucus which was present in both infected and uninfected animals. It varied in depth but this did not correlate with the length of time the animals had been in pseudo-oestrus.

An interesting feature of this model is the persistence of the infection; clearance is slow and many animals remain infected for several months provided the rats remain in the pseudo-oestrous stage. It is not clear whether the clearance is mediated by polymorpho-nuclear phagocytes, since their numbers are generally low. Studies currently in progress are investigating the involvement of antibodies in this clearance.

ACKNOWLEDGEMENT

The authors acknowledge with gratitude the contributions of Miss S.Smith in preparing the histological sections.

REFERENCES

1. A. Polak, Experimental Candida vaginitis (Vaginal Thrush), in : "Experimental Models in Antimicrobial Chemotherapy", Vol. 2, Zak and Sande, eds. (1986).

2. H.J. Scholer, Experimentalle Vaginal-Candidiasis der Ratte, Pathologia et Microbiologia, 23 : 62 (1960).

3. J.D. Sobel, G. Muller and J.F. McCormick, Experimental chronic vaginal candidosis in rats, Sabouraudia, 23 : 199 (1985).

4. R.J. McRipley, P.J. Erhard, R.A. Schwind and R.R. Whitney, Evaluation of vaginal antifungal formulations in vivo, Postgrad.Med.J., 55 : 648 (1979).

5. J.F. Reyley and S. McGregor, Quantification of vaginal Candida albicans infections in rodents, J.Med.Vet.Mycol., 24 : 455 (1986).

EFFECT OF BLOOD GROUP AND SECRETOR STATUS ON THE ADHESION

OF CANDIDA ALBICANS TO MUCOSAL SURFACES

F.D. Tosh   L.J. Douglas

Department of Microbiology, University of Glasgow
Glasgow G12 8QQ, The United Kingdom

INTRODUCTION

Adhesion of Candida albicans to epithelial cells is the first stage in the colonization of mucosal surfaces by this organism. Although the mechanism of adhesion is not fully established, there is now considerable evidence for the involvement of glycosides as epithelial receptors for the yeast [1,2]. A number of studies have demonstrated that, in vitro, C. albicans adheres more readily to exfoliated buccal or vaginal epithelial cells from some donors than from others [3,4], suggesting that some individuals may be more susceptible to colonization by the yeast. Recently, Burford-Mason et al. [5] reported that oral carriage of C. albicans in healthy subjects could be correlated with two host factors, namely blood group 0 and non-secretion of blood-group antigens, with the trend towards carriage being greatest in group 0 non-sectors. In the present study we have investigated the effect of donor blood group and secretor status on adhesion of C. albicans to buccal epithelial cells in vitro.

MATERIAL VE METHOD

Growth conditions and adhesion assays

Candida albicans GDH 2346 was used in adhesion assays with exfoliated buccal epithelial cells as described by Critchley and Douglas [1] in 1987. All buccal cell donors were healthy, non-smoking males in the age range 22-38 years. None had experienced recent antibiotic therapy or had apparent carriage of Candida. Adhesion assays were done in triplicate on at least three separate occasions. In some experiments, buccal cells were pretreated with anti-Lewis antiserum (a 1:2 dilution in 0.15M PBS, pH 7.2) at 37°C for 30 min. In other experiments, both yeasts and epithelial cells were suspended in a 50% solution (in PBS) of clarified, whole saliva during the assay.

Determination of ABO blood group and secretor status

The ABO and Lewis blood groupings of buccal cell donors were

Candida and Candidamycosis, Edited by E. Tümbay et al.
Plenum Press, New York, 1991

127

determined by conventional agglutination techniques. Secretor status was determined using the method of Burford-Mason et al. [5]

RESULTS

Effect of blood group and secretor status on yeast adhesion to human buccal epithelial cells

Initially, adhesion of C. albicans GDH 2346 to buccal epithelial cells from donors of two different blood-groups, A and O was compared. There was no significant difference between adhesion to group-A cells and that to group-O cells, either for secretors or non-secretors. However, within each blood group, adhesion to cells from secretor was significantly lower than that to cells from non-secretors ($p < 0.02$ for group A; $p < 0.05$ for group O). With group O donors, for example, 751 ± 31 yeasts (mean $\pm$ SE) adhered to 100 secretor cells as compared with 864 ± 41 yeasts which adhered to 100 non-secretor cells.

Secretors are defined as individuals capable of secreting antigens corresponding to their ABO blood group into saliva and other body fluids; non-secretors lack this capability. Secretors and non-secretors also differ in the expression of another closely related gene product, the Lewis (Le) antigen. This exists in two forms, Le^a and Le^b. Le^b is found in the body fluids of most secretors, whereas in non-secretors it is the Le^a form which predominates. Both secretors and non-secretors possess the appropriate blood-group antigen on their epithelial cell surfaces. When secretor saliva (blood group O) was added to adhesion assay mixtures containing buccal cells from the same individual, there was no effect on yeast adhesion. Addition of non-secretor saliva (group O), on the other hand, had a highly significant effect, enhancing adhesion by 22% ($p < 0.001$). With buccal cells from a non-secretor, self-saliva had no effect but secretor saliva inhibited adhesion by 21% ($p < 0.001$).

These results suggest that Le^b or H antigen in the secretor saliva somehow blocked adhesion to non-secretor cells, by binding either to the yeasts or to the epithelial cells. On the other had, Le^a antigen present in the non-secretor saliva may have enhanced adhesion to secretor cells by acting as a bridge to link yeast and uccal cell. Non-secretor cells would already possess Le^a antigen adsorbed on their surfaces. If this were to be blocked using the appropriate antiserum, then yeast adhesion should be inhibited. To test this possibility, buccal cells from a secretor and a non-secretor were pretreated at 37°C for 30 min with anti-Le^a and anti-Le^b antise-rum.Pretreatment of non-secretor buccal cells with anti-Le^a antiserum had a significant inhibitory effect on adhesion (24% inhibition; $p < 0.001$). None of the other pretreatments produced statistically significant differences.

Effect of sugars on the adhesion of C. albicans GDH 2346 to buccal epithelial cells

Sugar inhibition tests in which sugars - as potential receptor analogues - are added to adhesion assay mixtures are commonly employed in attempts to identify glycoside receptors for microorganisms. Such tests have been widley used already in work on Candida adhesion mechanisms[6]. In the present study, we examined the effects of different sugars on yeast adhesion to buccal cells from donors of two different blood groups. L-fucose inhibited adhesion to cells from donors of blood group A or 0 (both secretors), by approximately 25%. Similarly, N-acetylglucosamine inhibited adhesion to cells from a group A donor byş 24%. All of these differences were highly significant with p values of <0.001). However, when N-acetylglucosamine was used in assays with cells from a group 0 donor, there was no significant difference from the controls. When cells from donors of blood group A, but differing secretor status, were compared, L-fucose and N-acetylglucosamine gave a similiar inhibition of adhesion (approximately 20-25%). A mixture of these sugars produced significantly better inhibition, with adhesion values of 66% and 69% relative to controls for secretor and non-secretor cells, respectively. The most inhibitory sugar was N-acetylgalactosamine which allowed only 58-59% adhesion to group A buccal cells.

DISCUSSION

Correlations between ABO blood group or secretor status and vulnerability to disease have been documented for a number of bacterial infections[7]. With C. albicans, blood group 0 and non-secretion of blood-group antigens have been identified as possible risk factors for oral carriage of the yeast in healthy subjects[5]. The results of the present study indicate that Candida adhesion to the oral mucosa - the first step in the colonization process - is likely to be affected by the secretor status of the host. However, with our relatively small sample of buccal cell donors, there was no evidence that yeast adhesion was dependent on the ABO blood group.

The possible importance of secretor status was first suggested by Blackwell et al.[8] who briefly reported that Candida adhesion to buccal cells could be inhibited by secretor saliva but enhanced by non-secretor saliva. Our results confirm these findings and provide additional evidence that the Le^a antigen which is adsorbed on to the surface of the buccal cells in non-secretors could function as a receptor for the yeast. This conclusion would be consistent with previous work from this laboratory by Critchley and Douglas[1], which indicated that C. albicans GDH 2346 produces surface mannoproteins with a lectin-like affinity for L-fucose residues and that fucose-containing glycosides are likely to be the major epithelial cell receptors for this yeast strain. The Le^a antigen contains a terminal L-fucose residue. Fucose is also the immunodominant sugar of the H antigen of blood group 0 and it is possible that cell-bound H antigen could similarly function as a receptor although its presence in secretor saliva may effectively block adhesion. Our results do not support an analogous role for the Le^b antigen since treatment of buccal cells with anti-Le^b antiserum had no effect on adhesion.

In vitro sugar inhibition tests have been widely used with both bacteria and yeasts to predict the nature of the epithelial cell receptors. Our results indicate that such tests can be influenced by the ABO blood group of the buccal cell donor and this finding should be borne in mind in future work on Candida adhesion mechanisms. In experiments with buccal cells from a donor of blood group A, N-acetylgalactosamine gave greater inhibition of yeast adhesion than either L-fucose, N-acetylglucosamine or a mixture of these two sugars. Such a result might be expected if the blood group A antigen, in which N-acetylgalactosamine is the immunodominant sugar, could act as an epithelial cell receptor for C. albicans.

REFERENCES

1. I.A. Critchley and L.J. Douglas, Role of glycosides as epithelial cell receptors for <u>Candida albicans</u>, <u>J. Gen. Microbiol.</u>, 133: 637 (1987).
2. L.J.Douglas, Adhesion of <u>Candida albicans</u> to host surfaces, <u>F.E.M.S. Symposium on Candida and Candidamycosis, Proceedings</u>, E. Tümbay, H.P.R. Seeliger and Ö. Anǧ, eds., London, Plenum Publishing House (1990).
3. R.D. King, J.C. Lee and A.L. Morris, Adherence of <u>Candida albicans</u> and other <u>Candida</u> species to mucosal epithelial cells, <u>Infect. Immun.</u>, 27: 667 (1980).
4. J.D. Sobel, P.G. Myers, D. Kaye and M.E. Levison, Adherence of <u>Candida albicans</u> to human vaginal and buccal epithelial cells, <u>J. Infect. Dis.</u>, 143: 76 (1981).
5. A.P. Burford-Mason, J.C.P. Weber and J.M.T. Willoughby, Oral carriage of <u>Candida albicans</u>, ABO blood group and secretor status in healthy subjects, <u>J. Med. Vet. Mycol.</u>, 26: 49 (1988).
6. L.J. Douglas, Adhesion of <u>Candida</u> species to epithelial surfaces, <u>C.R.C. Crit. Rev. Microbiol.</u>, 15: 27 (1987).
7. A.K. Mourant, A.C. Kopec and K. Domaniewski-Sobczak, "Blood Groups and Diseases: A Study of Associations of Diseases with Blood Groups and Other Polymorphisms", Oxford University Press, Oxford (1978).
8. C.C. Blackwell, S.M. Thom, D.M. Weir, D.F. Kinane and F.D. Johnstone, Host-parasite interactions underlying non-secretion of blood group antigens and susceptibility to infections by <u>Candida albicans</u>, <u>in</u>: "Protein-Carbohydrate Interactions in Biological Systems", D.L. Lark, ed., pp. 231-233, Academic Press, London (1986).

CANDIDA ALBICANS - PLATELET INTERACTION : EVIDENCE FOR, IN VIVO AND

IN VITRO, CELL TO CELL ATTACHMENT

C.Mahaza[1]   R.Robert[1]   M.Miègeville[2]
G.Tronchin[1]   J.M.Senet[1]

[1]Laboratoire de'Immunologie-Parasitologie-Mycologie, UFR
des Sciences Médicales et Pharmaceutiques, Section
Pharmacie, 16 Bd. Daviers, 49100 Angers, France
[2]Laboratoire de Parasitologie, Faculté de Pharmacie, 1,
rue Gaston Veil, 44035 Nantes, France

INTRODUCTION

It is possible to elucidate the physiopathology of candidosis by
evaluating the different parameters involved in the relationship
between C.albicans and its hosts. The different steps of the host
colonization which can be evaluated are : adherence, multiplication
and dissemination through the organism. When this dissemination is
via the blood stream, the yeasts interact with blood components such
as fibrinogen[1], fibronectin[2], complement[3], leukocytes and platelets.
In vivo experiments show the attachment of C.albicans to platelet
aggregates. In vitro, electron microscopic observations demonstrate
that the platelets adhere, protrude spikes and finally spread and
flattened on the upper fibrillar layer of the fungus.

MATERIAL AND METHOD

Organisms and culture conditions

C.albicans 1066, (serotype A), originally isolated from a case of
septicemia was used throughout[1]. Blastospores were grown at 22°C for
48 h in Medium 199 (ph 6.4- Flow Laboratories). Germ tubes were
obtained in the same medium by incubating the blastospores at 37°C
for 3 h, in Petri dishes for tissue culture (35 mm 0-Greiner Labor-
technik).

Animals

Swiss female mice weighing 18 to 20 g (Centre d'élevage Depres
-France) were used for the in vivo studies.

Preparation of human platelets

As it was difficult to get a large amount of mice platelets for
in vitro experiments, human platelets were obtained from healthy

donors' blood collected in 2 mM EDTA. Platelets separated from other blood cells by differential centrifugation at 400 g for 5 min were submitted to a second centrifugation at 2,000 g for 15 min. The platelets were then washed in a Tris dextrose buffer saline (TDS) (Tris 25mM, dextrose ,5mM, sodium chloride 150 mM, pH 7.5). The pellets were then gathered and resuspended in the same buffer at a final concentration of 2.5 x 10 cells/ ml.

Study of C.albicans-platelet interaction in vivo

0.2 ml of a yeast suspension (2.5 x 10 blastospores or germ tubes per ml) was injected into the caudal vein of a mouse. After sectioning the tail, blood smears were made 5, 10, 30, 45, 60 and 90 min after the injection, stained with May Grunwald Giemsa (MGG) and examined under light microscopy.

Study of C.albicans-platelet interaction in vitro

Qualitative studies. Germ tubes obtained as described above, adhered to the Petri dishes. After two gentle washings with phosphate buffer saline (PBS) containing 1‰ sodium azide in order to stop germination, 3 ml of a 5% bovine serum albumin PBS was added for 1 h at 37°C. After washings the Petri dishes 3 times with TDS, 2 ml of platelets (2.5 x 10^8/ml) in TDS-EDTA 2mM were poured into the dishes. After incubation at 37°C for 5, 10, 15, 30, 45, 60 or 90 min, three gentle washings were made to remove nonfixed platelets, and the preparations were examined both under light and electron microscopy.

(i) The germ tube-platelet complexes were fixed with formal-dehyde (3.7% in TDS) for 1h at room temperature. After three washings with TDS, the preparation was stained with MGG and examined with light microscopy.

(ii) For scanning electron microscopy, the germ tube-platelet complexes were fixed, dehydrated, dried at the critical point, metal coated with gold-palladium and examined with a J.S.M. 35 JEOL microscope, according to the method described by Miégeville[4].

(iii) For transmission electron microscopy, germ tubes-platelet complexes were incubated for 1h at 37°C. After 3 washings with TDS, the cells were fixed for 1 h with 2.5% glutaraldehyde in sodium cacodylate (0.1 M, pH 7.4) and post-fixed for 1 h in OsO4 1% in the same buffer. Routine dehydration and Epon embedding procedures were used. Sections stained with uranyle acetate or uranyle acetate and lead citrate were examined using 100 CX JEOL microscope[5].

RESULTS AND DISCUSSION

In vivo, platelets rapidly became attached to blastospores of Candida albicans (Fig. 1a). This result could be observed 5 min after injection but was not found after 30 min. After inoculation of germ tubes, the fungi could only be detected in platelet aggregates 2 min after injection (Fig. 1b) and were no longer isolated from circulating blood. These results could be due to the eventual sequestration or the fungal-platelet complexes in the capillaries.

In vitro evidence of Candida albicans interactions with platelets. Light microscopy showed platelets fixed along the germ tubes whereas few or no attachments were observed on the blastospores (Fig.1c).

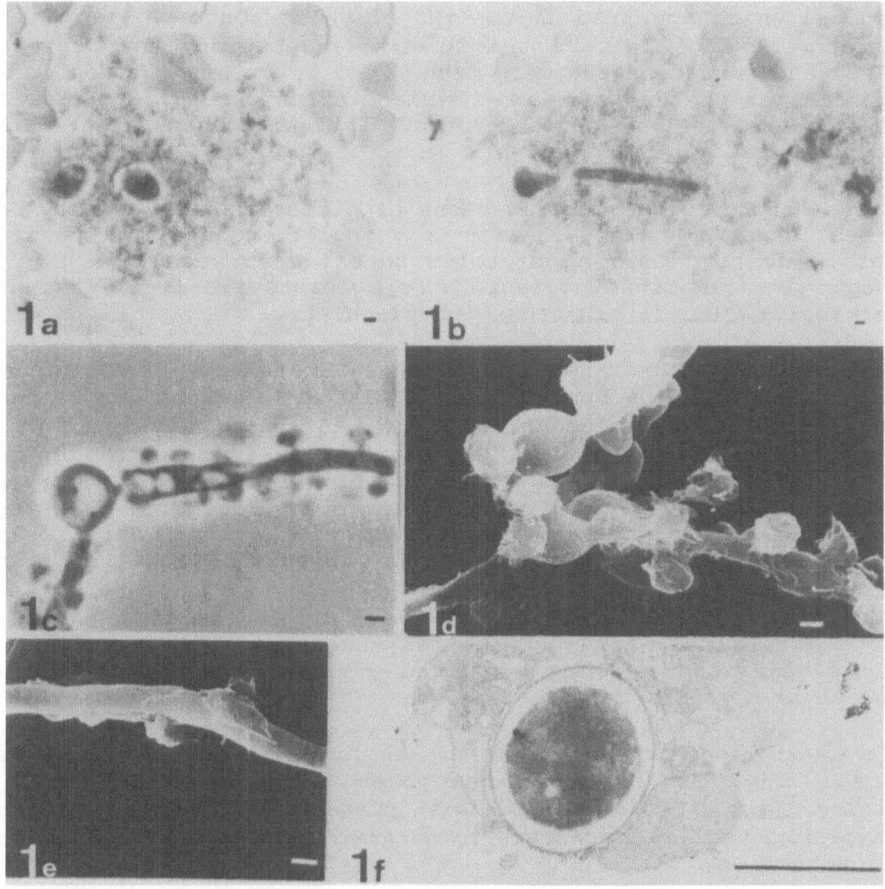

Fig. 1. Candida albicans-platelet interaction.
In vivo studies, smears stained with MGG :
a) Interaction with blastospores.
b) Interaction with germ tubes.
In vitro studies :
c) Observations with light microscopy.
d) and e) Observations with scanning electron microscopy.
f) Observations with transmission electron microscopy.

The kinetic study showed that platelet fixation was weak in the first 10 min. The number of fixed platelets increased to reach a maximum after 30 min of incubation at 37°C. Observations made after 90 min incubation showed a modification in the morphology of the platelets, with an apparent decrease in their number.

The same experiments were carried out again this time using scanning electron microscopy. The results were similar to those obtained by light microscopy, but more precise morphological changes were noted : the platelets lost their discoid shape and became globular. Spikes or pseudopodes which appeared to fix the platelets onto the germ tubes were formed (Fig. 1d). After 90 min incubation, the platelets flattened and spread on the germ tube (Fig. 1e). This flattening was responsible for the decrease in the platelet thickness, and explains why fewer platelets were observed under light microscopy after 90 min incubation.

These morphological modifications of platelets have already been described during platelet adhesion to vascular endothelium. Contrary to what is observed during platelet adhesion to vessel wall or to fibrinogen[7,8], no activating factors or agonists are involved in the interaction of Candida albicans with platelets.

Our results may be compared to those reported by Skerl et al.[6], who showed that Candida spp. mediate a platelet aggregation response through a complement-dependent mechanism, and that platelets promote the germination of Candida spp. Ultrastructural study with transmission electron microscopy shows that the fibrillar layer of the cell wall seems to be involved in this interaction (Fig. 1f). This layer is always present in the germ tube though only sometimes in blastospores[9,10] and this difference may be responsible for the greater fixation of platelets onto the germ tubes. The fibrillar network has already been described in its role in the adhesion of yeasts to buccal[11], intestinal, or endothelial mucosa as well as to plastic materials such as prosthesis, catheters, prosthetic valves or polystyrene Petri dishes [10,12].

This work shows that platelets can fix directly onto the germ tube and may be linked to the observations of Maisch et al.[13] concerning C.albicans interaction with platelet-fibrin clots. Further investigations will be necessary to determine the molecular basis of this interaction.

REFERENCES

1. A.Bouali, R.Robert, G. Tronchin, J.M.Senet, Characterization of binding of human fibrinogen to the surface of germ-tubes and mycelium of Candida albicans, J.Gen.Microbiol., 133 : 545 (1987).
2. R.A. Calderone and W.M.Scheld, Role of fibronectin in the pathogenesis of candidal infections, Rev.Infect.Dis., 9 : 400 (1987).
3. R.Calderone, L.Linehan, E.Wadsworth, A.Sandberg, Identificaton of C3d receptors on Candida albicans, Infect.Immun., 56 : 252 (1988).
4. M.Miegeville, O.Morin, Observations de différentes souches de levures et de leurs protoplastes en microscopie électronique a balayage, C.R.Acad.Sc.Paris, 283D : 417 (1976).
5. G.Tronchin, D.Poulain, J.Herbaut, J.Biguet, Cytochemical and ultrastructural studies of Candida albicans II. Evidence for a cell wall coat using concanavalin A, J.Ultrastr. Res., 75 : 50 (1981).
6. K.G. Skerl, R.A.Calderone, T. Sreevalsan, Platelet interaction with Candida albicans, Infect. Immun., 34 : 938 (1981).
7. J.S.Bennet, G.Vilaire, D.B. Cines, Identification of the fibrinogen receptor on human platelets by photoaffinity labeling, J.Biol.Chem., 257; 8049 (1982).
8. G.A.Marguerie, N.Thomas-Maison, M.J.Larrieu, E.F. Plow, The interaction of fibrinogen with human platelets in plasma milieu, Blood, 59 : 91 (1982).

9. M.J.Kennedy and R.L.Sandin, Influence of growth conditions on *Candida albicans* adhesion, hydrophobicity and cell wall ultra-structure, *J.Med.Vet.Mycol.*, 26 : 79 (1988).

10. G.Tronchin, J.P.Bouchara, R.Robert, J.M.Senet, Adherence of *Candida albicans* germ tubes to plastic : ultrastructural and molecular studies of fibrillar adhesins, *Infect. Immun.*, 56 : 1987 (1988).

11. L.J.Douglas, J.G.Houston, J.McCourtie, Adherence of *Candida albicans* to human buccal epithelial cells after growth on different carbon sources, *FEMS Microbiol. Letter*, 12 : 241 (1981).

12. D.Rostrosen, R.A.Calderone, J.E.Edwards, Adherence of *Candida* species to host tissues and plastic surfaces, *Rev.Infect.Dis.*, 1 : 75 (1986).

13. P.A.Maisch and R.A. Calderone, Adherence of *Candida albicans* to a fibrin-platelet-matrix formed *in vitro*, *Infect.Immun.*, 32 : 92 (1980).

CANDIDA ALBICANS - PLATELET INTERACTION: MOLECULES INVOLVED IN THE

ADHERENCE

C. Mahaza   R. Robert   J.M. Senet

Laoratoire d'Immunologie-Parasitologie-Mycologie, UFR
des Sciences Médicales et Pharmaceutiques, Section
Pharmacie, 16 Bd. Daviers, 49 100 Angers, France

INTRODUCTION

We have already described the Candida albicans interaction with platelet[1]. This interaction seems to be mediated by the fibrillar layer of the cell wall which is always present on the germ tube of Candida albicans. This layer is also present on blastospores when they are cultured on rich medium, but is not usually found after growth on Medium 199[2]. The specificity of this interaction was demonstrated by inhibition tests using platelet extracts and anti-C.albicans germ tube antibodies (C.Mahaza-unpublished data). This suggests the presence of a binding affinity between proteins or mannoproteins of the yeast and the surface molecules of platelets. In this study, affino-binding techniques carried out with radiolabelled components led to te identification of a 45-47 kDA molecule from Candida albicans surface layer capable of fixing the GP IIb-IIIa complex of the platelet membrane.

MATERIAL AND METHOD

Organisms and culture condition

Candida albicans 1066, serotype A, originally isolated from a case of septicemia was used throughout. Blastospores were grown at 22°C for 48h in brain heart medium (Difco) or in Medium 199 (Flow Laboratories). Germ tubes were cultured in Medium 199 by incubating blastospores at 37°C for 3h.

Preparation of yeast extracts

Dithiothreitol extracts (DTT-E) were obtained from germ tubes and blastospores using the methods described by Smail and Jones[3].

Radiolabelling of yeast extracts

The yeast extracts were iodinated using iodogen method (Pierce Chemicals) and according to the procedure described by Fraker and Speck[4].

Candida and Candidamycosis, Edited by E. Tümbay et al.
Plenum Press, New York, 1991

Preparation of human platelets

Blood was collected from healthy donors in EDTA. Platelets were separated by differential centrifugation at 400 g for 5 min. Platelet-rich plasma was submitted to a second centrifugation at 2,000 g for 15 min. The packed platelets were then washed four times in TDS (Tris 25mM, dextrose 5mM, sodium chloride 150 mM, pH=7.5) added with either 2mM or 10 mM EDTA. The pellets were then gathered and a suspension of 2.5×10^8 platelets/ml was prepared in the same buffer.

Radiolabelling and solubilization of platelets

Whole platelets were radiolabelled according to Fraker and Speck's method[4]. Radiolabelled platelets (2.5×10^8) were added to 500 ul of Nonidet NP 40 (Sigma Chemicals- 1 % in TDS-EDTA 2 or 10 mM). After 15 min of contact, the supernatant corresponding to the radio-labelled platelet extract was collected and dialysed overnight at 4°C against TDS-EDTA (2 or 10 mM).

Study of platelet factors involved in the interaction with Candida albicans

To 10^9 yeasts (germ tubes or blastospores) were added 100 ul of radiolabelled platelet extract. After 15 min of incubation, under continuous agitation, the yeasts were washed 5 times with 5 ml of TDS-EDTA (2 or 10 mM). 150 ul of sample buffer for SDS-PAGE, as described by Laemmli[5] were added to the last pellet so as to elute platelet factors fixed on the yeasts. The eluate was then electro-phoresed on 6 % SDS polyacrylamide gel. The gel was then dried and autoradiographed on Kodak X-OMAT/AR.

Study of the yeast factors involved in the interaction

This experiment was run throughout with a 2 mM EDTA final concentration. 66 ug of radiolabelled DTT yeast extract in 200 ul TDS-EDTA were mixed with 2.5×10^8 platelets and either 200 ul of TDS-EDTA or 200 ul of unlabelled but 50 times concentrated DTT yeast extract. After 15 min of incubation the suspension was centrifuged and the pellet washed 5 times with 5 ml of TDS-EDTA. The fungal factors attached to the platelets were then eluted with 150 ul of sample buffer for SDS-PAGE. The eluate was electrophoresed on 12.5 % SDS-PAGE. The gel was dried and autoradiographed on Kodak X-OMAT/AR.

RESULTS AND DISCUSSION

Electrophoresis of platelet factors involved in the attachment to C. albicans in 2mM TDS-EDTA (Fig.1) was carried out under reducing (2 mercaptoethanol) (lane a) and nonreducing conditions (lane b). Lanes 1 and 2 corresponded to the autoradiography of a total platelet extract after exposure for 4 h (lane 1) and 24 h (lane 2). These lines showed about 14 to 16 radiolabelled platelet fractions detected in the total extract. Lanes 3, 4 and 5 corresponded to the eluate of platelet factors fixed to the germ tube (lane 3), to blastospores grown in brain heart (lane 4) and to blastospores cultured on Medium 199 (lane 5). Lane 3 revealed that under nonreducing condition (b), the platelet factors involved in the interaction with C. albicans were composed of one major protein of 90-95 kDa, a lightly radiola-belled 140-145 kDa protein and in some experiments a protein of high molecular weight (450 kDa) was present under nonreducing conditions, thus indicating that the 450 kDa protein was a trimer molecule. For platelet factors fixed to blastospores cultured in brain heart (lane 4), or in Medium 199 (lane 5), autoradiography showed the presence of a lightly radiolabelled protein after 24 h of exposure. However, a similar profile to that observed with the germ tubes was obtained when the time of exposure was 7 days (lane 6), indicating that the difference between germ tubes and blastospores is only quantitative. Except for the 450 kDa protein, no platelet componenets were detected when experiments were made using the 10 mM EDTA buffer. This 450 kDA protein is probably thrombospondin, an endogenous protein. Its presence in some of our experiments is probably due to activation of the platelets during washings. The 90-95 kDA and the 140-145 kDa components were always absent when the 10 mM EDTA buffer was used., The patterns obtained with or without reducing conditions indicate that these fractions correspond to the GP IIb-IIIa which is a complex molecule linked by Ca++ and known to be a major adhesin of the platelet membrane. The total DTT germ tube extract was composed of 10 to 20 radiolabelled bands as shown in Fig.2 (lane 1). Platelet absorbed material, in both cases, from germ tube DTT-E (lane 2), on from blastospore DTT-E (lane 3) was represented only by one radiolabelled protein of 45/47 kDa. When an experiment was carried out with a mixture of radiolabelled germ tube extract and an unlabelled, 50 times concentrated solution of the same extract, no band was detected (lane 4), proving that <u>Candida albicans</u> paletelet interaction was specific. This component should be compared to the 47/48 kDa component identified by Fiss and Buckley [6-9] as a pathogenic marker of visceral candidosis. The mechanism of adhesion investigated here along with others described by our laboratory[10] must play a major role in the colonization process, which undoubtedly starts with an adherence mechanism.

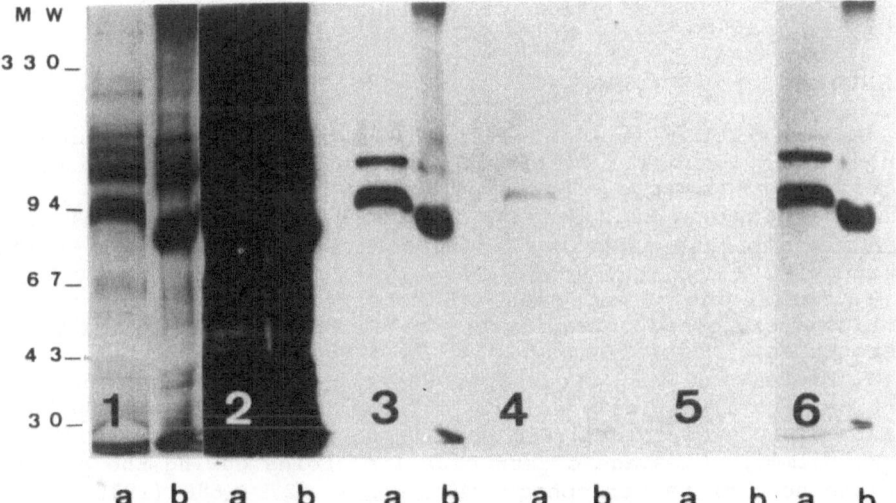

Fig. 1.Autoradiograph of SDS-PAGE with (lane a), without (lane b) reducer. Exposure during 4hrs (lane 1), 24 hrs (lane 2,3,4,5) and 7 days (lane 6). Lanes 1&2 correspond to the NP 40 solubilized radiolabelled platelet membrane; lane 3 to the platelet component adsorbed on <u>Candida albicans</u> germ tube;lane 4 on blastospores grown in a poor medium. Lane 6 corresponds to lane 4. Results obtained with 2 mM EDTA.

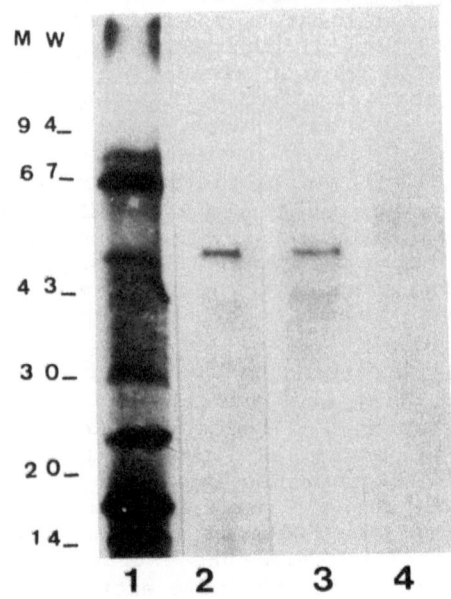

Fig. 2. Autoradiography of SDS-PAGE:
    Lane 1: _Candida albicans_ germ tube radiolabelled DTT-E
    Lane 2: germ tube DTT-E component adsorbed on platelet
    Lane 3: blastospore DTT-E component adsorbed on platelet
    Lane 4: same as 2 with 50 x unlabelled DTT-E

REFERENCES

1. C. Mahaza, Etude de l'interaction _Candida albicans_-plaquette. Thése de doctorat d'état Es-Scences Pharmaceutiques, N° 48, Faculté de Pharmacie - Angers (1989).
2. M.J. Kennedy and R.L. Sandin, Influence of growth conditions on _Candida albicans_ adhesion, hydrophobicity and cell wall ultrastructure, _J.Med.Vet.Mycol._, 26 : 79 (1988).
3. E.H. Smail and J.M. Jones, Demonstration and solibilization antigens expressed primarily on the surfaces of _Candida albicans_ germ tubes, _Infect. Immun._, 45 : 74 (1984).
4. P.J. Fraker and J.C. Speck Jr., Protein and cell membrane iodination with a sparingly soluble chloroamide, 1,3,4,6-tetrachloro-3a,6b-diphenylglycoluril, _Biochem.Biophys.Res.Com._, 80 : 849 (1978).
5. U.K. Laemmli, Cleavage of structural proteins during the assembly of the head of bacteriophage T4, _Nature_, 227 : 680 (1970).
6. E. Fiss, H.R.L. Buckley, E. Wadsworth, A. Sandberg, Purification of actin from _Candida albicans_ and comparison with the _Candida_ 48.000 - M protein, _Infect. Immun._, 55 : 2324 (1987).
7. R. Matthews, J.P. Burnie, S. Tabagchali, Immunoblot analysis of the serological response in systemic candidosis, _Lancet_, 22/29 : 1415 (1984)
8. N.A. Strockbine, M.T. Largen, H.R. Buckley, Production and characterization of three monoclonal antibodies to _Candida albicans_ proteins, _Infect.Immun._, 43 : 1012 (1984).

9. N.A. Strockbine, M.T. Largen, S.M. Zweibel, H.R. Buckley, Identification and molecular weight characterization of antigens from _Candida albicans_ that are recognized by human sera, _Infect.Immun._, 43 : 715 (1984).

10. G. Tronchin, J.P. Bouchara, R. Robert, J.M. Senet, Adherence of _Candida albicans_ germ-tubes to plastic : ultrasructural and molecular studies of fibrillar adhesins, _Infect.Immun._, 56 : 1987 (1988).

ADHESION OF CANDIDA TO MURINE GASTROINTESTINAL MUCOSA OF

ANIMALS TREATED WITH ANTI-CANCER THERAPY AND INHIBITION

BY A CHITIN DERIVATIVE

H. Sandovsky-Losica   E. Segal

Department of Human Microbiology, Sackler School of
Medicine, Tel-Aviv University, Tel-Aviv, Israel

INTRODUCTION

It is well established that Candida albicans is an inhabitant of
the gastrointestinal (GI) tract in a significant proportion of the
normal population 20-65%[1]. In debilitated individuals, following
various treatments such as broad spectrum antibiotics, irradiation or
anti-cancer cytotoxic drugs, an increased colonization of the GI
system is observed (56-75%)[2].

It is believed that massive colonization in the GI tract due to
antibiotic treatment, and damage to the GI mucosa due to irradiation
or anti-cancer chemotherapy, accompanied by impairment of the normal
host defense systems, may result in the GI tract serving as portal of
entry for Candida, leading to systemic candidiasis [2, 3, 4, 5].

Since attachment of microorganisms to epithelial mucosal
surfaces is generally considered the initiating point of colonization
and infection [6,7], the following questions can be asked : (1)
whether in debilitated hosts fungal adhesion to GI mucosa is
increased, and (2) whether the adhesion could be blocked and thereby
infection eventually prevented.

Chitin soluble extract (CSE) was recently found by us to be
effective in blocking the adherence of C. albicans to murine and
human vaginal cells and to human corneocytes [8,9]. This led us to
examine the effect of CSE on adherence in vitro of the fungi to GI
tissues from mice treated with anti-cancer therapy as well as from
untreated control mice.

MATERIAL AND METHOD

ICR female mice, 6 weeks old, grown under laminar flow, or conventional conditions were debilitated by exposure to 400R Cobalt
irradiation or by injection of the anti-cancer drugs Methotrexate
(I.P 3mg/mouse), or 5-Fluorouracil (I.V 200mg/kg). The effect of
these treatments was assessed by white blood cell counts and spleen
weights on various days post treatment.

Candida and Candidamycosis, Edited by E. Tümbay *et al.*
Plenum Press, New York, 1991

On various days post treatment, treated and untreated control mice were sacrificed, the small intestine was removed, and tissue disks (0.7 cm diameter) were cut using a metal punch. Tissue disks were exposed for 2 hrs (37°C) to 1ml of 2x108 radiolabelled <u>C. albicans</u> CBS 562; (<u>C. albicans</u> was labelled by a pulse of 20 ul of ^{3}H-leucine/10ml of yeasts, 30-50 C;/mMol, for 2 hrs at 37°C in phosphate buffered saline supplemented with CaCl2). Following incubation with the yeasts, the tissue disks were washed with PBS+CaCl2, digested (70% perchloric acid and 30% H_2O_2 at 90°C), and radioactivity was assayed. Percentage adherence was calculated by dividing the cpm of the tissue disk of the adhesion mixture by cpm of the <u>Candida</u> inoculum (1 ml of labelled yeasts).

CSE was prepared according to the procedure developed in our laboratory[8,9]. Briefly, a 20% suspension of chitin in distilled water was extracted by shaking, on a gyratory shaker for 8 hrs. The extract was concentrated and lyophilized.For inhibition experiments CSE was added to the adhesion mixture, and the test was performed as described above. Inhibition percentage was determined by comparison to the control (withoud CSE), which was considered 100%.

RESULTS AND DISCUSSION

1. <u>Treatment with anti-cancer therapy</u>

Irradiation, Methotrexate or 5-Fluorouracil treatment caused a decrease in the number of white blood cells (WBC). The decrease in the WBC reached its nadir on day 2 post irradiation, on day 3 post Methotrexate treatment and on day 5 post 5-Fluorouracil treatment (Fig. 1). Spleen weights decreased in Methotrexate and 5-Fluorouracil treated mice from the first day post treatment and remained low during the whole 7 day observation period.

2. <u>C. albicans adherence to tissues from animals treated with anti-cancer therapy</u>

As noted in Fig.2, in irradiated mice an increase in adherence level was observed, the most marked increase was founed on days 2-3 post treatment. In Methotrexate and 5-Fluorouracil treated mice an increase in adherence level was noted with a peak on day 3 or 4 post treatment, respectively. The peak values of adherence corresponded to the period in which irradiation, or anti-cancer chemotherapy was most effective, as evaluated by WBC decrease or reduction in spleen weights.

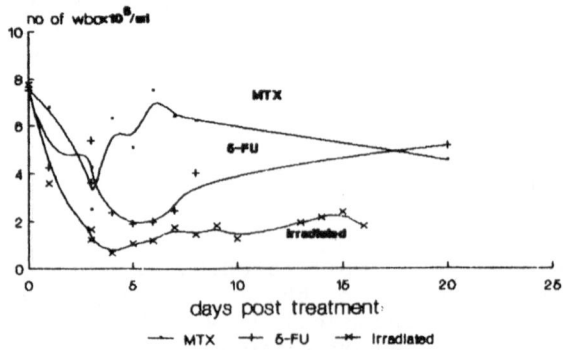

Fig. 1. White blood cell counts in debilitated mice.

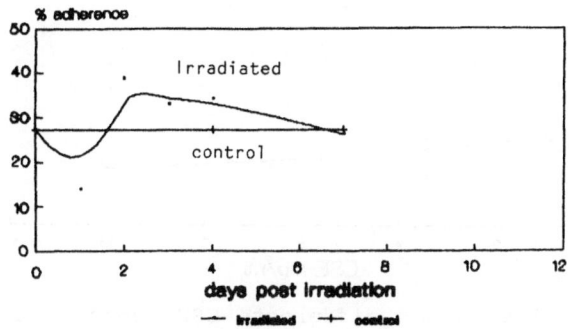

A: IRRADIATION

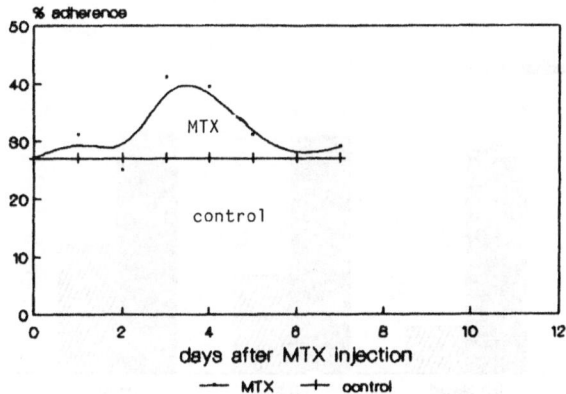

B: METHOTREXATE

C: 5-FLUOROURACIL

Fig. 2. Adherence of <u>C. albicans</u> to upper small
intestine from debilitated mice.

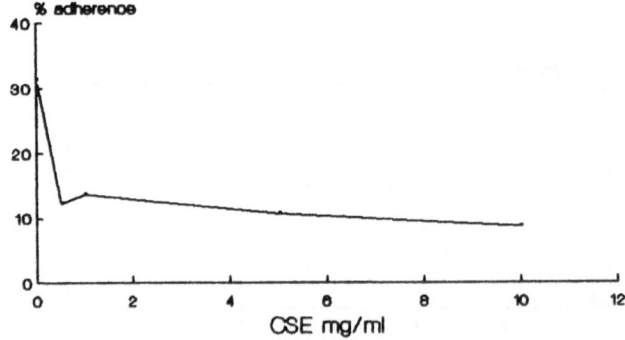

Fig. 3. Effect of different CSE concentrations
on adherence of C. albicans to GI mucosa.

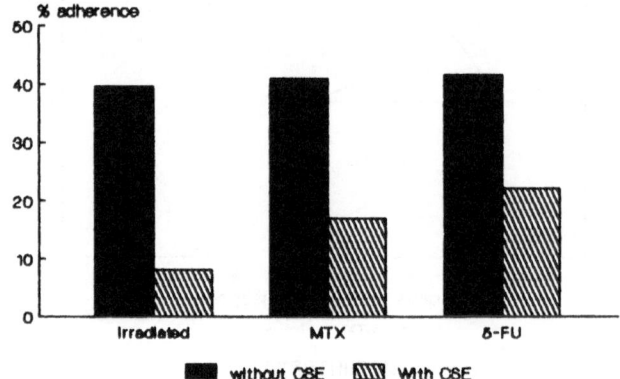

Fig. 4. Effect of CSE on adherence of C. albicans
to GI mucosa from debilitated mice.

3. Effect of CSE in vitro adherence to GI tissues:

Testing various CSE concentrations added to adhesion mixtures,
we found that CSE exhibits an inhibitory effect on the adhesion to
upper small intestine,from 10mg/ml down to 0.5 mg/ml (Fig.3). Testing
the effect of CSE (10 mg/ml) on adhesion of C. albicans to upper
small intestine from irradiated, Methotrexate or 5-Fluorouracil
treated mice at various day post anti-cancer treatment, we found that
adhesion was inhibited at all time intervals tested. As shown in
Fig.4, which represents data from time points with maximal adhesion
values, adhesion was blocked by CSE approximately by 50 percent.

In summary, we have shown in this study, that irradiation or
treatment with the anti-cancer drugs, Methotrexate or 5-Fluorouracil,
resulted in increased adherence to mucosal surfaces and that a chitin
derivative blocked the in vitro adhesion. The following step would
involve exploration of the possibility to use this approach in vivo,
attempting to block candidal colonization and/or dissemination.

REFERENCES

1. Stone, C.E. Geheber, L.D. Kolb, and W.R., Kitichens, Alimentary tract colonization by _Candida albicans_, J. Surg. Res., 14:273 (1973).
2. G. P. Body, Candidiasis in cancer patients, _Am. J. Med._, 77 (4D): 13 (1984).
3. P. Eras, M.J.Goldstein and P. Sherlock, _Candida_ infections of the gastrointestinal tract, _Medicine_ 51:367 (1972).
4. J.S. Trier and D.J. Bjorkman, Esophageal, gastric and intestinal candidiasis, _Am. J. Med._, 77 (AD): 39 (1984).
5. C. Hawkins and D. Armstrong, Fungal infections in the immuno-compromised host, _Clin. Hematol._, 13:599 (1984).
6. A. Ofek, and E.G. Beachey, General concepts and principles of bacterial adherence in animals and man, _in_: "Bacterial Adherence (Receptors and Recognition)", Series B:6, E.G.Reachey, ed., pp.3-29, Chapman and Hall, London (1980).
7. E. Segal, Pathogenesis of human mycoses: role of adhesion to host surfaces, _Microbiol. Sci._, 4: 344 (1987).
8. M. Kahane, E. Segal, M. Schewach-Millet and Y. Gov, In vitro adherence of _Candida albicans_ to corneocytes, inhibition by chitin soluble extract, _Acta Dermatol_. _Venerol._, 68:98 (1988).
9. N. Lehrer, E. Segal and L. Barr-Nea, _In vitro_ and _in vivo_ adherence of _Candida albicans_ to mucosal surfaces. _Ann. Microbiol._, 134B:293 (1983).

PHOSPHOLIPASE ACTIVITY IN <u>CANDIDA ALBICANS</u>, <u>CANDIDA</u> SPP

AND OTHER YEASTS

A. Rezusta[1]   M.C. Alejandre[2] J. Gill[2]   M.C. Rubio[2] M.S. Salvo[3]

[1] Hospital General "San Jorge" Huesca, Avda Martinez de
   Velasco, Huesca, Spain
[2] Hospital Clinico Universitario Zaragoza, Avda San
   Juan Bosco 15, 50009 Zaragoza, Spain
[3] Microbiologia, Ambulatorio General Solchaga, San
   Fermin 29, Pamplona, Spain

INTRODUCTION

The ability of <u>Candida albicans</u> to produce cytolytic enzymes
such as proteinases [1,2] and phospholipases [3,4] may be associated with
the pathogenicity of this fungus [5]. The phospholipases secreted by
<u>C. albicans</u> could play a part in the invasion of the host tissues in
lesions   of   candidosis [6] by   disrupting   the   epithelial   cell
membranes and allowing the hyphal tip to enter the cytoplasm [7].

There are five kinds of phospholipases: A,B,C,D and lysophospholipase. Phospholipase A was described by Costa et al.[3], Price and
Cawson [8], Hugh and Cawson [9] ; phospholipase B by Banno et al.[6],
phospholipase C by Costa et al.[3]; lysophosholipase by Hugh and
Cawson [7,] Price at al [8] and Banno et al.[6] Phospholipase D has not been
found in <u>C. albicans</u>.

All phospholipase activities in mycelial form cells were
extremely low as compared to those of yeast form cells [6].

In   this   study,   the   authors   investigated the phospholipase
activity in different yeast strains isolated from several clinical
specimens.

MATERIAL AND METHOD

<u>Clinical isolates</u>

Four hundred and ninety nine isolates of <u>Candida</u> species
(<u>C.albicans</u>, 441; <u>C. tropicalis</u>, 21; <u>C. parapsilosis</u>, 15;
<u>C. glabrata</u>, 15; <u>C. krusei</u>, 3; <u>Candida</u> spp., 3; <u>C. guilliermondii</u>, 1;
<u>Saccharomyces cerevisiae</u> 1 and <u>Trichosporon beigelii</u> 1 from the
Microbiology Service of the University Hospital of Zaragoza and one
<u>C. albicans</u> ATCC 13053 were tested. All isolates were identified by
standard procedures [10, 11].

Preparation, inoculation and incubation of plates

The method of Price et al.[13] modified by Samaranayake et al.[12] was used. The test medium comprised Sabouraud Dextrose Agar (SDA), supplemented with 1 M sodium chloride, 0.005 M calcium chloride and 8 % sterile egg yolk (Oxoid). The egg yolk was centrifuged at 500 g for 15 min and the supernatant made up to its original volume in sterile distilled water and incorporated into the sterile medium. Equal volumes of citric acid and di-sodium hydrogen phosphate buffer and the test medium were mixed aseptically and plates with final pH value of 4.3 were prepared.

The yeast inocula for the assay were prepared by sub-culturing the stock cultures on SDA for 18 hours. A final yeast suspension of 1-2 x 10^6 yeasts/ml was perapared by addition of an appropiate volume of sterile saline. A Steer's multipoint inoculator was then used to transfer the samples to the test plates (6 samples per plate). The plates were incubated at 37°C for 4 days, in a humid chamber [12].

Measurement of phospholipase activity

Phospholipase activity (Pz value) was determined by the ratio of the diameter of the colony to the total diameter of colony plus precipitation zone. Thus, Pz = 1.00 meant that the isolate tested was phospholipase negative, and Pz= 0.70, for instance, meant that the test strain was releasing considerable amount of phospholipase [10].

RESULTS

Of the 441 isolates of C. albicans tested, 353 (80.04%) were phospholipase-positive.All C. tropicalis, C. parapsilosis, C. glabrata, C. krusei, Candida spp, C. guilliermondii, T. beigelii and S. cerevisiae isolates were negative for phospholipase activity (Table 1). These results did not include one strain of C. albicans (ATCC 13053), which is positive.

Only 19.06 % of the C. albicans isolates did not produce phospholipase, while the remainder (80.04 %) demonstrated varying levels of activity, Pz ranging from 0.25 to 0.9. The distribution of Pz values of the 441 isolates is shown as a histogram in Figure 1. Statistical results are shown in Table 2.

Relationship between phospholipase activity and infection sites is demonstrated in Tablo 3.

Table 1. Phospholipase activity in 501 yeast isolates.

Microorganisms	Number of Strains	Positive
C.albicans	441	353 (80.04%)
C.tropicalis	21	0
C.parapsilosis	15	0
C.glabrata	15	0
C.krusei	3	0
Candida spp.	3	0
C.guilliermondii	1	0
Saccharomyces cerevisiae	1	0
Trichosporon beigelii	1	0

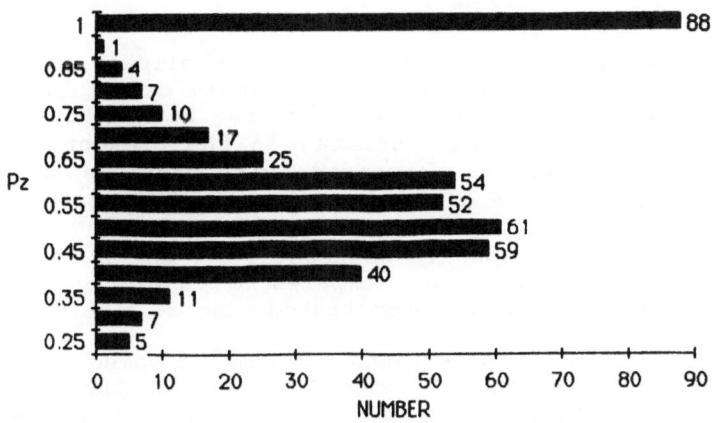

Fig. 1. Histogram showing the distribution of phospholipase activity in C. albicans.

Table 2. Descriptive statistics of phospholipase activity in 353 C. albicans isolates.

Minimum	0.900
Maximum	0.250
Range	0.650
Median	0.500

Table 3. Correlation of phospholipase activity of C. albicans and infection site.

Specimens	Number	Phospholipase positive	Phospholipase negative
Urine	71	58 (81.69%)	13 (18.13%)
Vaginal excretions	59	45 (76.27%)	14 (23.72%)
Nail	60	44 (73.33%)	16 (26.66%)
Urethral excretions	8	6 (75.00%)	2 (25.00%)
Skin	32	25 (78.12%)	7 (21.87%)
Folliculitis	7	7 (100%)	0
Wound pus	31	28 (90.32%)	3 (9.67%)
Respiratory tract	167	135 (80.83%)	32 (19.16%)
Others	6	5 (83.33%)	1 (16.66%)

DISCUSSION

Perhaps the presence or absence of phospholipase activity could be employed as a more useful discriminatory criterion for strain differentiation. It is noteworthy that the percentage of phospholipase-negative C. albicans isolates in the present study was a little low (19.06 %) as compared with previous studies which demonstated that 23 % (6 of 28) oral isolates (14), 45 % (5 of 11) blood isolates, 50 % (14 of 28) wound isolates and 70 % (9 of 13) urine isolates were phospholipase negative[13]. On the other hand, phospholipase activity was lower compared to studies of oral isolates by Williamson et al. [14] which demonstrated that 94 % were positive.

In two cases the authors found the same Pz value in C. albicans isolates from three different infection sites in the same patient. This finding supports Price et al.[13] who say that it is now possible to use the Pz value as a marker to identify and compare C. albicans strains isolated from multiple infection sites in the same patient. Appropriate clinical symptoms together with the same Pz value in isolates from blood and other sites in the same patient would be strong evidence that systemic infection exists.

The results of this study indicate that C. tropicalis and C. glabrata and C. parapsilopsis do not produce phospholipase detectable by this method as indicated by Samaranayake et al.[12] Barret et al. [15] found that S. cerevisiae and C. parapsilopsis showed by radioenzymeassay low phospholipase activity.

C. guillermondii, T. beigelii and C. krusei do not produce phospholipase which assay has not been documented previously.

REFERENCES

1. F.Macdonald and F.C.Odds, Inducible proteinase of Candida albicans in diagnostic serology and the pathogenesis of systemic candidosis, J. Med. Microbiol., 13: 423 (1980).
2. R.Rüchel, R. Tegeler and M. Trost, A comparison of secretory proteinases from different strains of Candida albicans, Sabouraudia, 20: 233 (1982).
3. A.L. Costa, C.Costa, A.Misefari and A.Amato, On the enzymatic activity of certain fungi, VII. Phosphotidase activity on media containing sheep's blood of pathogenic strains of Candida albicans, Atti societa peloritana discienze fisiche mathematische naturali, XIV: 93 (1968).
4. D.B. Louria, R.G. Brayton and G. Finkel, Studies on the pathogenesis in experimental Candida albicans infections in mice, Sabouraudia, 2: 271 (1963).
5. F.C. Odds, "Candida and Candidosis", Leicester University Press, Leicester (1979).
6. Y.Banno, T.Yamada and Y.Nozawa, Secreted phospholipases of the dimorphic fungus, Candida albicans. Separation of three enzymes and some biological properties, Sabouraudia 23: 47 (1985).
7. D.Hugh and R.A. Cawson, The cytochemical localization of phospholipase A and lysophospholipase in Candida albicans, Sabouraudia, 13: 110 (1975).
8. M.F.Price and R.A.Cawson, Phospholipase activity in Candida albicans, Sabouraudia, 15: 179 (1977).
9. D.Hugh and R.A.Cawson, The cytochemical localization of phospholipase A and lysophospholipase in Candida albicans, Sabouraudia, 13: 110 (1975).

10. J.Lodder, "The Yeasts: A Taxonomic Study", 2nd Ed., Elsevier, Amsterdam (1970).

11. D.W.R. Mackenzie, Serum tube identification of _Candida albicans_, _J. Clin. Pathol._, 15: 563 (1962).

12. L.P. Samaranayake, J.M. Raeside and T.W. Macfarlane, Factors affecting the phospholipase activity of _Candida_ species _in vitro_, _Sabouraudia_, 22: 210 (1984).

13. M.F. Price, I.D. Wilkinson and L.O. Gentry, Plate method for detection of phospholipase activity in _Candida albicans_, _Sabouraudia_, 20: 7 (1982).

14. M.I. Williamson, L.P. Samaranayake and T.W. Macfarlane, Phospholipase activity as a criterion for biotyping _Candida albicans_, _Sabouraudia_, 24: 415 (1986).

CANDIDA ALBICANS IN CELL CULTURE SYSTEMS: A MODEL FOR PATHOGENICITY

STUDIES

H. Hänel

Hoechst A.G., Department for Chemotherapy, Mycology/
Parasitology, Postfach 80 03 20, 6230 Frankfurt am Main
80, The Federal Rebuplic of Germany

INTRODUCTION

A number of factors influence the penetration process of
C. albicans into mammalian cells. One of them is the production of a
phospholipase by the yeast. Histochemical investigations confirm an
increased production of this enzyme at the site of penetration. The
combined action of proteases and phospholipases probably facilitate
the penetration not only of germ tubes but also of blastospores of
C. albicans. In an in vitro test system basing on the plate method of
Price et al[1]. various compounds were investigated for their
phospholipase inhibiting activity. Using this system the β-adrenergic
receptor blocker propnanolol proved to be active [2]; therefore, the
author had to develop a cell culture system to test the influence of
propranolol on the adhesion and penetration of C. albicans to vaginal
fibroblasts. The isomer R-propranolol was selected, for it lacks the
β-blocking activity but exhibits the full phospholipase inhibition.

MATERIAL AND METHOD

A) Cells

Wistar female rats were killed with CO_2, the vaginal tissue dis-
sected, cut and rinsed with PBS, submersed on gauze with 0.25% tryp-
sine + 0.5% EDTA solution in PBS for 30 minutes. There after the
epithelial layers were carefully scraped off, transfered to Williams
E cell culture medium (Flow Lab.) + 10% fetal calf serum (Gibco
Corp.), washed twice with the medium and centrifuged 10 minutes at
61g (4°C). Resuspended cells in the same medium were transferred to
culture flasks at 5% CO_2 95% relative humidity and 37°C. The cells
were co-cultured every week.

B) Treatment

The cells were grown on glass slide chambers (Nunc) with remov-
able frame until an incomplete monolayer was formed.

20 mg R-propranolol was dissolved in 1 ml DMSO. 20 ml of this

solution dissolved in 4 ml Williams E medium results in a concentration of 100 mg/ml. Lower concentrations were prepared accordingly. After treatment 40 ml of a <u>C. albicancs</u> suspension was added to each 4 ml chambers. The suspension was made of overnight grown <u>C. albicans</u> cultures (serotype A, strain 175) on maltpeptone agar at 37°C. At 536 nm the suspension had an extinction of 1.0. Therefore the final concentration was 105 colony forming units <u>C. albicans</u> per ml Williams E medium. The slide chambers with the cell cultures and yeasts were incubated for 6 h at 37°C, 5% CO_2 and 95% relative humidity.

C) Scanning electron microscopy (SEM)

Cell cultures were fixed in Bouin solution for 24 h at 4°C, washed twice for 4 h each in PBS and then transferred through a dehydrating row 20% ethanol, 40%, 60%, 80% and 90% ethanol for 12 h each. Then they were submerged in 96% and absolute isopropanol for 24 h each. Alcohol was removed by ethylacetate during 48 h. The wet sample was then critically point dried (Polaron, Waterford) and sputtered with gold (Polaron). The samples were scanned in a Leitz 1600T-SEM.

RESULTS

R-propranolol had a strong effect on several parameters of the <u>Candida</u>-fibroblast interaction. It inhibited the adhesion in a dose dependent manner from 25 mg/ml up to 100 mg/ml. At 50 mg/ml and 100 mg/ml R-propranolol 1-3 blastospores were present at each mm^2 of the cell culture, whereas the control cultures harboured from 50 to 200 blastospores per mm^2 (Fig. 1). Additionally, the hyphae and pseudohyphae grew short when R-propranolol was applied. The penetration in the controls was obvious (Fig. 2) whereas the penetration was not usual under the influence of R-propranolol (Fig. 1). At 100 mg/ml a slight damage of the fibroblasts became obvious by formation of vacuoles.

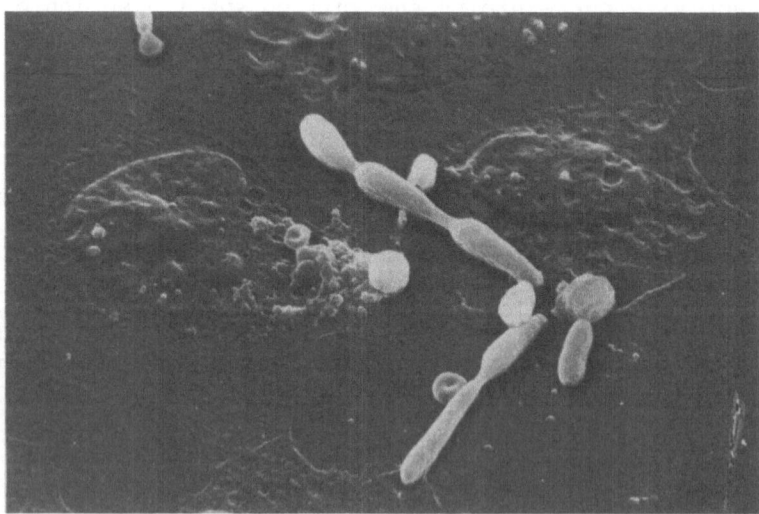

Fig. 1. Scanning electron microspcope picture of a cell culture with fibroblasts. After infection with <u>C.albicans</u> cultures were treated with 50 mg/ml R-propranolol.

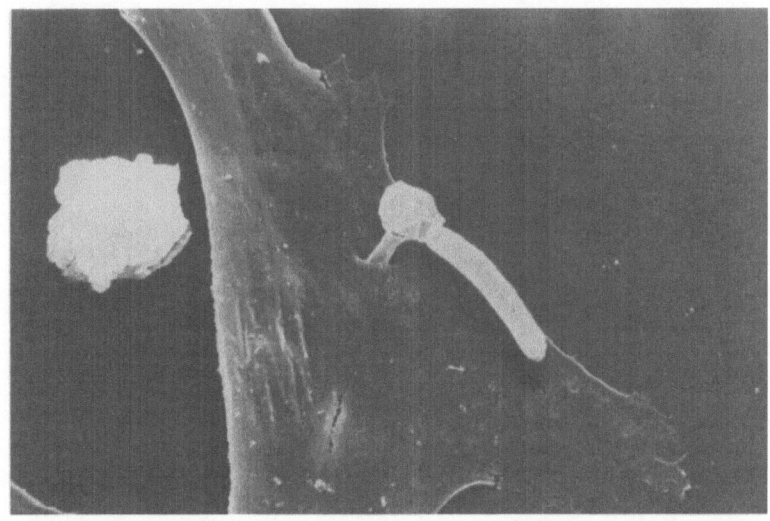

Fig. 2. Cell culture in Figure 1 treated with the solvent. Note penetration of the hyphal tip into the fibroblast.

DISCUSSION

Lipophilic β-blocking agents like propranolol are known to inhibit phospholipase A1 from mammals as shown by Pappu et al[2]. It is assumed that they inhibit by binding to the substrate. Yet it has been described that the phospholipases are extremely variable. The phospholipase C from Clostridium spp. differs extremely from the human phospholipase C which itself has only few DNA similarities to phospholipase C from Trypanosoma brucei. It is highly likely that the phospholipases from Candida albicans are not identical to the human phospholipases which would enable a specific inhibition. The role of phospholipase in the attachment was described by Barrett-Bee et al[3]. It also has been discussed as a promoter of cell fusion[4] which would explain the intimate contact between the hyphae and the host cells. The author assumes that it is one of the multiple adhesion factors of Candida which are far from being fully understood.

REFERENCES

1. M.F. Price, I.D. Wilkinson and L.O. Gentry, Plate method for detection of phospholipase activity in Candida albicans, Sabouraudia, 20: 7 (1982).
2. A.S. Pappu, P.J. Yazaki and K.Y. Hostetler, Inhibition of purified lysosomal phospholipase A1 by beta-adrenoceptor blockers, Biochem. Pharmacol., 34: 521 (1985).
3. K. Barrett-Bee, Y. Hayes, R.G. Wilson and J.F. Ryley, A comparison of phospholipase activity, cellular adherence and pathogenicity in yeasts, J. Gen. Microbiol., 131: 1217 (1985).
4. E.A. Dennis, Phospholipases, in: "The Enzymes", Vol. 16, 3rd Ed., P.D. Boyer, ed., Academic Press, New York (1983).

VIRULENCE OF PROTEINASE-POSITIVE AND PROTEINASE-NEGATIVE CANDIDA ALBICANS TO MOUSE AND KILLING OF THE YEAST BY NORMAL HUMAN LEUKOCYTES*

S. Kuştimur[1]   H. El-Nahi[1]   N. Altan[2]

Faculty of Medicine, Gazi University, Ankara, Turkey
[1]Department of Microbiology
[2]Department of Biochemistry

INTRODUCTION

Candida infections, which are of opportunistic type are prevalently seen all over the world and C.albicans is the most frequently isolated species in human beings [1-4]. Candida species may lead to a wide spectrum of diseases due to physiological changes that occur in healthy persons and to traumatic, hematologic,endocrinal and other predisposing factors [2,5,6]. Species apart from C.albicans are responsible only for 10% of human candidiasis [7]. There are several factors that have been suggested as contributing to the pathogenicity of C.albicans [8-18]. Among them the secretory acid proteinase enzyme has been accepted lately by many investigators to play an important role in the pathogenicity of this microorganism [12-21]. This enzyme is produced primarily by C.albicans. However, some less pathogenic species like C.tropicalis and C.parapsilosis also manufacture the enzyme [13-15].

The virulence of proteinase-producing strains and of proteinase-deficient mutant strains has been investigated under experimental conditions in mice, and it has been shown that the proteinase-deficient mutant strains are less pathogenic for mice. Also, it has been demonstrated that the proteinase releasing strains are more virulent due to their effective resistance to phagocytosis and the intracellular killing [13,19,22,23] . In this connection, we aimed to assess the role of secretory acid proteinase in the virulence of C.albicans by means of in-vivo mouse experiments along with in-vitro phagocytosis and intracelluler killing experiments.

MATERIAL AND METHOD

Candida strains : In this study, 100 yeasts were isolated from various samples received by Microbiology Department from clinics and out-patient clinics. Isolates of Candida species were classified according to classical methods.

*This study was supported in part by The Scientific and Technical Research Council of Turkey (TUBITAK, TAG-612) .

Control Strains : Two C. albicans strains, known to produce secretional proteinase, coded 73/079 and 83/008, were taken from Dr. F.C.Odds, Leicester University, England.

The test for acid proteinase enzyme : The modification of the method described by Kwon-Chung et al [22] was applied.

The experiment of virulence in the mouse : Six groups were formed, each containing 10 mice weighing 20-25 gm. One group received C.albicans 83/008 as positive control, three groups three different C.albicans strains with known proteinase activity, and one group a proteinase negative strain as negative control. The last group received physiological saline injection. The animals which were injected 0.5 ml of Candida suspension (10^6 yeast/ml) through their lateral tail veins, were observed everyday for 30 days. The samples obtained from the mice that had been injected with Candida species were examimined from mycological and histopathological aspects as proposed by Mac Donald and Odds[23].

Granulocyte candidacidal experiment : The study was done according to the methods of Lehrer and Cline[24].

RESULTS AND DISCUSSION

We have classified many Candida strains isolated from various clinical specimens into their species. By modifying Kwon-Chung et al.'s method[22] we have found the enzymatic activity in eleven different C.albicans isolates (Table 1). The strains showed proteolytic activity after different incubation periods. We found the maximal activity at the 18th hour in 8 strains, at the 22nd hour in 2 strains and at the 32nd hour in 1 strain. We saw a decrease down to the initial activity after 40th hour in all strains. Also the control strain showed maximal activity at the 18th hour (Figure 1). Kwon-Chung et al. [22] have also have found this activity around the 22nd hour. Our findings support the results of many studis which demonstrated that different C.albicans strains have an enzymatic activity varying from strong to weak proteolytic effect [12-15,22,27]. In the study, we also examined the activities of one C.tropicalis, one C.pseudotropicalis and two C.krusei strains. No activity was observed. Budtz-Jörgensen [12] and Mac Donald [15] have found proteinase activity in various C.tropicalis strains but not in C.pseudotropicalis and C.krusei which support our findings.

The second part of our study included an in-vivo virulence assay of the proteinase-positive C.albicans. 50% of the four groups of mice (40 in number), which had been injected with proteinase-positive C.albicans died within the first 72 hours. Within 10 days 80% of them died. The death rate on the 16th day was 100%. In the proteinase-negative group only a single mice died within 10 days. On the 30th day 70% of them were alive (Figure 2). In the organ cultures prepared from the mice, that were either dead or killed by cervical dislocation, proteinase-positive C.albicans grew at a rate of 90% in the kidney and 40% in the other organs. In-vivo experimental studies have shown taht the kidneys are target organs in disseminated Candida infections[23,25]. On the other hand, the proteinase-negative strains showed a growth rate of 10% in the kidney and no growth in the other organs. In the histopathological examinations of the organs, especially in the kidneys there were multiplication of the yeast blastospores and hyphae formation along with severe tissue changes (Figure 3). Our findings correlate with other works and also indicate that proteinase increases the virulence of C.albicans and is an important factor in the pathogenesis [22,23,25].

Table 1. Absorbance values of secretory acid proteinase activity at 280 nm in different time periods of C. albicans, C. tropicalis, C. parapsilosis, C. krusei and control strains isolated from various specimens.

Strain Number		2nd hr	12th hr	16th hr	18th hr	22nd hr	32nd hr	36th hr	40th hr
C. albicans	1	0.224	0.450	0.380	0.740	0.460	0.350	0.300	0.260
"	2	0.210	0.240	0.250	0.780	0.300	0.260	0.250	0.230
"	3	0.250	0.290	0.310	0.340	0.420	0.455	0.380	0.290
"	4	0.217	0.380	0.390	0.420	0.480	0.420	0.350	0.260
"	5	0.140	0.310	0.650	0.850	0.820	0.710	0.515	0.300
"	6	0.170	0.200	0.595	0.798	0.770	0.730	0.470	0.235
"	7	0.199	0.234	0.615	0.740	0.780	0.300	0.225	0.160
"	8	0.255	0.370	0.862	0.775	0.722	0.600	0.459	0.340
"	9*	-	-	0.715	0.925	0.737	-	-	-
"	10*	-	-	0.827	0.870	0.842	-	-	-
"	11*	-	-	0.460	0.990	0.708	-	-	-
C. tropicalis	1	0.000	0.000	0.000	0.000	0.000	0.000	0.000	0.000
C. pseudotro-picalis	1	0.000	0.000	0.000	0.000	0.000	0.000	0.000	0.000
C. krusei	1,2	0.000	0.000	0.000	0.000	0.000	0.000	0.000	0.000
Control 83/008		0.240	0.360	0.460	0.980	0.920	0.590	0.460	0.270
Control 73/079		0.135	0.180	0.210	0.600	0.522	0.330	0.270	0.210

* : The proteinase activities were determined when the other strain enzyme activities were at their highest levels.

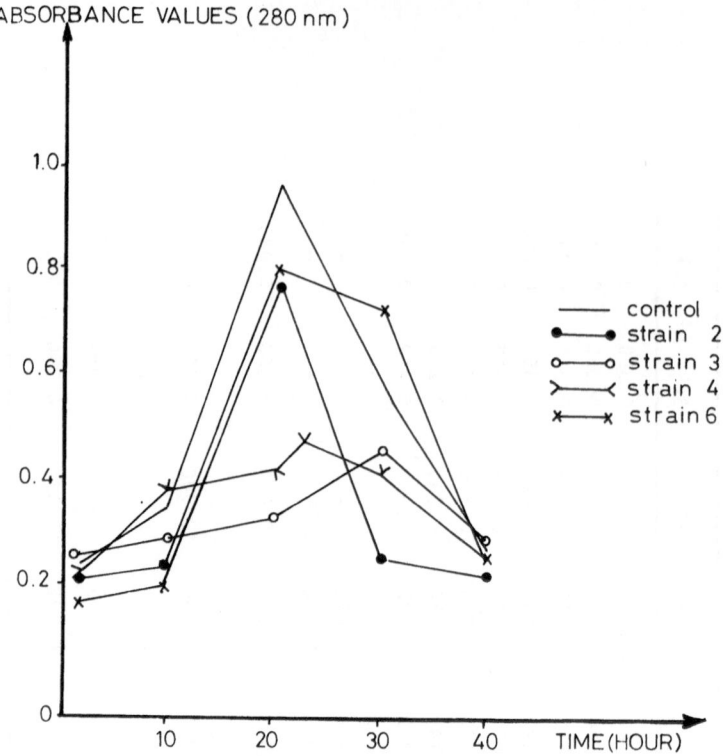

Fig. 1 . Proteinase activities of various C. albicans strains and control strain coded 83/008.

The most important defense mechanism of the body against Candida is the phagocytosis function of the granulocytes. According to the researchhers, the virulent C.albicans strains that produce significant amounts of proteinase are more resistant to intracellular death compared to the attenuated strains [22,23]. In our study we have detected the fate of three strongly virulent proteiniase-positive C.albicans strains, the control strain 83/008 and two proteinase-negative strains in the presence of normal healthy human granulocytes. Granulocytes showed no difference between the proteinase positive and proteinase negative strains in terms of phagocytosis. However, one hour after the inoculation proteinase-positive blastospores were alive at a higher rate compared to the proteinase-negative strains. The difference between the intracellular death rates was found to be significant ($p<0.05$) (Table 2). Walther et al[26]. have shown that the rate of intracellular death is related to the yeast/leukocyte ratio; the death rate is 66% at a ratio of 1/5, where as it goes down to 12% at 7/1. The difference among the death rate of proteinase-positive and negative strains at rates of 1/1, 1/3, 1/5 was statistically significant. The yeast/leukocyte ratio in our experiment is 1/1.

Also we studied the effect of the granulocytes and the autologous sera of two patients with chronic mucocutaneus candidiasis (CMC) on proteinase-positive C.albicans. The intracellular death rates of the

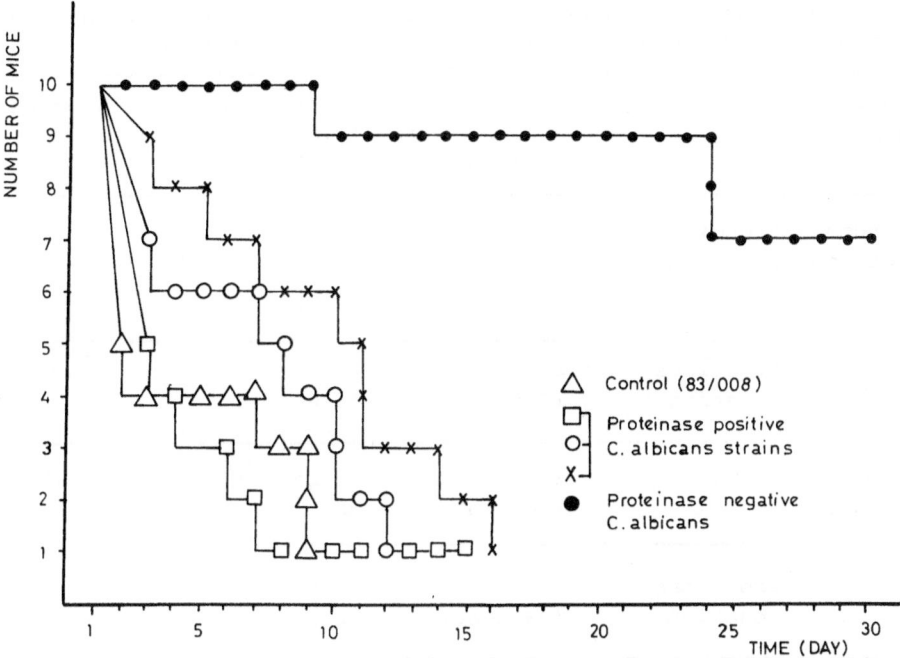

Fig. 2. Death ratio of mice by proteinase-positive and proteinase-negative *C. albicans* strains.

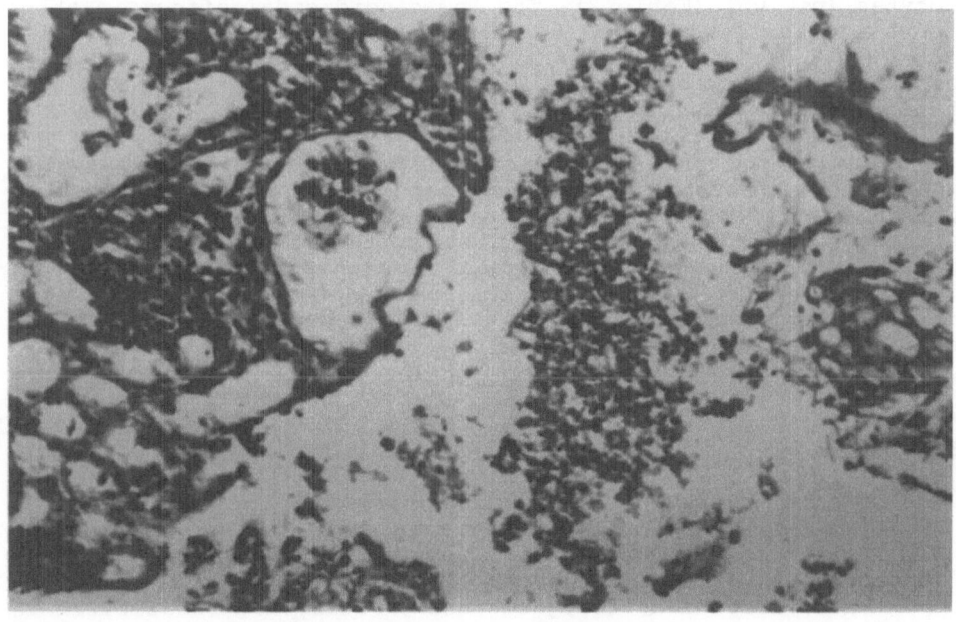

Fig. 3. Multiplication of the yeast hypha formation along with severe tissue changes in the kidney.

Table 2. Death ratios of proteinase-positive and negative C. albicans strains by normal human and patients' granulocytes.

Leukocyte (10⁷/ml)	C. albicans (10⁷/ml)	Number of strains and patients	% intracellular death*	Average **
Normal human granulocytes	Proteinase positive	4 (Strain)	23.41 26.83(Control) 31.25 45.50	31.75$^{\pm}$4.85
	Proteinase negative	2 (Strain)	48.75 56.66	52.70$^{\pm}$3.96
Granulocytes of patients with CMC	Proteinase positive	2 (Patient)	25.25 35.75	30.5$^{\pm}$5.26
	Proteinase negative	2 (Patient)	39.4 42.00	40.70$^{\pm}$1.29

* Averages are based on experiments done at least four times.
** t-Student test.

patients' leukocytes did not show any significant difference when compared with the rate of the normal leukocytes (p>0.05) (Table 2). The candidacidal effect of the same patients' leukocytes against the proteinase-negative strains were found to be low, relative to normal leukocytes (p<0.05) (Table 2). The low candidacidal effect in these patients does not seem to be related to the yeast itself. In these patients, as some researchers have also shown [27,28], especially the serologic factors are effective on the leukocyte functions, but this mechanism is not known yet.

In conclusion, our data correlates with other findings which suggest that the proteinase in an important factor in the pathogenesis of C.albicans. It shows that the strains which produce proteinase can escape from the host defense mechanisms much more easily, leading to formation of disseminated candidiasis. It emphasizes that the proteinase should be accepted among the virulence factors such as the formation of germinal tubes, adhesion to endothelial cells and mucosal surfaces, the production of chlamydospores and the endotoxin.

REFERENCES

1. P.A.Ketchum,"Microbiology-Introduction For Health Profesionals", 1st ed., John Wiley and Sons, New York (1984).
2. W.K. Joklik et al, "Zinsser Microbiology", 80th ed., Appleton-Century-Crofts, Connecticut (1984).
3. U.K. Unat, "Tıp Parazitolojisi-İnsanın Ökaryonlu Parazitleri ve Bunlarla Oluşan Hastalıkları (Medical Parasitology)", 3. Baskı, İstanbul Üniversitesi Cerrahpaşa Tıp Fakültesi Yayınları, İstanbul (1982).

4. A. Yücel, Tıp bakımından önemli Candida türlerinin mikolojisi (Mycology of important Candida species in medicine), Türk Mikrobiol.Cem.Derg., 1-2 : 45 (1987).

5. C.H. Kirk Patrick et al, Chronic mucocutaneous candidiasis, model-building in cellular immunity, Ann. Intern. Med., 74 : 955 (1971).

6. E. Jawetz, J.L. Melnick, and E.A. Adelberg, "Review of Medical Microbiology", 17th ed., Lange Medical Publications, Connecticut (1987).

7. J.P. Duguid, B.P. Marmion and R.H.A. Swain, "Mackie and Mac Cartney Medical Microbiology", 13th ed., The English Language Book Society and Churchill Livingstone, Hong Kong (1983).

8. E. Segal et al, Correlative relationship between adherence of C.albicans to human vaginal epithelial cells in vitro and candidal vaginitis, Sabouraudia, 22 : 191 (1984).

9. J. Mc Courtie and L.J. Douglas, Relation between cell surface composition, adherence, and virulence of Candida albicans, Infect. Immun., 45 : 6 (1984).

10. F.C. Odds, Commensalism and pathogenicity of yeasts on skin, J.Med.Microbiol., 17 : 3 (1984).

11. M.F. Price and R.A. Cawson, Phospholipase activity in Candida albicans, Sabouraudia, 15 : 179 (1977).

12. E. Budtz-Jörgensen, Proteolytic activity of Candida albicans spp. as related to the pathogenesis of denture stomatitis, Sabouraudia, 12 : 266 (1971).

13. R. Rüchel et al., Secretion of acid proteinases by different species of the genus Candida, Zbl.Bakt.Hyg., I.Abt.Orig.A., 255 : 537 (1983).

14. R. Rüchel, A variety of Candida proteinases and their possible target of proteolytic attack in the host, Zbl.Bakt.Hyg.A., 257 : 266 (1984).

15. F. Mac Donald, Secretion of inducible proteinase by pathogenic Candida species, Sabouraudia, 22 : 79 (1984).

16. R. Rüchel et al., Identification and partial characterization of two proteinases from the cell envelope of C.albicans blasto-spores, Zbl.Bakt.Hyg.A, 260 : 523 (1985).

17. H.Remold et al., Purification and characterization of a proteolytic enzyme from C.albicans, Biochim. Biophys. Acta, 658 : 399 (1968).

18. S. Kuştimur, Kandidalarda proteinaz aktivitesinin virulans ile ilişkisi (The relationship between proteinase activity and virulance in Candida), Gazi Üniv. Tıp Fak.Derg., 3 : 221 (1987).

19. F. Mac Donald et al., Inducible proteinase of C.albicans. Diagnostic serology and in the pathogenesis of systemic candidosis, J.Med.Microbiol., 13 : 423 (1980).

20. R.Rüchel, Properties of a purified proteinase from the yeast C.albicans, Biochim.Biophys. Acta, 659 : 99 (1981).

21. F.Mac Donald et al., Purified C.albicans proteinase in the serological diagnosis of systemic candidosis, JAMA, 20 : 2409 (1980).

22. K.J. Kwon-Chung et al., Genetic evidence of role of extra-cellular proteinase in virulence of Candida albicans, J.Infect.Immun., 49 : 571 (1985).

23. F. Mac Donald and F.C. Odds, Virulence for mice of a proteinase-secreting strain of Candida albicans and a proteinase-deficient mutant. J.Gen.Microbiol., 129 : 431 (1983).

24. R.I. Lehrer, and M.J. Cline, Interaction of Candida albicans with human leukocytes and serum, J.Bacteriol., 95 : 996 (1969).

25. D.B. Louria and R.G. Brayton, Behavior of Candida cells within leukocytes, Proc.Soc.Exp.Biol.Med., 115 : 93 (1964).

26. T. Walther, M. Rytter, C. Schönborn et al., Differences in the intracellular killing of proteinase-positive and proteinase-negative Candida albicans strains by granulocytes, Mykosen, 29 : 10 (1986).

27. F.M. Laforce, D.M. Mills, K. Inverson et al., Inhibition of leucocyte candidacidal activity by serum from patients with disseminated candidiasis, J.Lab.Clin.Med., 86 : 657 (1974).

28. D. Djawari, O.P. Hornstein, J. Gross and W.Meinhow, Defect of phagocytosis and intracellular killing of Candida albicans by granulocytes in patients with familiar and non-familiar chronic mucocutaneous candidosis, Arch.Derm.Res., 260 : 159 (1977).

THE INVESTIGATION OF PATHOGENITY AND VIRULENCE OF CANDIDA

E.Dalkılıç T.Aksebzeci İ.Kocatürk N.Aydın B.Koçulu

Department of Microbiology, Faculty of Medicine,
Erciyes University, Kayseri, Turkey

INTRODUCTION

Candidiasis is the most frequently encountered opportunistic
fungal infection. It is caused by a number of species of Candida with
Candida albicans being the most frequent etiologic agent [1-3].

We investigated the relationship between the resistance to
digestion and tube germ formation of the phagocytosed Candida cells
and also their virulence.

MATERIAL and METHOD

Candida spp.

A total of 39 strains, isolated from various clinical specimens
were used.

They were 23 Candida albicans, 2 C. tropicalis, 2 C. pseudotropicalis, 4 C. stellatoidea, 1 C. guilliermondii, 1 C. parapsilosis, 2
C. krusei, 2 Geotrichum and 2 Torulopsis strains.

Phagocytosis Experiment

In this experiment we injected 50 ml 10/50 machine-oil into
peritoneal cavity of 3 kg weight Albino rabbits. Ten days after
injection, we took all the blood of the animals by cutting the femoral
artery and added anticoagulant to the blood. The plasma was separated
and dispersed into sterile tubes in 0.5 ml volume. For phagocytosis
experiment,1 drop of Candida suspension was put in the plasma. Giemsa
stainings were made from these specimens at 0, 2, 4, 12 and 24 hours.

The number of macrophages were adjusted as 4 times of Candida
number and Candida cells were put in macrophage-buffer suspension[4].

We took samples at 0, 2, 4, 12 and 24 hours and stained with
Giemsa. We examined them under light microscope with high
magnification. To determine the effect of phagocytosis on Candida,
samples were inoculated into Sabouraud-dextrose agar (SDA).

RESULTS

1) The number of *Candida* cells which have no resistance to digestion by macrophages decreased after 24 hours (Fig. 1 and 2).

2) The number of virulent *Candida* cells resistant to digestion by macrophages increased in number after 24 hours (Fig. 3 and 4).

3) In contrast to avirulent forms, virulent *Candida* cells multiplied within and then broke down the macrophages, budded and formed pseudomycelia (Fig. 5-9).

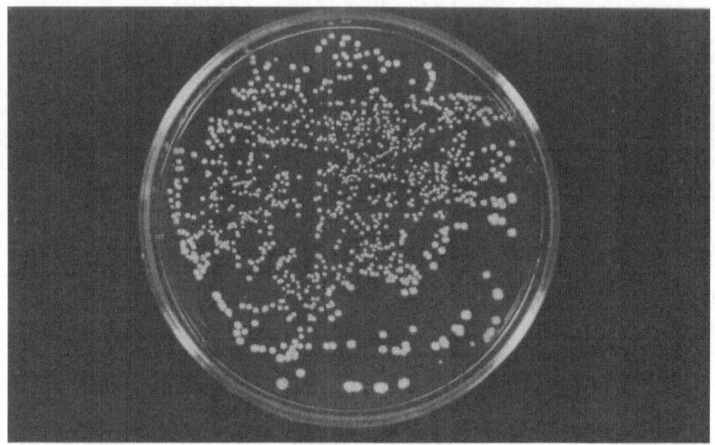

Fig. 1. The concentration of avirulent *Candida* cells at the beginning of phagocytosis experiment.

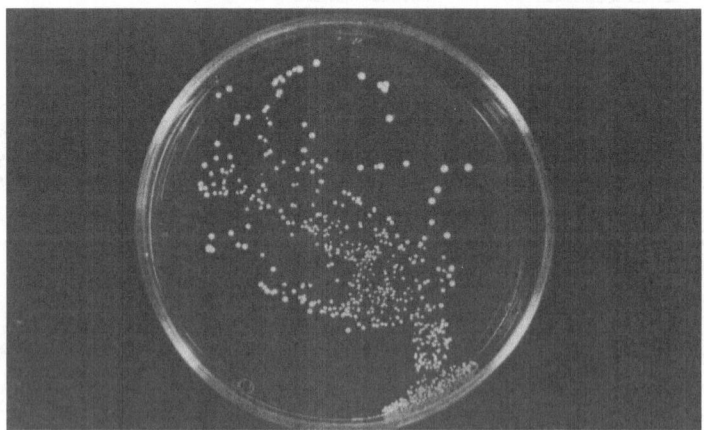

Fig. 2. The concentration of avirulent *Candida* cells 24 hours after dealing with macrophages.

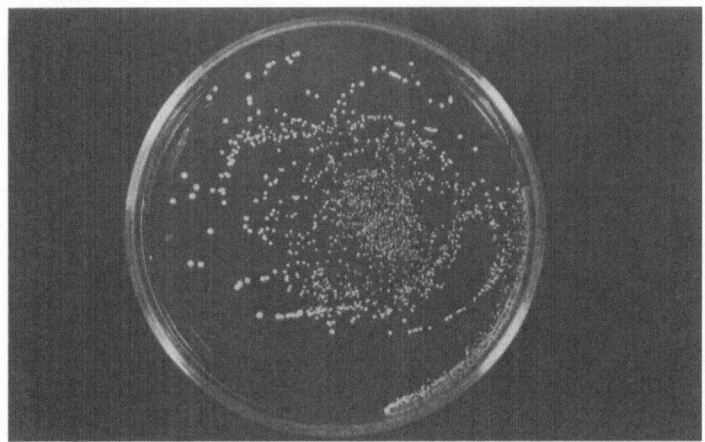

Fig. 3. The concentration of virulent <u>Candida</u> cells at the
beginning of phagocytosis experiment.

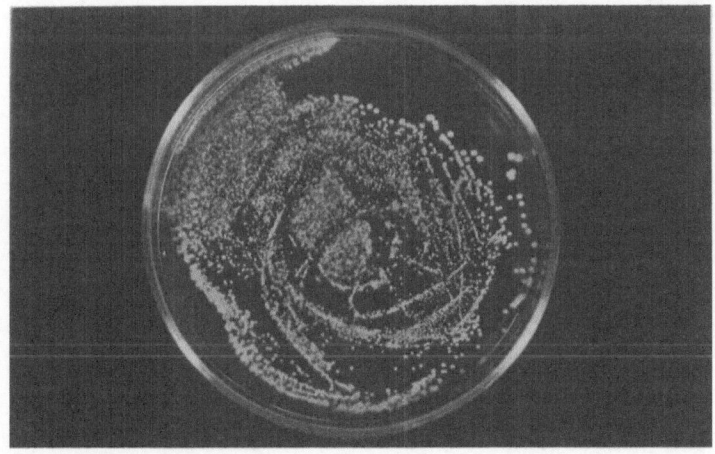

Fig. 4. The concentration of virulent <u>Candida</u> cells 24 hours
after dealing with macrophages.

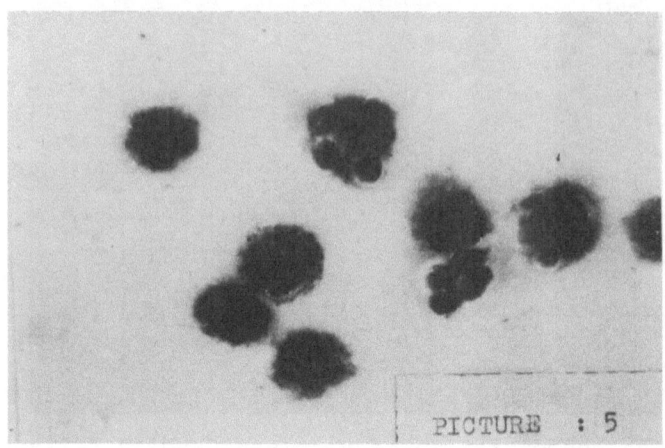

Fig. 5. The phogocytosis of avirulent <u>Candida</u> cells by
macrophages.

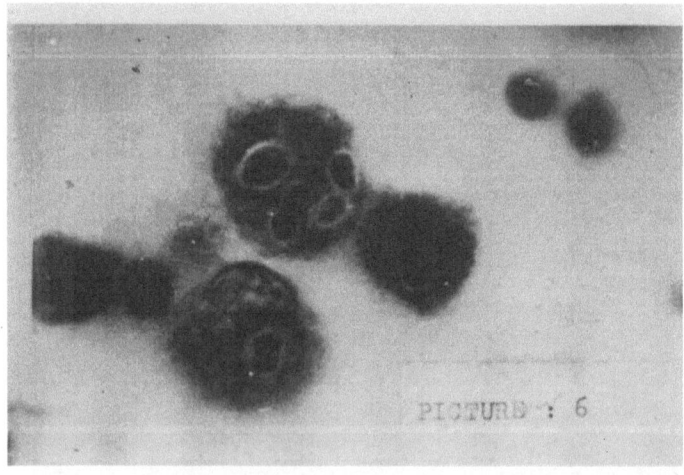

Fig. 6. The digestion of avirulent <u>Candida</u> cells inside
macrophages.

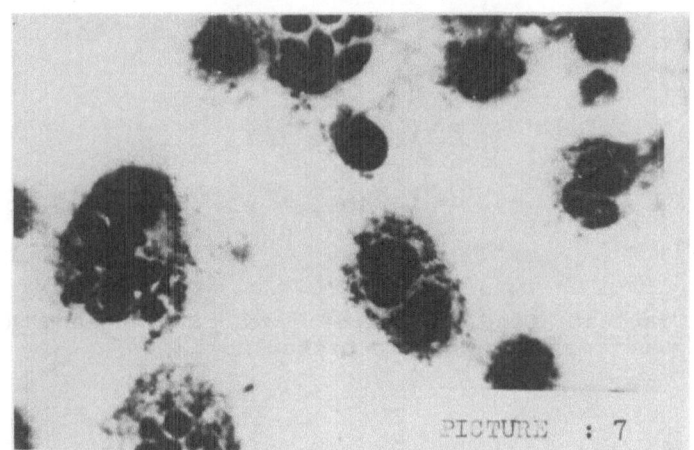

Fig. 7. The engulfed virulent _Candida_ cells inside
phagocytic macrophages.

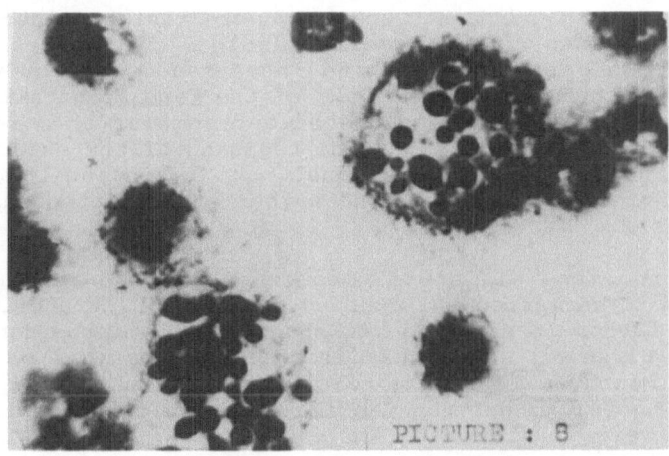

Fig. 8. The multiplication of virulent _Candida_ cells
with budding inside macrophages.

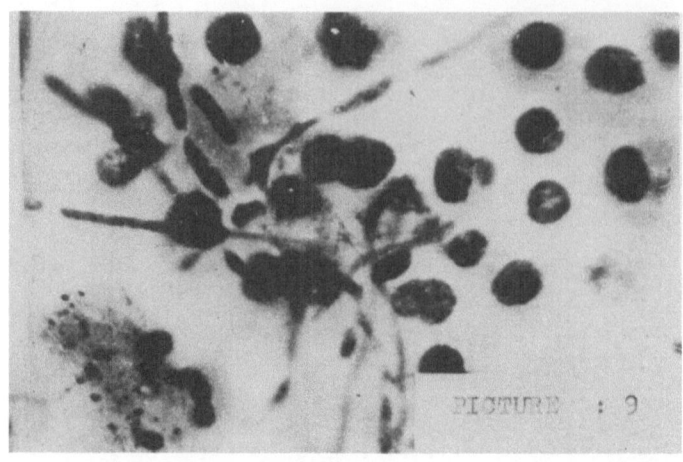

Fig. 9. Virulent <u>Candida</u> cells developing pseudomycelia after macrophages are broken down.

DISCUSSION

According to the investigation of Ekmen [4] and Dalkılıç [5], there is no difference between the pathogenic and non-pathogenic <u>Candida</u> cells in phagocytosis experiments.

The pathogenic <u>Candida</u> has scarcely more resistance to digestion by macrophages than non-pathogenic <u>Candida</u> [4,5] (Figures 5-9). The virulent <u>Candida</u> cells survived and increased in number within and then broke down the macrophages. Some of the <u>Candida</u> strains which we used in our experiments were digested and some of them resisted to digestion. As seen in Table 1, at the begining of the experiments the number of <u>Candida</u> cells were equal, but resistant <u>Candida</u> cells increased in number in culture. The number of non-pathogenic <u>Candida</u> cells decreased.

There are many different objections to the resistance of a microorganism to digestion [6,9]. Turning to acid of intracellular pH is an important factor in digestion of phagocytosed microorganisms [4,5,7]. According to Hirsh [10], some internal cell enzymes must be secreted. In our experiments, all <u>Candida</u> cels were faced with intraphagocyte enzymes, but we think that the <u>Candida</u> cells which showed resistance to digestion were virulent.

As seen in Table 1, there is a difference in the number of colonies of virulent and avirulent strains after 24 hours.

This difference is also observed in avirulent and virulent strains of some other micro-organisms like <u>Mycobacterium tuberculosis</u>, <u>Streptococcus</u>, <u>Staphylococcus</u> and <u>Cryptococcus neoformans</u>[11,12,13].

Table 1. Digestion within macrophage and germ tube formation of yeasts studied.

Yeast studied	Number of strains	Digestion in macrophage		The results of germ tube formation in 150 minutes	
		(−)	(+)	(+)	(−)
Candida albicans	23	20	3	20	3
Candida pseudotropicalis	2	1	1	1	1
Candida stellatoidea	4	3	1	3	1
Candida tropicalis	2	−	2	2	−
Candida krusei	1	1	−	−	−
Candida guilliermondii	1	−	1	−	1
Candida parapsilosis	1	1	−	1	−
Geotrichum candidum	2	1	1	−	2
Torulopsis glabrata	1	−	1	−	1
Total	38	29	9	28	9

(+) : Positive   (−) : Negative

We also investigated the relationship between the virulence and tube germ formation. In some strains the tube germ formation occured in 2 hours. But any relationship between the tube germ formation and virulence could not be determined. In conclusion, the tube germ formation occurs in many Candida cells in 2 hours. The virulent Candida cells show resistance to digestion by macrophages and continue to survive, but avirulent cells are digested.

REFERENCES

1. G.J.Platenkamp, Application of serological tests in the diagnosis of invasive candidiasis, Mycoses 31 (Suppl.2): 27 (1988).
2. M.H.Weiner and W.J.Yount, Mannan, antigenemia in the diagnosis Candida infections, J.Clin Invest., 58: 1045 (1976).
3. K.Higashide, R.Aman, O.Yamamuro, Clinical characteristics correlated with different fungi causing vulvovaginal mycosis, Mycoses 31: 213 (1988).
4. H.Ekmen, Invitro phagocytosis of Candida albicans, Türk Hijyen ve Tecrübi Biyolojisi Dergisi, XXIII: 71 (1963).
5. E.A.Dalkılıç, Determination of Candida virulence using various tests, Mikrobiyoloji Bülteni, 6: 313 (1972).
6. M.Holliday, A modified cytoplasmic antigen of candidiasis, J.Immunol.Methods, 31: 71 (1979).
7. C.B.Smith, Candidiasis; pathogenesis, host resistance and predisposing factors, in: "Candidiasis" pp. 53-70, Rowen Pess, New York (1985).
8. R.Monellie and L.T.Rosenberg, The role of complement in the phagocytosis of Candida albicans by mouse peripheral blood leukocytes, J.Immunol., 107: 476 (1971).
9. M.A.Ghannoum, Mechanism potentiating Candida infections, A review, Mycoses 31: 543 (1988).

10. J.G.Hirsch, Studies on the bactericidal action of phagocytic cell, *Bacteriol.Rev.*, 24: 133 (1960).
11. G.Mehdi, and M.Stuart, Role of serum in the intracellular killing of staphylococci in rabbit monocytes, *J.Bacteriol.*, 91: 1393 (1966).
12. E.Suter, The multiplication of tuberculer bacilli within normal phagocytes in tissue culture, *J.Exp.Med.*, 96: 137 (1952).
13. E.S.Bulmer, The *Cryptococcus neoformans* II. Phagocytosis by human leucocytes, *J.Bacteriol.*, 94: 1480 (1967).

INHIBITION OF CANDIDA PSEUDOTROPICALIS BY CERTAIN GRAM-NEGATIVE BACTERIA OF VETERINARY IMPORTANCE

K.S. Diker   F. Aydın   M. Arda

Department of Microbiology, Faculty of Veterinary
Medicine, Ankara University, Ankara, Turkey

INTRODUCTION

Candida species commonly colonize the oral cavity and intestinal canal of man and various animals [1]. There are data which suggest that the endogenous bacterial flora may suppress the colonization of Candida [2,3,4]. Experiments performed by different investigators indicate that both in vivo and in vitro some E. coli, Salmonella and Streptococcus strains inhibit C. albicans [5,6,7]. At least one report (Jeffries and Schileru, Ann. Meett. Am. Soc. Microbiol. 1975, Abstract D52, p.60) suggest that the production of bacteriocins may be responsible for this inhibition. It has been also reported that in vitro inhibiton may be due to pH change and nutrient depletion of media [8].

A number of bacteria have been shown to inferfere with the in vitro growth of C. albicans, C. krusei and C. tropicalis [4,8]. No previous work has determined the interactions between Gran-negative organisms and C. pseudotropicalis.

This study was conducted to determine the possible in vitro interactions of C. pseudotropicalis and organisms frequently encountered in animal infections.

MATERIAL AND METHOD

The following organisms originally isolated from animal clinical sources were used for their anticandidal activity: Escherichia coli, Salmonella typhimurium, Pseudomonas aeruginosa, Aeromonas hydrophila, Moraxella bovis, Campylobacter jejuni and C. coli. Number of each species tested is shown in Table 1. For demonstrating anti-candidal activity, Candida pseudotropicalis and C.albicans originally isolated from the liver of a cat and urine of a dog, respectively, were used as target strains.

All tests were carried out on brain-heart infusion (BHI) agar (Difco) and Mueller-Hinton (MH) agar (Oxoid). In cross streaking method, subcultures of bacteria were transferred to solid test medium by making a streak across the center of the plate. After incubation period, the bacterial growth was scraped off with glass slide and the plates were exposed to chloroform vapour for 60 min to kill residual cells. After aeration 2 h, diluted suspension of Candida (10^5–10^6 cell per ml) were streaked at right angles across the area of bacterial streak. The plates were examined after 24–48 h of incubation. Inhibition was recorded when colonies were absent in the area of cross streak than beyond.

In macro colony method, diluted suspensions of Candida were streaked over solid test medium. After allowing the Candida to growth for 2 h, bacteria were spot inoculated and plates were incubated for 24–48 h. Inhibitory activity was recorded when there was no growth of the target strain on and around the spot of bacterial growth.

To test the effect of nutrient depletion on growth, each experiment was carried out by overlaying the test medium with melted fresh medium after Candida growth. After each test, pH changes of media were also recorded. Each experiment was carried out in duplicate. The effect of Candida on the growth of bacteria was also studied by the same methods.

RESULTS

The inhibitory activity of bacteria against C. pseudotropicalis and C. albicans is shown in Table 1. All bacteria tested, except Campylobacter, inhibited the growth of C. pseudotropicalis in at least one of experiments, and many of these strains inhibited the growth of C. albicans also. The zone of inhibiton was clear on BHI agar plates, as compared to MH agar plates. The final pH values of the surfaces of the media varied from 5.3 to 7.4. The growth inhibition patterns did not differ whether or not the test media were overlaid with fresh media. Candida spp. did not produce any inhibitory activity against bacteria.

Table 1. Inhibition of Candida growth by Gram-negative organisms.

Bacteria	No. of strains tested	C.pseudotropicalis		C.albicans	
		CS[a]	MC[b]	CS	MC
P.aeruginosa	2	++[c]	++	++	+
A.hydrophila	2	++	++	+	+
S.typhimurium	3	+	+	+	-[d]
E.coli	1	++	++	+	+
E.coli	1	+	+	-	-
M.bovis	1	+	+	+	-
M.bovis	1	+	-	-	-
C.jejuni	2	-	-	-	-
C.coli	2	-	-	-	-

a: cross-streaking method   b: Macro-colony method
c: inhibition of Candida growth   d: no inhibition

DISCUSSION

Studies based on the inhibition of in vitro grown C. albicans by several bacteria led to conlusion that these microorganisms play a role in the resistance to candidiasis [4]. It is generally accepted that Gram-negative bacilli are more effective in inhibiting C. albicans, C. tropicalis and C. krusei [4,8]. Based on the assays described in this paper, the most strains of Gram-negative bacteria inhibited in vitro growth of C. pseudotropicalis. This inhibition of candidal growth by Gram-negatives was not an all-or-none phenomenon as there were variations between genera and within species regarding inhibitory activity. These observations are in good agreement with those reported for C. albicans [4,6].

Previous investigators have suggested that bacteria may suppress the growth of Candida by either competing for nutrients, changing pH of medium or by secreting antifungalsubstances [8]. Nutrient depletion as a possible mechanism of the observed in vitro inhibition of C. pseudotropicalis could be ruled out as, on one hand, the growth inhibition patterns did not differ whether or not the test media were overlaid with fresh media and, on the other hand, some strains belonging to the same species were unequally effective on Candida. In addition, as tested Candida strains have been shown to grow well at pH values between 5.3 and 7.4, the changes in pH could not be the cause of the observed inhibitions. Besides, the degrees of inhibition were so great in some cases as to suggest that other factors were operating to contribute to an inhibition of Candida growth.

C. albicans is accepted as the most pathogenic species of the genus whereas C. pseudotropicalis may very rarely be involved in pathological lesions of animals. The finding that C. pseudotropicalis was more sensitive to the inhibitory activity of bacteria than was C. albicans may be one of the interpretations of this condition.

REFERENCES

1. F.C. Odds, "Candida and Candidosis". Leicester University Press, Leicester (1979).
2. E. Balish and A. W. Phillips, Growth and virulence of Candida albicans after oral inoculation in the chick with a monoflora of either Escherichia coli or Streptococcus faecalis, J. Bacteriol., 91:1744 (1966).
3. D. Gale and B. Sandoval, Response of mice to the inculations of both Candida albicans and Escherichia coli. I. The enhancement phenomenon, J. Bacteriol., 73:616 (1957).
4. H.D. Isenberg, M.A. Pisano, S.L. Carito and J.I. Berkman, Factors leading to overt monilial disease. I. Preliminary studies of the ecological relationship between Candida albicans and intestinal bacteria, Antibiot. Chemother., 10:353 (1960).
5. J.M. Caves, J.A. Carpenter, and M.K. Hamdy, Interaction between Salmonella enteritidis and Candida albicans, Proc. Soc. Expl. Biol. Med., 143:433 (1973).
6. R.P. Hummel, E.J. Oestreicher, M.P. Maley and B.C. MacMillan, Inhibition of Candida albicans by Escherichia coli in vitro and in the germfree mouse. J. Surg. Res., 15:53 (1973).
7. W.F. Liljemark and R.J. Gibbons, Suppresion of Candida albicans by human oral streptococci in gnotobiotic mice, Infect. Immun., 8:846 (1973).
8. T.F. Paine. The inhibitory actions of bacteria on Candida growth, Antibiot. Chemother., 8:273 (1958).

ISOLATION OF <u>CANDIDA ALBICANS</u> HISTIDINE AUXOTROPH (HIS 4) AND

CHARACTERIZATION OF ITS ETIOLOGIC GENE

S.Gottlieb[1] E.Segal[1] G.Lebens[2] I.Polacheck[2] Z.Altboum[3]

[1]Department of Human Microbiology, Sackler School of
Medicine, Tel Aviv University, Tel Aviv, Israel
[2]Department of Clinical Microbiology, Hadassah Medical
Center, Jerusalem, Israel
[3]Department of Microbiology, Israel Institute of
Biological Research, Ness-Ziona, Israel

INTRODUCTION

The clinical importance of <u>C. albicans</u> requires the development
of a genetic system in this organism for better understanding of its
mechanism(s) of pathogenesis and to eventually develop improved
therapies for candidal infection. Our previous studies focused on the
attachment of the fungus to host mammalian tissues [1-3]. Those studies
involved a specific <u>C. albicans</u> isolate-CBS562. We considered it
important to develop a genetic system in this strain. Consequently,
we describe herewith the isolation of an histidine auxotroph (his4),
and isolation, characterization and chromosomal location of the
homologous HIS4 gene.

MATERIAL, METHOD AND RESULTS

Isolation of his4 mutant

<u>C. albicans</u> CBS 562 was mutagenized with NTG (final concent-
ration of 150 µg/ml). Out of 1050 colonies screened, a single His⁻
strain (SAG5) was isolated. SAG5 had an absolute requirement for
L-histidine, as shown in Table 1. SAG5 was defective in histidinol
dehydrogenase activity, the enzyme that converts L-histidinol to
L-histidine in the histidine biosynthetic pathway. In analogy to
<u>S.cerevisiae</u> this mutation was characterized as his 4C [4] and therefore
SAG5 genotype was designated his4. SAG5 was a stable mutant and did
not revert to prototrophy spontaneously or following mutagenesis with
NTG or UV irradiation. Therefore, further studies were undertaken to
isolate the corresponding HIS4 gene for construction of an expression
vector in SAG5.

Isolation of C. albicans HIS4 gene

Based on the observation that <u>C. albicans</u> genes can complement
<u>S. cerevisiae</u> genetic lesions [5], we used this approach to clone the

Candida and Candidamycosis, Edited by E. Tümbay *et al.*
Plenum Press, New York, 1991

Table 1. Growth characteristics of <u>C. albicans</u> CBS562 and
SAG5 strains.

| Media[1] | Ability to grow | |
	CBS562	SAG5
SD	+	−
SD+L-histidinol phosphate	+	−
SD+L-histidinol	+	−
SD+L-histidine	+	+

<u>C.albicans</u> HIS4 gene. A genemic library from the wild type
<u>C. albicans</u> CBS 562 was constructed in the shuttle vector YEp24, that
contains the URA3 gene as a selective marker. The gene bank was
constructed by the following method: genomic DNA of strain CBS 562
was partially and totally digested with the restriction endonuclease
BamHI and the DNA fragments were size fractionated. Fractions
containing fragments in the 5-15 kbp size range were pooled and
ligated with a linearized YEp24 DNA, that was cleaved at the unique
BamHI site (at its tetracycline resistance gene). The ligated mixture
was used to transform <u>E.coli</u> HB101 to ampicillin resistance.
Approximately 3000 ampicillin resistant tetracylcline sensitive
colonies were isolated, pooled, grown and their plasmid DNA was
isolated. <u>S. cerevisiae</u> 8534-15C: Matα, his4-34, uru3-52, leu2-112
(whose his4-34 marker carries two point mutations) was transformed
with the <u>C. albicans</u> gene bank by the protoplast and lithium acetate
techniques [6]. Numerous Ura+ transformants were obtained from which
two Ura+His+ (STC5, STC7) cotransformants were isolated. These
strains carry a cloned <u>C. albicans</u> HIS4 gene, as shown by plasmid
curing experiments, detailed below: Strain STC7 was grown under
nonselective conditions (rich medium) and individual colonies were
replica plated onto selective media. Out of 25 colonies tested, six
colonies failed to grow on minimal media without uracil and
histidine. Strains that grew on the minimal media contained a plasmid
(pSTC7), while the strains which failed to grow lost their plasmid.
Transformation of <u>S. cerevisiae</u> 8534-15C strain with either pSTC5 or
pSTC7 resulted in high frequency of Ura+His+ transformation.

<u>Restriction endonuclease map of the HIS4 gene</u>

Plasmids pSTC5 and pSTC7 were isolated and digested with several
restriction enzymes. The DNA fragments were separated in agarose
gels. The results presented in Fig.1 indicated that pSTC5 and pSTC7
are similar plasmids. Their total length was 21.92 kbp and they contained a 14.15 kbp DNA insert. A linear restriction cleavage map of
the plasmids is presented in Fig.2.

<u>Hydridization of HIS4 gene with C. albicans genome</u>

Confirmation that the cloned HIS4 gene originated from
<u>C. albicans</u> was achieved by hybridization experiments with
<u>C. albicans</u> genomic DNA. <u>C. albicans</u> CBS562 and <u>S. cerevisiae</u> S288C
genomic DNA's were digested with the restriction enzyme HindIII, and
the Southern blot of those DNA fragments was hybridized to a nick
translated ^{32}P-labelled HIS4 gene. The results presented in Fig.3
showed that the HIS4 gene hybridized only to the <u>C. albicans</u> genome.

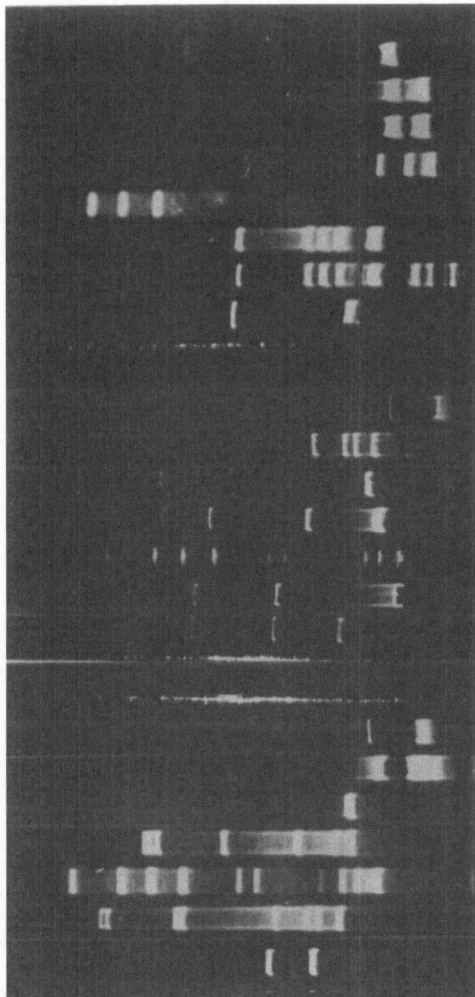

YEp24-BamHI
pSTC5-BamHI
pSTC7-BamHI
λ-HindiIII
ØX174-HaeIII
pSTC5-EcoRI
pSTC7-EcoRI
YEp24-EcoRI

YEp24-PvuII
pSTC5-PvuII
YEp24-EcoRV
pSTC5-EcoRV
λ+ØX174
pSTC5-HindIII
YEp24-HindIII

YEp24-KpnI
pSTC5-KpnI
YEp24-NcoI
pSTC5-NcoI
λ+ØX174
pSTC5-PstI
YEp24-PstI

Fig. 1. A restriction endonuclease pattern of pSTC5 and pSTC7 DNA.
Following digestion with the indicated restriction enzymes,
the DNA fragments were separated on 1% agarose gel in TBE
buffer.

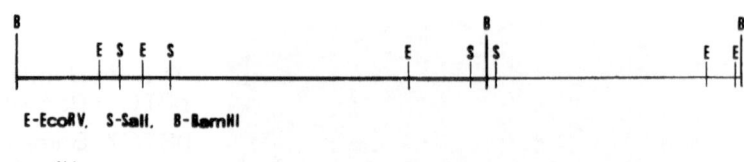

E-EcoRV, S-SalI, B-BamHI

├──┤1kb

Fig. 2 . A linear restriction cleavage map of pSTC7. The thick line
contains the cloned C. albicans HIS4 gene. The thin line
shows the vector YEp24.

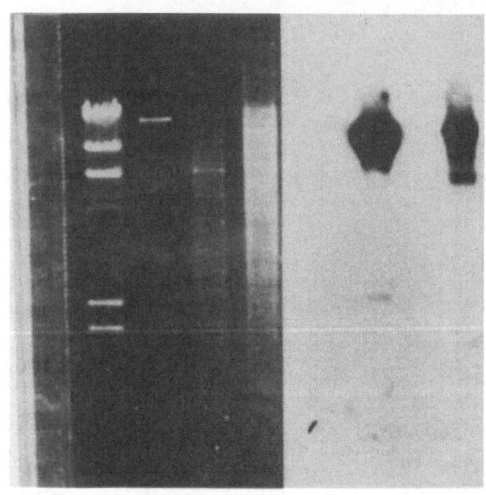

Fig. 3 . Hybridiazation of HIS4 gene to C. albicans genomic DNA.
A: Agarose gel electrophoresis of HindIII restriction
endonuclease-digest of pSTC7 (Lane 1), S. cerevisiae
S288C genomic DNA (Lane 2) and C. albicans CBS562
genomic DNA(Lane 3).
B: Hybridization of ^{32}P-labelled C. albicans HIS4 gene to
a Southern blot of the gel in Fig. 3A.

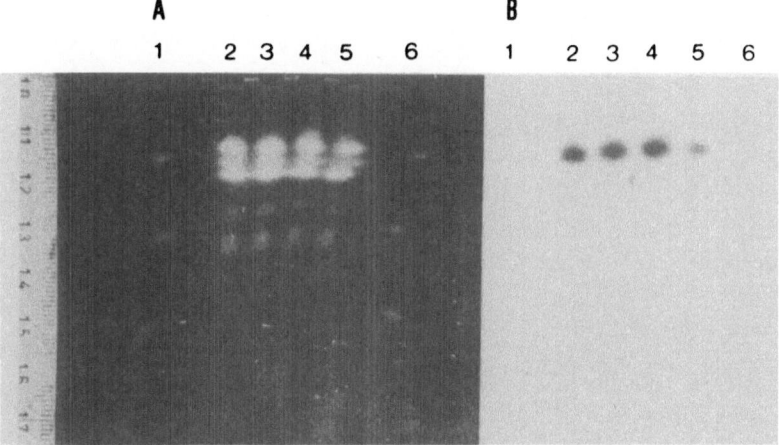

Fig. 4. Chromosomal location of HIS4 gene.
        A : Electrophoretic karyotype of C. albicans CBS 562 (Lanes
            2-5) and S. cerevisiae 2012 (Lanes 1 and 6) resolved
            by OFAGE.
        B : Hybridization of ^{32}P-labeled C.albicans HIS4 gene to a
            Southern blot of the chromosomes in Fig. 4A.

Chromosomal location of HIS4 gene

    Chromosomes of C. albicans CBS562 were resolved by orthogonal-
field alteration-gel-electrophoresis (OFAGE)[7]. The electrokaryotyping
revealed six separate chromosomal bands, three of which seem to be
doublets (Fig.4A). Hybridization of Southern blot of those
chromosomal bands with ^{32}P-labelled HIS4 gene located the HIS4 gene
on the largest chromosomal band (Fig. 4B).

SUMMARY AND CONCLUSION

    The present work describes the isolation of an histidine (his4)
auxothroph from C. albicans CBS 562 and isolation of its etiological
HIS4 gene. This genetic system will be tested for construction of an
expression vector in C. albicans.

REFERENCES

1. E. Segal, CRC Crit. Rev. Microbiol., 14, 229 (1987).
2. E. Segal, Microbiol. Sci., 4: 344 (1987).
3. H. Sandovsky-Losica, E. Segal in: "Proceedings of FEMS-Symposium
   on Candida and Candidamycosis", E.Tümbay, H.P.R. Seeliger and Ö.
   Ang, eds., Plenum Publishing House, London (1990).
4. F.R. Fink, Genetics, 53: 445 (1966).
5. A.Rosenbluh et al., Mol. Gen. Genet., 200: 500 (1985).
6. F. Sherman, " Methods in Yeast Genetics ", Gold Spring Harbor
   Laboratory, Cold Spring Harbor, N.Y. (1972).
7. I. Polacheck and G.A. Lebens, J. Gen. Microbiol., 135: 65 (1989).

INCIDENCE OF YEAST-LIKE ORGANISMS AND ESPECIALLY CANDIDA ALBICANS IN

HUMAN ENVIRONMENT

O. Marcelou-Kinti   E.Sarpakis   A.Samanidau
E. Kakepis   E.Marcelos

Department of Parasitology, Entomology and Tropical
Diseases; Athens School of Hygiene, 196 Alexandres
Avenue, 104 42 Athens, Greece

INTRODUCTION

Candida spp. are the most important fung; among the group
referred as opportunistic. Candida albicans, which is the most
frequently isolated species from candidiasis, comprises one of the
normal flora members of human mucosa.

Although Candida isolation from the human environment is
considered rare, recently its prevalence in the environment is
being continuously increased.

This work presents the results from the investigation of
different materials from human environment for the pesence of Candida
spp.

MATERIAL AND METHOD

Material

The material of this study consists of 318 animal droppings, 68
floor samples from shower bathrooms of sport installations, 31 floor
samples from dressing rooms of sport installations, 276 coastal sea
water samples, 20 sand samples, 14 samples from swimming pools, 50
samples from the body surface and intestines of flies, 135 pigeon
guano samples for the isolation of C. neoformans and 50 sand samples
for the isolation of Pityrosporum spp.

Method

For the isolation of fungi from animal droppings a quantity of
about 0.5 gr is diluted in 1 ml normal saline solution containing 0.5
gr/lt chloramphenicol. Two hours later, 2 petri dishes containing
Sabouraud dextrose agar with chloromphenicol and 2 petri dishes of
SGA with 0.5 gr chloramphenicol and 0.5 gr actidione were inoculated
with 0.1 ml of dropping dilution each.

Candida and Candidamycosis, Edited by E. Tümbay *et al.*
Plenum Press, New York, 1991

For the isolation of fungi from liquid material 100 ml of sea water and swimming pool water were filtered, and the filter was placed on the surface of SGA with chloramphenicol, SGA with chloramphenicol and actidione.

The examination of the floor for the presence of fungi is based on the method of carpet (moquette).

The presence of C.neoformans in pigeon droppings is realised by culturing on a medium containing Guizotia abyssinica.

RESULTS AND DISCUSSION

The results of this study revealed that C. albicans is isolated from the human environment although this yeast-like fungus is endogenous, living as a saprophyte on healthy mucous membranes.

The occurence of yeast-like organisms in animal intestine is presented in Table 1.

Candida albicans was more frequently isolated from rabbit droppings (15%) followed by bat guano, field rats, rats. Other Candida species were isolated in a lower frequency.

The presence of C. albicans was considerable in sport installations. This fact proves the role of man in environmental pollution.

Candida albicans was isolated from coastal sea water and sand samples in a percentage of 0.8 % to 11.2 % depending on the number of swimmers.

Candida albicans was also isolated from swimming pools. Its presence depends on the correct chlorination of water because C.albicans as well as other yeast-like fungi are sensitive to the antifungal action of chlorine.

Cryptococcus neoformans was in a percentage of about 25 % from dry pigeon droppings (Table 2).

Table 1. Yeast-like fungi isolation from animal droppings.

Origin	C. albicans		C.krusei		C.pelliculosa		C.stellatoidea		C.tropicalis		Total
	No	%	No	%	No	%	No	%	No	%	
Rabbits	3	15.0	1	5.0	1	5.0	–	–	–	–	20
Mice	11	8.2	1	0.7	–	–	1	0.7	2	1.4	133
Rats (lab.)	3	9.6	1	0.7	1	0.7	–	–	–	–	31
Guinea pigs	2	5.8	1	2.9	–	–	1	2.9	–	–	34
Bat guano	5	10.0	1	2.0	1	2.0	2	4.0	1	2.0	50
Field rats	3	12.0	2	8.0	–	–	1	4.0	1	4.0	25
Singing birds	2	8.0	1	4.0	1	4.0	1	4.0	–	–	25

Table 2. Occurrence of _Cryptococcus neoformans_ in pigeon droppings.

Source	No of samples	No of positive samples	%
Fresh droppings	76	0	-
Dry droppings	135	34	25.1

Table 3. _Pityrosporum_, dermatophytes and other fungi in beach sand*.

Species	% of positive samples
Pityrosporum	21.0
M.gypseum	21.6
Chr.keratinophilum	13.2
Candida sp.	5.1
Rhodotorula	5.1
Geotrichum	2.5
Other fungi	9.9

* A total of 190 samples were examined and
  41 (21.6 %) of them were negative for fungi.

Pitrosporum spp. were isolated in a frequency of 35 % and 18 % from sand and sea-water, respectively (Table 3).

All these findings show the main role of man in the spreading of anthropophilic fungi.

REFERENCES

For references please contact O. Marcelou-Kinti, the first auther of this paper.

INVESTIGATION OF CANDIDA PATHOGENICITY AND OTHER

ETIOLOGIC FACTORS IN DIAPER DERMATITIS

K. Gücüyener[1]   A. Gülekon[2]   S. Kuştimur[3]

Faculty of Medicine, Gazi University, Ankara, Turkey
[1] Department of Pediatrics
[2] Department of Dermatology
[3] Department of Microbiology

INTRODUCTION

Diaper dermatitis (DD) is one of the most common cutaneous disorders of children below two years of age. The inflammation occurs on the lower parts of the abdomen, genitalia, buttocks and upper portion of the thighs [1,2]. Recent studies refuse the ancient dogma, the role of ammonia and urea splitting bacteria in the etiology of this disorder, and incriminate a combination of wetness, impervious diaper coverings leading to a primary irritant dermatitis (PID) and Candida albicans as the primary factor in the initiation of DD [3].

Specimens from children with DD and from a series of normal infants were examined for the presence of C. albicans and other microorganisms.

MATERIAL AND METHOD

Twenty patients diagnosed as DD and sixteen control infants were seen in the Outpatient Department of Pediatrics and Dermatology, Medical Faculty, Gazi University, between September 1988 and February 1989.

Skin scrapings from the affected sites were taken from twenty patients with DD using a sterile tenotomy knife and in the control group from normal skin in the napkin area. Scrapings were directly examined by microscope and cultured on Sabouraud dextrose, blood and endo agars. Fresh stool specimens were obtained from diaper and cultured onto agar medium. The growing C. albicans strains were identified by germ tube formation, demonstration of chlamydospore and urease production, and carbohydrate fermentation tests [4,5]. Each child's medical history was taken including information on past and present use of antibiotics and/or steroids, recent diarrhea, diapering practices, hygenic status, duration of lesion, and type of feeding. Skin pathologies and also oral mucosa were examined.

Candida and Candidamycosis, Edited by E. Tümbay *et al.*
Plenum Press, New York, 1991

Table 1. The clinical and laboratory characteristics of the patients.

No	Age	Oral candi- diasis	Duration of Lesi- on/Day	Clinical diagnosis	Direct microscopy	Sabouraud	Blood agar Endo agar
1	16/365	−	12	PID	−	−	Diphtheroid
2	15/365	−	5	PID	−	−	Staph. coag.(−) Diphtheroid
3	20/365	+	5	Seborrheic dermatitis	−	−	−
4	23/365	−	7	Candidiasis	+	C.albicans	−
5	2.5/12	+	15	Candidiasis	+	C.albicans	−
6	1/12	+	15	PID	−	C.albicans	Staph. coag.(+) Diphtheroid
7	1/12	+	15	Candidiasis	+	C.albicans	Gr(−) bacillus Diphtheroid
8	16/365	−	6	Candidiasis	+	C.albicans	Staph. coag.(−)
9	17/365	+	10	Candidiasis	+	C.albicans	−
10	21/365	+	5	Candidiasis	+	C.albicans	Staph. coag.(+)
11	25/365	+	10	Candidiasis	+	C.albicans	−
12	15/365	+	3	Candidiasis	+	C.albicans	Staph. coag.(−)
13	16/365	+	10	PID	+	C.albicans	Gr(−) bacillus
14	1.5/12	+	20	PID	+	C.albicans	Staph. coag.(−)
15	7/12	−	30	Seborrheic dermatitis	−	−	−
16	1	−	20	PID	−	−	Diphtheroid Staph. coag.(−)
17	40/365	−	20	Candidiasis	+	C.albicans	−
18	9/12	−	7	PID	−	C.albicans	−
19	2.5/12	−	15	Candidiasis	+	C.albicans	−
20	7/12	−	30	PID	−	−	−

RESULTS

The characteristics of the patients, presence of oral Candida, duration of lesions, clinical diagnosis, and presence of microorganisms and Candida are shown in Table 1.

The mycologic results and clinical diagnosis are shown in Table 2.

The average age of the patients with DD was 2.54 months, of the controls 1.52 months. The mean duration of the lesions was 12.35 days.

The isolated microorganisms from the lesion sites in the DD group were as follows : coagulase-positive staphylococci in 2 (10%) patients, coagulase-negative staphylocci in 5 (25%) patients, diphtheroids in 5 (25%) patients, and Gram-negative bacilli in 2 (10%) patients.

Table 2. The mycologic results and clinical diagnosis.

Clinical Diagnosis	No of Patients	%
PID	8	40
Sebor.dermatitis	2	10
Candidiasis	10	50
C.albicans	14	70

E.coli was isolated in 2 (10%), and Candida in 5 (25%) patients. In the control patients; C. albicans was isolated only from 1 (7.15%) patient, coagulase-negative Staphylococcus from 8(50%) patients, diphtheroids from 4 (25%) patients, Gram-negative bacilli from 4(25%) patients. The stool cultures showed no growth of any kind of pathogenic bacteria.

Four (40%) patients out of 8 were diagnosed as PID. Two patients had diarrhea. Thirteen patients were fed with human milk while the remaining 7 with formula and cows's milk. The socio-economic status of all patients were generally low.

DISCUSSION

The pathogenesis of DD is not well defined. Numerous factors such as friction and irritation, urine and ammonia feces, infections of Candida and other bacteria may play a role [6,7]. Primary dermatologic disorders as psoriasis, atopic and seborrheic dermatitis can also involve the diaper area, but they are not primary etiological factors. In this study, we did not have any atopic patients, but one had seborrheic dermatitis (10%). This is inconsistent with the current literature.

Recent evidence indicates that ammonia and Bacillus ammoniagenes are not important etiological factors in DD [8]. Insufficient cleaning of the infant, infrequent changes of diapers and irritant chemicals used in cleaning and sterilizing diapers, occlusive rubber and plastic pants may be the factors often incriminated. None of our patients were using disposable diapers, but all were using occlusive rubber or plastic pants; and the mean number of diaper change per day was 5.72. Those factors listed above may contribute. However, in 50% of those to finding 40% PID among our patients.8 patients, C. albicans was also isolated from the diaper area.

Microbial factors either fungal or bacterial have long been viewed as important etiologic factor s. Since bacteria obtained from the skin of infants with and without various forms of DD show no significant difference in type or quantity from microbial flora, there is no firm proof that bacteria account for the dermatitis [9,11].

We also have isolated coagulase-negative Staphylococcus, Gram-negative bacilli and diphteroids which constitute the normal permenant skin flora in our patients.

However, in 2 patients the isolation of coagulase-positive Staphylococcus was interpreted as a secondary event due to the inflammation [6,7].

In recent years most authorities agree that C. albicans is the invader in 41-85% of infants having active DD [10-12]. However, it is still controverial whether C. albicans initiates DD or merely aggrevates a dermatitis already caused. In an experimental study, C. albicans was applied to the skin and covered with an occlusive dressing for 24 hours, resulting in a dermatitis characterised with erythematous papules and pustules in 95% of the cases [13]. Contrarily when high concentrations of ammonia under an occlusive dressing were applied for 24 hours, there appeared no reaction [8]. The gastrointestinal tractus is an important reservoire for C. albicans which can spread onto the skin and induce dermatitis [14].

Our results are inconsistent with the previous reports, indicating that C. albicans plays an important primary instigating role in DD either by gaining pathogenicity in the masserated diaper region or/and by reaching the lesion site from its natural reservoire, the alimentary tract.

All of our patients ranked in the low socio-economic status, and there was no correlation between the isolation of C. albicans and the type of feeding of the baby.

REFERENCES

1. S. Hurwitz, "Clinical Pediatric Dermatology", pp.27-29, W.B. Saunders Company, Philadelphia (1981).
2. A. Rook et al, "Textbook of Dermatology", pp.239-244, Blackwell Scientific Publications, Oxford (1986).
3. P.J. Honig, Diaper dermatitis, Postgrad Med., 74:79 (1983).
4. J.D. Baver, Mycology in Clinical Laboratory Methods", The C. V. Mosby Company, St. Louis (1982).
5. G.S. Kobayashi et al, The mycosis, in: "Gradwohl's Clinical Laboratory Methods and Diagnosis", The C. V. Mosby Company, St. Louis (1980).
6. J.J. Leyden, Diaper dermatitis, Dermatol. Clin., 4:23 (1986).
7. L.W. Wetson et al, Diaper dermatitis: Current concepts, Pediatrics, 66:532 (1980).
8. J.J. Leyden et al, Urinary ammonia and ammonia producing organisms in infants with and without diaper dermatitis, Arch. Dermatol., 113:1678 (1977).
9. P.N. Dixon et al, Role of Candida albicans infection in napkin rashes, Br. Med. J., 2:23 (1969).
10. J.J. Leyden et al, The role of microorganisms in diaper dermatitis, Arch. Dermatol., 114:56 (1978).
11. L.F. Montes et al, Microbial flora of infant skin, Arch. Dermatol., 103:640 (1971).
12. R.M. Brookes et al, Skin flora of infants with napkin rash, Br. J. Dermatol., 85:250 (1971).
13. A. Rebora et al, Experimental infection with Candida albicans, Arch. Dermatol., 108:69 (1973).
14. A. Rebora et al, Napkin (diaper) dermatitis and gastrointestinal carriage of Candida albicans, Br. J. Dermatol., 105:551 (1981).

THE PATHOGENETIC ROLE OF CANDIDA IN PSORIASIS

Ü. Soyuer   Ö. Aşçıoğlu   E. Aktaş

Department of Dermatology, Faculty of Medicine
Erciyes University, Kayseri, Turkey

INTRODUCTION

Psoriasis is a chronic, recurrent and inflammatory skin disease. It is characterized by erythematous scaly patches on knees, elbows, sacral area and scalp. The etiopathogenesis of psoriasis is still unknown. Recent reports have suggested a role for the alternative complement pathway [1]. Psoriasis may result from the interaction of yeasts and other microorganisms with an abnormally responsive alternative complement pathway [1,2]. C. albicans is a potent activator of alternative complement pathway. This suggested that the yeast flora of gastrointestinal tract may play an important role in the etiopathogenesis of psoriasis.

In this study, yeasts residing in the gut of psoriatic patients were compared with those of healthy controls.

MATERIAL AND METHOD

The stool specimens of 39 patients with psoriasis vulgaris (28 females and 11 males aged 18-60) and 20 healthy controls (10 females and 10 males aged 20-48) were examined.None of the patients had taken systemic corticosteroids, antibotics or immunosupressive drugs before and during the mycological examinations.

About 200 mg of stool from each subject was diluted with 0.1 M saline as to give a 10% suspension. The suspension was homogenized, ten-fold serial dilutions were prepared in sterile saline, and yeast counts were made.

Stool specimens of two groups were examined by direct microscopy, and cultures were made on Sabouraud's dextrose agar with chloramphenicol. Candida strains isolated from stool samples were first tested for germ-tube production in serum. Germ tube positive isolates were assumed to be C. albicans. Identifications of germ-tube negative isolates was based on biochemical tests.

RESULTS

Psoriatic patients sohwed a significantly increased yeast colonization of gut (Fig. 1).

Candida and Candidamycosis, Edited by E. Tümbay *et al.*
Plenum Press, New York, 1991

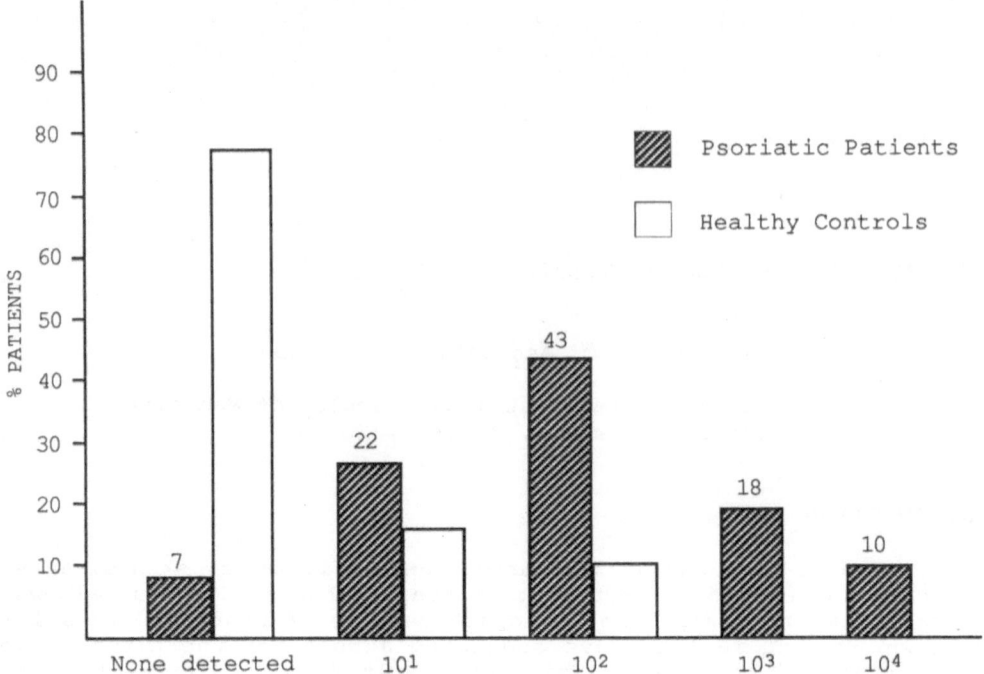

Fig. 1. Total faecal count of yeasts in psoriatic patients and
healthy controls.

Mycological examinations showed that psoriatic patients had 93%
positive and controls had 25% yeast positive stool samples.
C. albicans was frequent in both groups (Table 1).

Table 1. Direct microscopical examination of stool samples and
Candida species isolated.

	Direct microscopical examination of stool samples			
	Blastospores	Blastospores and pseudohyphae	Positive	Negative
Patients n=39	34 (%87)	2 (%5)	36 (%92)	3 (%7)
Controls n=20	5 (%25)	–	5 (%25)	15 (%75)

		Candida species isolated from stool samples				
	No Growth	C.albicans	C.krusei	C.para-psilosis	C.pseudo-tropicalis	C.tropi-calis	Total
Patients	5	26	3	2	2	1	34
Controls	16	3	–	1	–	–	20

DISCUSSION

Candida species form part of the normal microbial flora of the human intestinal tract [3]. Increase of intestinal yeast colonization may trigger the eruption of psoriasis on the skin [4]. However, recently it was proposed that prosiasis is ordinarily result of interaction of various microbial products with an abnormally responsive alternative complement pathway. C. albicans is one of the potent activators of alternative complement pathway [1]. In addition to this hypothesis, high phospholipase A activity of C. albicans isolated from the intestines of psoriatic patients[5] supports the suggestion about the possible pathogenetic role of Candida in psoriasis.

This study has shown that intestinal Candida colonization is significantly increased in psoriasis. None of the patients showed gastrointestinal symptoms, and there was no predisposing factor for candidosis except skin disorder.

In conclusion, we suggest that Candida colonization of gut may be one of the etiopathogenetic factors of psoriasis.

REFERENCES

1. E.W.Rosenberg and P.W. Belew, Microbial factors in psoriasis, Arch. Dermatol., 118:143 (1982).
2. N.Crutcher, E.W. Rosenberg, P.W. Bele et al., Oral nystatin in the treatmnet of psoriasis, Arch. Dermatol., 120:435 (1984).
3. D.W. Warnock, D.C.E. Speller, P.J. Finan et al., Antibodies to Candida species after operations o the large intestine: Observations on the association with the oral and faecal colonization, Sabouraudia, 17:405 (1979).
4. W. Wachowiak, G.V. Stryker, J. Marr et al., The occurence of monilia in relaton to psoriasis, Arch. Dermatol., 19:713 (1929).
5. H. Hänel, I. Menzel and H. Holzmann, Hohe Phospholipase A-Aktivität von Candida albicans aus Darm von Psoriatikern, Mykosen, 31:451 (1988).

CANDIDA INCIDENCE IN INFANTS WITH NAPKIN DERMATITIS IN THE SOUTH-EAST REGION OF TURKEY

M.Derici [1]   İ.Mevlitoğlu [1]   M.Mete [2]

Faculty of Medicine, Dicle University
Diyarbakır, Turkey
[1] Department of Dermatology
[2] Department of Microbiology

INTRODUCTION

Napkin dermatitis, which generally occurs on the convex side of perianal region and is recovered by antiseptic application, is a regional dermatitis [1]. Although the etiopathogenesis is unknown, it is believed that Candida plays an important role in napkin dermatitis. Candida grows in the diaper area in wet and airless conditions [2]. It is claimed that Candida originates from the gastrointestinal system (GIS), a reservoire [3].

In this study, the relation between Candida incidence, genera and the ratio of Candida in faeces of infants with napkin dermatitis in the south-east part of Turkey was investigated.

MATERIAL AND METHOD

In the last 6 months we examined 40 patients of 0-2 group who visited the Dermatology Department, Medical School, Dicle University. The selected cases had not been treated with systemic or topical antimycotics for at least a week. Materials were taken from inguinal region, labia major, scrotum, perianal region, the front part of hips or thighs, in which the lesion was most active. Before taking the specimens, the region was cleaned with 70% ethyl alcohol. One of the specimens was used for direct microscopic examination and the other for culture. The material was cultured on Sabouraud's agar with antibiotics 37°C for 3 days. Growing yeast colonies were identified by conventional methods [4].

The faecal specimens of cases were examined in the Parasitology Laboratory. Twenty-five cases with other dermatosis were accepted as the control group.

Napkin dermatitis was classified in 5 categories. Grade 1: Erythema, Grade II: Erythema and papules, Grade III: Erythema and papulopustules, Grade IV: Erosions, Grade V: Crater like ulcers, nodules.

Candida and Candidamycosis, Edited by E. Tümbay et al.
Plenum Press, New York, 1991

197

Table 1. Candida presence in skin and faeces of patients with napkin dermatitis.

Clinical Picture (Napkin Dermatitis)	Number	Candida (+)			
		Skin	%	Faeces	%
Erythema (Ery.)	12	10	83.3	2	16.6
Ery. and papules	5	5	100	1	20
Ery. and papulopustules	3	2	66.6	2	66.6
Erosions	15	13	86.6	13	86.6
Ulcers, nodules	5	2	40	2	40

RESULTS

Forty cases with napkin dermatitis and 25 patients as control group were evaluated. Candida was found on the skin in 80% of the cases. In 93.8% of these cases C. albicans was found. Candida guilliermondii was observed in only two cases (Table 1).

In cases with erosion the positiveness of Candida was found higher than that of the others (Table 1).

In the skin of 2 cases in the control group C. albicans was observed. These two cases were newborns. In the faeces of the same group Candida was not encountered.

DISCUSSION

The 80% Candida positiveness which was observed in infants with napkin dermatitis was compared with other studies.Leyden and Kligman[5] have found C. albicans in 80%of cases with napkin dermatitis and stated that 1/3 of this percentage is id reaction. Dixon et al [6] after reviewing 117 cases, have found this ratio as 41%. Desmons and Dligny [7] examining 53 infants with napkin dermatitis, have found C. albicans incidence as 62%. Maleville et al. [8] have found 50% and Ekşioğlu et al. [9] have found 52.5%. Grigoriou et al. [10], between 1976 and 1980, have made a statistical evaluation and found Candida incidence as 98% in infants with napkin dermatitis.

The presence of Candida in two control infants can be explained by transmission to infants from their mothers during birth and due to bad hygiene. Candida was found in the stool of 50% of test patients. Rebora and Leyden [3] found this positiveness as 34% and Desmons and Dligny [7] as 8%. Rebora and Leyden [3] could not observe Candida in the faeces of patients with mild napkin dermatitis and thought that Candida which arrives from GIS is the cause of severe napkin dermatitis.

In the faeces of the control group, Candida was not observed as was reported by Rebora and Leyden [3].

It is believed that Candida plays an important role in the etiopathogenesis of napkin dermatitis and passes from GIS to the diaper area.

As a result, the high incidence of _Candida_ in napkin dermatitis can be explained by low socio-economic conditions and bad hygiene in the South-East Anatolia Region. In future studies, _Candida_ presence in mothers of infants with napkin dermatitis will be investigated.

REFERENCES

1. M.Larregue, P.Gallet, J.P.Rat et al., W-shaped napkin rash in infants, _Poitiers-Dermatologica_ (Basel), 151: 104 (1975).
2. C.Gezen, Deri ve mukozaların _Candida_ infeksiyonları, _in_: "Candida ve İnfeksiyonları", E.Tümbay, ed., pp. 29-34, Bilgehan Basımevi, İzmir (1986).
3. A.Rebora and J.J.Leyden, Napkin (Diaper) dermatitis and gastrointestinal carriage of _C. albicans_, _Br.J.Dermatol._, 105: 551 (1981).
4. E.Tümbay, "Pratik Tıp Mikolojisi", 1.Baskı, pp.45-47, Bilgehan Basımevi, İzmir (1983).
5. J.J.Leyden and A.M.Kligman, The role of microorganisms in DD. _Arch.Dermatol._, 114: 56 (1978).
6. P.N.Dixon, R.P.Warin, M.P.English, Role of _C. albicans_ infection in napkin rashes, _Br.Med.J._, 2: 23 (1969).
7. F.Desmons and J.Y.Dligny, Study of the role of _Candida albicans_ infection in different infant conditions including a diaper rash (64 cases), _Franc.Ann.Dermatol.Venereol._, 104: 185 (1977).
8. J.Maleville, M.Capbeen, D.Boineau, Cutaneous microbial flora in 206 children with DD and pyodermitis, _Franc.Ann.Dermatol. Venereol._, 104: 701 (1977).
9. M.Ekşioğlu, T.Akan, N.Kürkçüoğlu, Diaper dermatitte mikroorganizmaların rolü, _in_: "XI Ulusal Dermatoloji Kongresi, Samsun, 1986", pp. 123-130, (1986).
10. D.Grigoriou, J.Delacretaz, D.Borelli, "Medical Mycology", pp. 229-231, Roche, Basle (1987).

AN INVESTIGATION OF <u>CANDIDA</u> SPECIES IN PATIENTS WITH ORAL LEUKOPLAKIA

İ.Kökçam[1]   M.Yılmaz[2]   M.Bolat[1]   S.S.Kılıç[2]   S.Bakır[1]

Faculty of Medicine, Fırat University, Elazığ, Turkey
[1] Department of Dermatology
[2] Department of Microbiology

INTRODUCTION

In 1966, Cawson [1] described the distinctive clinical and histo-logical features of a form of oral candidosis presenting itself as a leukoplakia. This was termed chronic hyperplastic candidosis or candidal leukoplakia [2].

It is known that the most common cause of oral mucous membrane keratosis is tobacco smoking [3,4]. Hyperkeratosis is a prerequisite for <u>Candida</u> infection as <u>Candida</u> is a keratophilic microorganism [5].

The present study was carried out in patients with oral leuko-plakia. The aim was to compare the occurence rates of <u>Candida</u> species in patients with and without oral leukoplakia.

MATERIAL AND METHOD

Patients

The patients in this study were seen at the Mental Diseases Hospital of Elazığ, eas Anatolia. We have found 69 (17.60 %) cases of leukoplakia out of 392 patients. All of the patients with leukoplakia were smokers. In this group (Group 1) there were 64 men (92.75 %) and 5 women (5.25 %). Their average age was 32 years with a range from 17 years to 60 years.

Control group (Group 2) consisted of 44 patients with no clini-cally detectable diseases of the oral mucous membrane. In this group all of the patients were men. The patients had an average age of 35 years, with range between 22 and 65 years.

Microbiologic tests

The samples were taken by swabbing the keratotic patches with sterile cotton applicators in the first group. In the control group, the mucosa of the left cheek and right cheek were swabbed.

The swabs from the mucous membranes were streaked onto Sabouraud Glucose Agar (SGA) with chloramphenicol and incubated for 72 hours at

37°C. Representative colonies presumed to be C. albicans were tested for their ability to form chlamydospores on cornmeal-agar and germ tubes within 4 hours in human serum. Chlamydospores and rapid germ tube development confirmed the identity of C. albicans. The other Candida species were identified by conventional methods.

Statistical analysis

The candidal growth rates in patients with and without leukoplakia were compared by using the Z test. A "p" value below 0.05 was accepted as significant.

RESULTS

Group 1. Candida species were cultured from the oral keratotic patches of 14 (20.28 %) of the 69 cases with leukoplakia. The frequencies of the species were as follows: Candida albicans 7(50 %), C. krusei 3 (21.42 %), C. tropicalis 2 (14.28 %), C. guillermondii 1 and Rhodotorula 1 (This patient had onychophagia).

Group 2. Of the mucosal swabs from the 44 patients in this group, 6 (13.64 %) gave positive cultures for Candida species. Candida albicans was isolated from 4 (66.6 %), C. krusei from 1 (16.6 %), and C. guillermondii from 1 patient.

The difference between the rates of Candida growth in the groups was statistically significant (p<0.001).

DISCUSSION

Candida albicans is the most commonly isolated yeast from leukoplakia, but also other Candida species may be encountered [6,7]. Also in our study, C. albicans was the dominating species in both groups, constituting 50 % of all yeasts in the leukoplakia lesions. In addition the following species were identified in the patients with leukoplakia: Candida krusei, C. tropicalis, C. guillermondii and Rhodotorula. In 6 patients (13.64 %) among the control group, various Candida species were isolated. The Candida species found in the control group and in patients wih leukoplakia in our study are compatible with the results of the other workers [6,8].

The statistically significant difference in the prevalence of Candida species encountered in Group 1, compared to the control group, supports the view that their growth is facilitated in leukokeratotic sites [9].

Various authors have reported the frequency of Candida albicans and other Candida species to be between 30-75 % in leukoplakia cases and control groups [3,9,10].

According to Budtz-Jörgersen[11], the ability to detect C. albicans in oral cavity is dependent on the sampling techiques employed, and this has been emphasized by work of Arendorf and Walker[3]. In our study, the prevalence of Candida species rates in both groups are not parallel to those of previous studies.

We concluded that Candida species are more common microorganisms in patients with leukoplakia than in control group.

REFERENCES

1. R.A.Cawson, Chronic oral candidiasis and leukoplakia, <u>Oral Surg.</u>, 22: 582 (1966).
2. R.A.Cawson and T.Lehner, Chronic hyperplastic candidiasis-Candidal leukoplakia. <u>Br.J.Dermatol.</u>, 80: 9 (1968).
3. T.M.Arendrorf and D.M.Walker, The prevalence and intra-oral distribution of <u>Candida albicans</u> in man, <u>Arch.Oral Biol.</u>, 25: 1 (1980).
4. B.W.Neville, Leukoplakia, <u>in</u>: "Clinical Dermatology", Vol.4, D.J.Denis, ed., Unit 21-13, Harper and Row Pub., New York (1987).
5. J.J.Pindborg, Disorders of oral cavity and lips, <u>in</u>: "Textbook of Dermatology", A.Rook, ed., pp. 2108-2110, Blackwell Scientific Publications, London (1986).
6. P.Krogh et al., Yeast organisms associated with human oral leuko-plakia. <u>Acta Derm.Venerol.</u> (Suppl.), 121: 51 (1986).
7. J.Reinholdt, P.Krogh, P.Holmstrup, Degradation of IgA1, IgA2, and S-IgA by <u>Candida</u> and <u>Torulopsis</u> species, <u>Acta Pathol.Microbiol. Immunol.Scand.</u>, 95: 265 (1987).
8. P.Krogh et al., Yeast species and biotypes associeted with oral leukoplakia and lichen planus, <u>Oral Surg. Oral Med. Oral Pathol.</u>, 63: 48 (1987).
9. O.P.R.Hornstein and E.Schirner, Prevalence rates of candidosis in leukoplakias and carcinomas of oral cavity, <u>Arch.Dermatol.Res.</u>, 266: 99 (1979).
10. R.J.Bastian and P.C.Reade, The prevalence of <u>Candida albicans</u> in the mouths of tobacco smokers with and without oral mucous membrane keratoses, <u>Oral Surg. Oral Med. Oral Pathol.</u>, 53: 148 (1982).
11. E.Butz-Jörgensen, The significans of <u>Candida albicans</u> in denture stomatitis, <u>Scand.J.Dent. Res.</u>, 82: 151 (1974).

THE MINOR SALIVARY GLAND NETWORK IN EXPERIMENTAL ORAL CANDIDOSIS

M.Lacasse[1]   C.Fortier[1]   L.Trudel[1]
A.J.Collet[2]   N.Deslauriers[1]

Laval University, Quebec, Canada
[1] GREB, Dental School
[2] Department of Anatomy, Faculty of Medicine

INTRODUCTION

Candida albicans is an opportunistic pathogen capable of infecting many mucous membranes in humans including the oral mucosa. In order to achieve successful colonization, C. albicans must overcome both non-specific and specific defense mechanisms present in the saliva and at mucosal surfaces. Since the incidence of local infection and systemic dissemination is lower than expected from the ubiquity of the microorganism in the oral flora, immuno-surveillance mechanisms can therefore be expected to have a synergistic effect with oral bacterial competition in the normal handling of Candida [1]. Cell-mediated immunity may play a role in host resistance to mucosal candidal infections as suggested by Marra and Balish [2] and Budtz-Jörgensen [3] who correlated the clearing of the infection with the development of cellular hypersensitivity. In mucocutaneous candidosis, indirect evidence suggests that a major host defense mechanism against Candida could involve T lymphocytes [4]. It is also possible to attribute a role to innate resistance in host defense against Candida. Some reports have demonstrated the in vitro effectiveness of peripheral NK cells against fungal pathogens [5-7], and polymorphonuclear cells are known to be of major importance in systemic candidosis (rev. 4).

In our experimental model of oral candidosis it is clear that dramatic alterations of the local immune system are brought about by candidal colonization of the buccal epithelium[8]. Nair & Schroeder [9] suggested that the minor salivary glands (MSG) could also be accessible to oral antigens via retrograde passage through salivary ducts. We thus looked at the possibility that de novo colonization of the murine oral mucosa by C. albicans was related to a retrograde passage of microbial antigens through salivary ducts and if infection induced the proliferation and/or the recruitment of immunocytes into the MSG.

MATERIAL AND METHOD

Animals and microorganisms

CD-1 male mice were 17-23 weeks old. C. albicans (LAM1) is a clinical isolate. Eosine methylene blue agar (EMB) modified by addition

Candida and Candidamycosis, Edited by E. Tümbay et al.
Plenum Press, New York, 1991

205

of chlortetracycline (100 µg/ml) [10] was used for the enumeration of C. albicans in saliva samples.

Experimental protocol

Mice were inoculated with 10^8 viable yeast cells in sterile saline dispensed with a precision micropipet (50 µl) or with a Calgiswab® (pelleted cells). At different times after intraoral challenge, mice were either sacrificed for histological preparations (sections were stained in series with H & E, PAS and toluidine blue) or saliva was collected after cholinergic stimulation [10] for microbiological determinations.

RESULTS

1. Kinetics of oral colonization by C. albicans

C. albicans is not found to be part of the normal oral flora in the CD-1 mouse since it could not be detected in samples from 16 individuals. Intra-oral inoculation of the microorganism results in colonization of every normal adult CD-1 mouse tested. From the enumeration of CFUs in the saliva it appears that the swabbing procedure induces a higher rate of colonization than micropipet inoculation of the same number of blastospores (Fig 1). Oral proliferation is also delayed after micropipet inoculation (day 4) as compared to Calgiswab application (day 2). At the peak of colonization, C. albicans reaches the level of some indigenous bacterial species found in the murine oral cavity and then sharply decreases.

2. Sequential histological study of the inflammatory reaction during candidal infection of the MSG network

Micropipet inoculation produces a mild inflammation of the MSG, generally limited to a few secretory ducts and contiguous connective tissue. Inflammation, predominantly mononuclear, is more extensive near the colonization peak. Following Calgiswab inoculation, inflammation of the main secretory ducts is regularly observed throughout candidal colonization with a maximum on day 4 of the kinetics (Fig.2). At that time, infiltration of the ductal wall and surrounding connective tissue appears to involve a large portion of the ducts as we could detect the infiltrate over many successive sections. Mononuclear cells generally predominate but numerous PMNs are found in localized areas. Mononuclear cells are found in the deep layers of the ductal epithelium while PMNs are located more superficially, often accumulating in glandular lumen. Large numbers of mast cells accumulate in the periglandular connective tissue of the soft palate, surrounding some secretory ducts on day 4 (Fig. 3 and 4).

Interacinar inflammation was most extensive on day 4 following swab inoculation. While involvement of secretory end pieces occurs occasionally on days 2 and 6, day 4 reveals an intense infiltration of many acini by inflammatory cells. Lesions show a high rate of infiltration by mononuclear cells. PMNs vary in number from one site to another but are often found in the glandular lumen. Inflammation of the interacinar connective tissue occurs only in discrete glundular lobules (Fig. 5, 6 and 7).

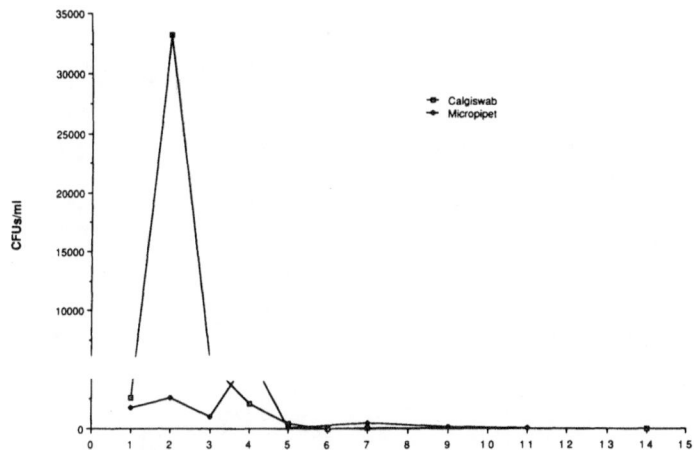

Fig. 1. Oral colonization by <u>Candida albicans</u> was more successful
using the Calgiswab inoculation technique, oral proliferation being delayed after micropipet inoculation.
Results are expressed as mean CFU counts per ml of saliva.

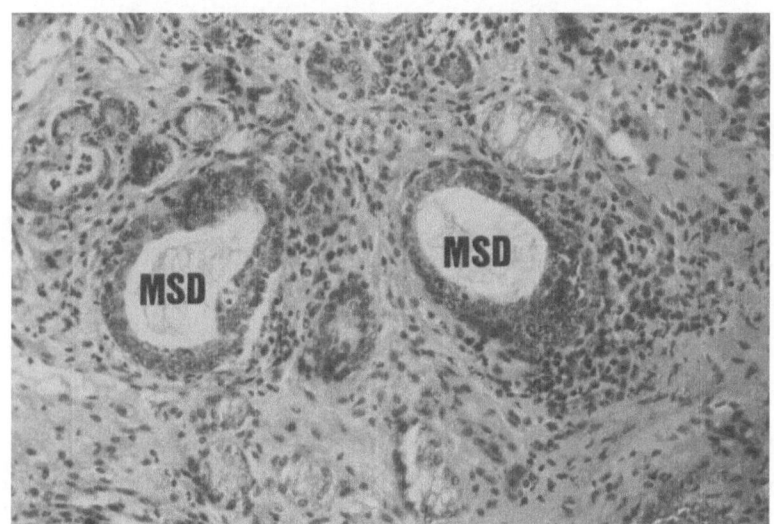

Fig. 2. Inflammation of the main secretory ducts (MSD) was
maximal on day 4 of the kinetics. Mononuclear and
polymorphonuclear cells were located around ducts and
infiltrated the ductal epithelium. Calgiswab inoculation,
soft palate, H & E (x230).

DISCUSSION

Inflammation of the main secretory ducts represents the most
striking feature of MSG involvement in experimental candidosis.
Secretory ducts are the first target of the inflammatory reaction
(starting on day 2) and remain affected until day 6 after intra-oral
(swab) colonization. At the peak of colonization by <u>C. albicans</u> (day
2), mononuclear cells infiltrated deep ductal epithelial layers while

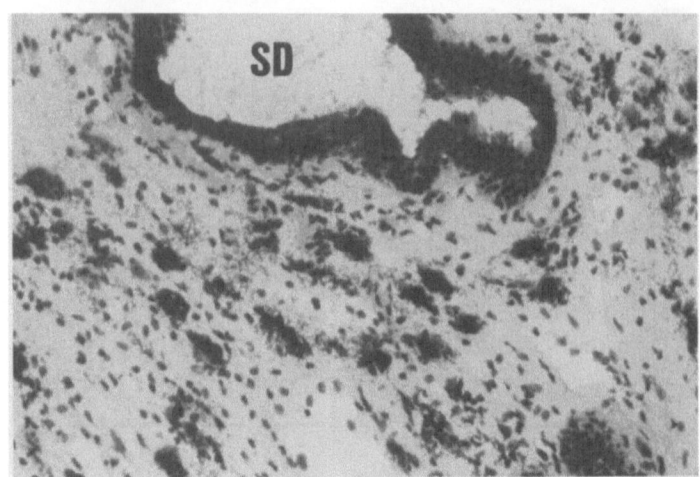

Fig.3. High numbers of mast cells surrounded some secretory
ducts (SD) on day 4. Calgiswab inoculation, soft palate,
toluidine blue (x350).

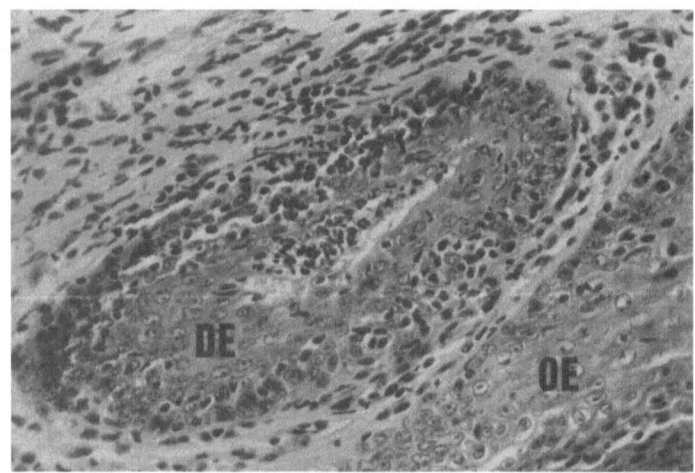

Fig.4. Inflammatory cells infiltrated the walls of many secretory ducts on day 4 after calgiswab inoculation: mononuclear cells were found in the deeper layers of the ductal
epithelium (DE) while PMNs were located more superficially and accumulated in the glandular lumen. Buccal mucosa,
PAS (x350) (OE= oral epithelium).

PMNs were mostly located in the superficial layers and accumulated in
the lumen. At that time, about 20-30% of the glandular tissue is
involved. The inflammation decreased rapidly after day 6. The
kinetics of this transient inflammatory process indicate a rapid
recovery from candidal infection in our model.

A similar kinetics and distribution of inflammatory cells was
observed following oral epithelial infection by C. albicans [8]. This
suggests that the micrrorganism can rapidly trigger a powerful defense

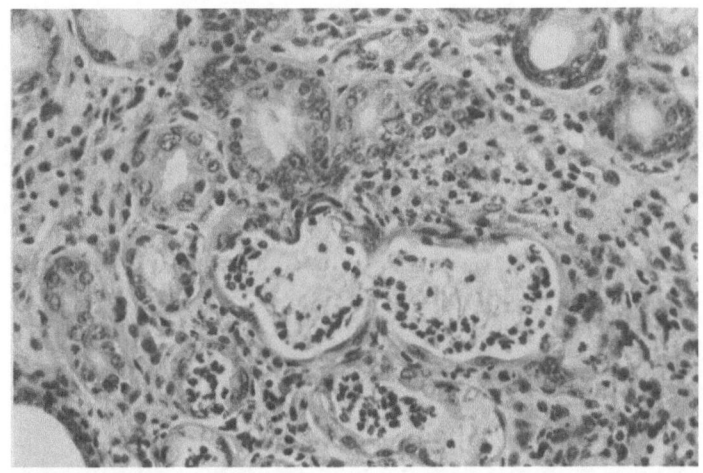

Fig. 5. Inflammatory lesions involving glandular acini showed a steady infiltration by mononuclear cells while PMNs varied in numbers. Nevertheless, PMNs constituted the large majority of cells in the glandular lumen. Day 4 after Calgiswab inoculation, soft palate, H & E (x230).

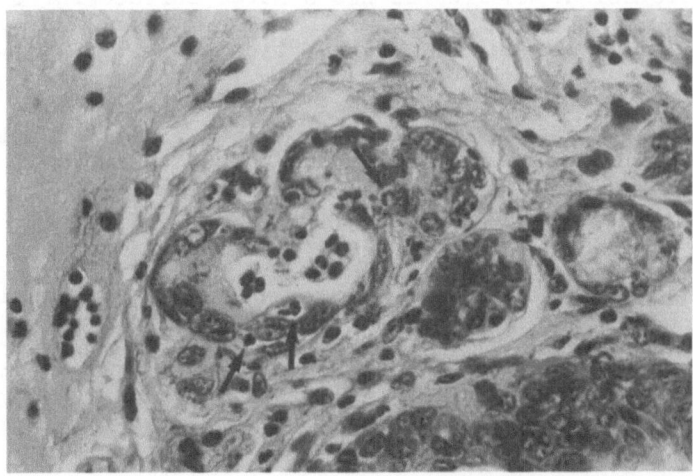

Fig.6. PMNs (arrows) infiltrated the acinar epithelium and reached the lumen on day 4 post inoculation. Calgiswab challenge, soft palate, H & E (x600).

reaction upon colonization of oral surfaces, mobilizing multiple components which then hamper microbial invasion and dissemination. Mast cells accumulated around some main secretory ducts and in the periglandular connective tissue. Healthy oral mucosa does not show such focal accumulations of mast cells around secretory ducts [11]. This suggsets that mast cells could be involved in the initiation and/or the development of the inflammatory reaction. The sequential recruitment of mast cells, macrophages, lymphocytes and neutrophils in oral candidal infection is being investigated.

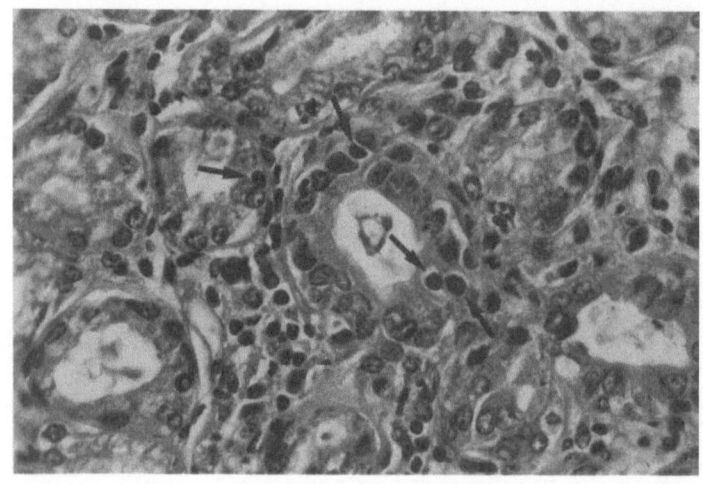

Fig. 7. Mononuclear cells (arrows) penetrated the acinar epithelium but were rarely seen inside the glandular lumen.
Day 4 after Calgiswab inoculation, soft palate, H & E
(x 600).

The tubulo-acinar MSG represents a large surface for C. albicans
to invade after gaining access to duct openings. In fact, duct
openings were shown to be the first portion of the MSG network to be
colonized by C. albicans, while infection of the deeper recretory
units depends on both glandular secretory cycles [9] and microbial
adaptation to this special microenvironment. A longitudinal study of
candidal invasion of the epithelium and the MSG network is under way.

ACKNOWLEDGEMENTS

The authors greatly appreciated the technical assistance of
A. Pusterla (Dept.Pathology). We also wish to thank Dr.F.A.Auger for
providing the strain of C. albicans and Gene Bourgeau for revising
the style. This study was supported by MRC (Canada) Grant MA-9943.

REFERENCES

1. Epstein, Truelove and Izutzu, Rev.Infect.Dis., 6: 96 (1984).
2. Marra and Balish, Infect.Immun., 10: 72 (1974).
3. Budtz-Jörgensen, Scand.J.Dent.Res., 81: 360 (1973).
4. Rogers and Balish, Microbiol.Rev. 44: 660 (1980).
5. Jimenez and Murphy, Infect.Immun., 46: 552 (1984).
6. Murphy and McDaniel, J.Immunol., 128: 1577 (1982).
7. Nabavi and Murphy, Infect.Immun., 50: 50 (1985).
8. Lacasse, Collet, Chandad and Deslauriers, (submitted) (1989).
9. Nair and Schroeder, Arch.Oral Biol., 28: 145 (1983).
10. Deslauriers, Oudghiri, Séguin and Trudel, Immunol.Invest., 15:
    339 (1986).
11. Lacasse, Collet, Mourad and Deslauriers, Adv. Exp. Med. Biol.
    216 A: 375 (1987).

ISOLATION OF <u>CANDIDA</u> FROM SPUTUM SPECIMENS OF PATIENTS WITH PULMONARY

TUBERCULOSIS AND BRONCHIAL ASTHMA AND INVESTIGATION OF <u>CANDIDA</u>

ANTIGENS IN THEIR SERA

G. Mutlu   D. Çolak   M. Pamukçu

Department of Microbiology, Faculty of Medicine
Akdeniz University, Antalya, Turkey

INTRODUCTION

Candidiasis is the most frequently encountered opportunistic fungal infection. It is caused by a number of <u>Candida</u> species with <u>Candida albicans</u> being the most frequent etiologic agent. <u>Candida</u> species are responsible for a number of different types of infections in normal and immunocompromised patients. The repeated recovery of yeasts from multiple specimens from the same patient usually indicates colonization. The simultaneous recovery of the same species of yeast from several body sites including urine is a good indicator of disseminated infection and subsequent development of fungemia [1,2].

In this study we examined whether <u>C. albicans</u> causes systemic disease in colonized patients. Sera from 32 colonized patients and 2 immunocompromised patients with systemic candidiasis were evaluated for both circulating antigens of <u>C. albicans</u> serotypes A and B by Ouchterlony immunodiffusion method.

MATERIAL and METHOD

We evaluated sera and sputum specimens from 39 patients who had chronic pulmonary tuberculosis, 11 patients with bronchial asthma and 2 patients at risk for systemic candidiasis. Sputums were inoculated onto Sabouraud's medium (Difco Lab. No: 708107) and incubated at 37°C for 3 days. The growing yeasts were identified as <u>C. albicans</u> according to chlamydospore or germ tube formation.

<u>Cultures</u>

<u>C. albicans</u> serotype A MIV-244 and <u>C. albicans</u> serotype B MIV-243 were used in preparing antigens.

<u>Antigen preparation</u>

(a) <u>Preparation of antigen that was used for immunization of rabbits</u>. Yeasts were inoculated onto Sabouraud's medium and incubated

Candida and Candidamycosis, Edited by E. Tümbay *et al.*
Plenum Press, New York, 1991

at 25°C for 18 h. Then cells were collected in % 0.9 NaCl solution containing % 0.5 formalin and washed three times in % 0.9 NaCl. Antigen suspensions were diluted to the turbidity of MacFarland nephelometer standards number 2. Antigens were killed by heating at 56°C for 2 h.

b) Prepatarion of antigen used for immunodiffusion test. 48 h C. albicans serotypes A and B cultures on Sabouraud's medium in Roux boites were used to prepare stock suspensions in physiological saline. The cells were killed by heating at 60°C for 1 h.

All the C. albicans strains were disintegrated by sonic vibrations at the power of 60 watts for 2 h and centrifuged. The supernatants were used as antigens. Protein concentrations were measured by biuret method. Concentrations were 1.07 gr/dl for C. albicans serotype A antigen and 1.02 gr/l for C. albicans serotype B antigen.

Antisera production

Male rabbits with a weight of 3 and 3.5 kg were used. Two rabbits were immunized by iv route with 1.0 ml of antigen suspension three times per week for five weeks. Blood samples were drawn from rabbits before and on the fifteenth day of immunization and one week drawn from last injection. We tested these sera for C. albicans specific antibodies by immunodiffusion and agglutination tests.

Investigation of antigens in patients' sera

We investigated circulating protein antigens of C. albicans serotypes A and B by Ouchterlony immunodiffusion test. The test medium consisted of 1 % agarose (Bio Meriux, no: 7 381 1) solution in calcium-magnesium-veronal buffer (pH: 7.2). Immune sera were dropped into the central wells and patients sera into the peripheral wells. Then the plates were let stand at room temperature for 72 h under conditions suitable for immunodiffusion.

RESULTS

C. albicans were isolated from sputum specimens of 24 patients with pulmonary tuberculosis and 8 patients with bronchial asthma. It was also isolated from urine, pus and throat cultures of 2 patients with systemic candidiasis.

The rabbit antisera gave precipitin lines with C. albicans serotypes A and B antigens which were disintegrated by sonic vibration. We also tested C. albicans antigens that were used for immunization with rabbit immune sera by tube agglutination test. The C. albicans specific antibody titer was 1: 1024.

DISCUSSION

The application of serological methods in diagnosis of mycotic disease developed during recent years. It concerns also candidiasis.

Numerous serological procedures have been developed to assess levels of C. albicans specific antibody in patients' sera. The detection of C. albicans antibodies can not help diagnosis, because most of the individuals are exposed to <u>C. albicans</u> and develop anti-bodies early in life.

Serological tests for specific antibody are frequently false-negative because many patients with candidiasis are immunocompromised and produce few, if any, antibodies. Therefore, several laboratories are currently investigating methods to detect circulating protein antigens or metabolites of <u>C. albicans</u> such as cell wall mannan [3,7], mannose [8], whole cell proteins [9], arabinitol [10], extracelluler prote-inases [11] and cytoplasmic proteins [13]. But standardization of <u>C. albicans</u> antigens suitable for routine serodiagnostic use has not been established yet.

In this study, we used whole cell proteins as antigens because they cantain all surface antigens such as cell wall mannan and they are easy to prepare. Although there is no detectable antigen in colo-nized patients' sera, we determined <u>C. albicans</u> serotype B antigen in the serum of one of the patients with systemic candidiasis [14]. Simul-taneous positive cultures from different body sites, like the case mentioned, indicate systemic candidiasis, therefore, we suggest that the antigen we used to immunize rabbits is specific.

The fact that we could not determine any antigen in sera of colonized patients with chronic pulmonary disease shows that there is no risk of disseminated candidiasis in those patients unless there are other damaging factors.

REFERENCES

1. S.M.Finegold and E.J.Baron, "Diagnostic Microbiology", Seventh edition, pp. 678-774, C.V. Mosby Company, St.Louis (1987).
2. J.W.Rippon, "Medical Mycology", pp. 177-204, W.B. Saunders Company, Philadelphia (1974).
3. L.de Repentigny, E.Reiss, Current trends on immunodiagnosis of candidiasis and aspergillosis, <u>Rev.Infect.Dis.</u>, 6: 301 (1984).
4. J.H. Ellsworth, E. Reiss, R.L. Bradley et al., Comparative serological and cutaneous reactivity of candidal cytoplasmic proteins and mannan separated by affinity of concanavalin. <u>Am.J. Clin. Microbiol.</u>, 5: 91 (1977).
5. L.O.Gentry, I.D.Wilkinson, A.S.Lae et al., Latex agglutination test for detection of <u>Candida</u> antigen in patients with dissemi-nated disease, <u>Eur.J.Clin.Microbiol.</u>, 2: 122 (1983).
6. J.M.Jones, Quantitation of antibody against cell wall mannan and major cytoplasmic antigen of <u>Candida</u> in rabbits, mice and humans, <u>Infect.Immun.</u>, 30: 78 (1980).
7. M.H.Weiner and M.Coats-Stephen, Immunodiagnosis of systemic candidosis: Mannan antigenemia detected by radioimmunoassay in experimental and human infection, <u>J.Infect.Dis.</u>, 140: 989 (1979).
8. T.P.Monson, K.P.Wilkinson, Mannose in body fluids as an indicator of invasive candidiasis, <u>J.Clin.Microbiol.</u>, 14: 557 (1981).
9. M.Manning-Zweerink, C.S.Maloney, T.G.Mitchell et al., Immunoblot analyses of <u>Candida albicans</u> associated antigens and antibobies in human sera, <u>J.Clin.Microbiol.</u>, 23: 46 (1986).
10. J.W.M.Gold, B.Wong, E.M.Bernad et al., Serum arabinitol concent-rations and arabinitol/creatinine ratios in invasive candidiasis, <u>J.Infect.Dis.</u>, 147: 504 (1983).
11. F.Mac Donald F and F.C.Odds, Inducible proteinase of <u>Candida albicans</u> in diagnostic serology and in the pathogenesis of systemic candidosis, <u>J.Med.Microbiol.</u>, 13: 423 (1980),

12. G.F.Araj, R.L.Hopfer, S.Chesnut, Diagnostic value of the enzyme-linked immunosorbent assay for detection of Candida albicans cytoplasmic antigen in sera of cancer patients, J.Clin.Microbiol., 16: 46 (1982).
13. P.S.Stevens, S.Huang, L.S.Young et al., Detection of Candida antigenemia in human invasive candidiasis by a new solid phase immunoassy, Infection, 8: 334 (1980).
14. R.C.Matthews, J.P.Burnie, S.Tabaqchali, Isolation of immunodominant antigens from sera of patients with systemic candidiasis and characterization of serological response to Candida albicans, J.Clin.Microbial., 25: 230 (1987).

CANDIDA GUILLIERMONDII VAR.GUILLIERMONDII INFECTION IN

INFERTILE COUPLES

B.Nagy[1] P.Sutka[2] M.Ziwe-Abidine[2] I.Kovács[1] V.Forgács[3]
T.Pulay[1] R.Gimes[1] F.Paulin[1] J.Csépli[3] S.Csömör[1]

[1]1st Department of Obstetrics and Gynaecology, Semmelweis Medical University, Baross u.27, 1088 Budapest,
[2]Mycological Department, Institute for Agricultural Control, Kereti K., 1024 Budapest,
[3]Department of Obstetrics and Gynaecology, City Hospital, Köves u.2, 1204 Budapest, Hungary

INTRODUCTION

Candida guilliermondii (C.g.) occurs in animals, but it can also affect humans. Morris and Morris[1] have detected C.g. in human vaginal smear, Utley et al.[2] in the case of endocarditis, Dick et al.[3] in the case of fatal disseminated candidiasis, and Santha et al.[4] have found out that 25 % of blood donors had C.g. precipitins and agglutinins. Sutka and Sutka[5,6] have produced by biotechnological method a preparation which may make rapid diagnosis possible. They have proved that in cows and heifers C.g. causes serious genital changes, eg. thickening of the uterus, reduction in the excretion of sexual hormones, asymptomatic infertility, reduced functional sperm index, and have treated these animals succesfully[7,8].

We applied Sutka's antigen and method in the routine infertility examination of couples.

MATERIAL AND METHOD

All patients with a previous history of infertility and to undergo C.g. screening turned to the Infertility Section of our clinic. The sera of 373 infertile persons (128 men and 245 women) and 56 fertile (10 men and 46 women) by the following tests.

Precipitation test

We performed the precipitation reaction by a special extracellular antigen, produced by a biotechnological method (Hung. Pat.No. 183115, US. Pat. No. 4678748).

Agglutination reaction

The whole cell agglutination antigens were prepared according to

Candida and Candidamycosis, Edited by E. Tümbay *et al.*
Plenum Press, New York, 1991

Hung. Pat. No. 183115. We used the classical method with <u>Candida albicans</u> and <u>Candida guilliermondii</u> 4922 strains. The titer higher than 1/80 was considered a sign of <u>Candida</u> infection.

Indirect immunofluorescence

Indirect immunofluorescence was employed (in 14 cases) with slides previously coated with the suspension of C.g. 4922 strain.

Histological examinations

Tissue samples were taken from the ovary of some patients (3 cases) participating in our IVF programme. We placed the samples into 10 % formaldehyde solution. After embedding in paraffin and slicing, deparaffinated sections were impregnated with Gömöri-Grocott reagent[7].

Antimycotic treatment

Nizoral (Ketoconazole. Janssen - Richter G. Hungary) was administered to C.g. positive patients, one tablet (200 mg) pro day for one month.Control studies were carried out following the treatment and one month later.

Statistical analysis

Statistical analysis was performed by a BMDP programme, the level of significance (P) was 0.05[9].

RESULTS

Out of 245 infertile women 131 (53.3 %) and out of 128 men 75 (61.7 %) proved C.g. positive, but in the female control group in 46 pregnants (between 6 and 15 pregnancy week) we found 23.9 % positive. In semen donors we detected two infected with C.g.. Patients having an aqqlutination titer higher than 1/80 were taken as C.g. positive. The precipitation reaction showed positivity in each case. Simultaneous C.a. and C.g. agglutination reactions resulted in differing titers in most cases, in some they coincided, leaving the diagnosis dependent upon the precipitation results. By the time of the first control, examination titers reached 1/2560, while in the second control they diminished to or went under 1/80. In a few cases (14) this form of determination, was also performed, the titers doubled as compared with those of the agglutination reaction.

By Gömöri-Grocott's reaction we succeeded in confirming the presence of yeast cells in the stroma ovary. In the semen of C.g. infected men we detected yeast cells. After three months of antimycotic treatment the semen parameters changed to normal.

DISCUSSION

Veterinarians have been successfully applying the antigen produced by Sutka and Sutka's method for the diagnosis of <u>Candida guilliermondii var. guilliermondii</u> infection and achieved considerable results in treating infertility in the cattle stock.

In certain cases the reason for infertility has been left unexplained in humans as well.

We tested the sera of 245 infertile women, 128 infertile men and 46 pregnant women (between 6 and 15 weeks). In the infertile group 131 (53.4 %), in the fertile 11 (23.9 %) women were found C.g. positive, the level of significance being considerable (P=0.0001) between those two groups. In the infertile men 75 (61.7 %) and in the semen donors 2 (20 %) proved C.g. positive.

C.g. positive infertile patients were subjected to a one-month ketoconazole treatment. We performed control studies to check the efficacy of the treatment.

By the second control examination approximately 60 % of the patients' precipitation titers went down to zero, while the agglutination titers were less than 1/80. The antigen used for the precipitation reaction turned out to be suitable for the serological detection of human systemic mycoses caused by C.g. and in our opinion, improvement - EIA, ELISA[3], RIA, DELFIA forms - could considerably contribute to the detection of human systemic mycoses.

As yeasts produce toxins, antibiotics and hormones as well as hormone receptors[10,11] in reproductive biology, a greater emphasis should be placed on their detection and on clarifying their function.

REFERENCES

1. P.D. Morris and D.F. Morris, Normal vaginal microbiology of women of childbearing age. Relations to the use of oral contraceptives and vaginal tampons, J. Clin. Pathol. 20 : 636 (1967).
2. J.R. Utley, J. Milis, J.C. Hutchinson, Valve replacement of bacterial and fungal endocarditis, Circulation 48 : 42 (1973).
3. J.D. Dick, B.R. Rosengard, W.G. Merz, R.K. Stuart, G.M. Hutchins, R. Saral, Fatal disseminated candidiasis due to amphotericin-B-resistant Candida guilliermondii, Ann. Intern. Med., 102: 67 (1985).
4. J. Santha, E. Karuczka and P. Sutka, Screening blood donors infected with Candida guilliermondii var. guilliermondii, in: "Abstracts- ISHAM Congress" Revista Iberica de Micologia 5, Suppl. 1 (1988).
5. P. Sutka and K. Sutka, Hungarian Patent No. 183115 (1980).
6. P. Sutka and K. Sutka, Process for the production of immunobiological preparations applicable in the diagnosis, prevention and treatment of Candida guilliermondii infection. US. Patent No. 4678748, in: "Biotechnology Res. Abstr. 4, p.5, 3254-W4", (1987).
7. P. Sutka and I. Mészáros, Breeding bull's infection with Candida guilliermondii, Magy. Állatorvosok Lapja, 33 : 151 (1978).
8. P. Sutka and K. Sutka, Antigen for detection of Candida guilliermondii var. guilliermondii infection in ruminants (Abstr.), in : "Inst. Pasteur First Int. Smposium Nov. 17-20, 1986"; Paris (1986).
9. W.J. Dixon, "BMDP. Statistical Software Manual", University of California Press, Berkeley, CA. 733 (1983).
10. A. El-Sokkary, M. Abou-Gabal, E.K. Nafie, A.A. Farid, Isolation and purification of a new toxin from Candida guilliermondii, Bull. Fac. Pharm. Cairo Univ., 12 : 201 (1973).
11. G. Kolata, Steroid hormone systems found in yeasts, Science, Aug 31: 913 (1984).

THE DIAGNOSIS OF CANDIDA INFECTIONS AND THE RAPID IDENTIFICATION OF

CANDIDA ALBICANS

O. Marcelou-Kinti   A.Voyadjoglou-Samanidou   E.Marcelos

Department of Parasitology, Entomology and Tropical Diseases; Athens School of Hygiene, 196 Alexandres Avenue, 104 42 Athens, Greece

INTRODUCTION

The role of yeast-like organisms of the genus Candida in human diseases makes necessary the application of simple, rapid and trustworthy methods for the prompt diagnosis of a severe, generalized disease.

In this paper a series of experiments are described, carried out at the Mycology Laboratory of the Athens School of Public Health, for the diagnosis of Candida mycosis.

MATERIAL AND METHOD

Material

        250 strains of yeast-like organisms isolated from vaginitis
        150 strains isolated from oral mucosa
        100 strains from different skin and nail disorders
         50 strains from patients in ICU
        250 strains isolated from the environment

        Total number of strains: 800

Method

For the identification of the isolated strains the consequent steps described below were followed:

1. Chlamydospore production in PCB (Potato, Carrot, Bile)
2. Germ tube test after 2-4 hours at 37°C, in blood serum
3. Sugar fermentation (glucose, maltose, saccharose, lactose, galactose, raffinose, inositole, etc.)
4. Sugar assimilation
5. Reduction of TTC (doubtful method)
6. Actidione resistance
7. Application of API 20 C for identification

Candida and Candidamycosis, Edited by E. Tümbay *et al.*
Plenum Press, New York, 1991

Table 1. Characteristics of yeast-like organisms useful for their taxonomy.

Fungus species	Fermentation				Assimilation									
	Gluc.	Malt	Sacch.	Lact	Gluc	Malt	Sacch.	Galact.	Raff.	Trehal.	Inos.	Lact.	Sel.	Xyl.
C. albicans	AG	AG	-	-	+	-	+	+	-	-	-	-	-	-
C. tropicalis	AG	AG	AG	-	+	+	+	+	-	+	-	-	±	+
C. pseudotropicalis	AG	-	AG	AG	+	+	+	+	-	+	-	+	+	-
C. krusei	AG	-	-	-	+	-	-	-	-	-	-	-	-	-
C. parapsilosis	AG/-	A/-	A/-	-	+	-	+	-	-	-	-	-	-	-
C. guillermondii	-/AG	-	-/AG	-	+	+	+	+	+	±	-	-	+	+
C. stellatoidea	AG	AG	-	-	+	+	-	+	-	+	-	-	-	+
T. glabrata	-	-	AG	-	+	-	-	-	-	-	-	-	+	-
T. cutaneum	AG/-	AG/-	AG/-	AG/-	+	+	+	+	±	±	+	-	+	+
R. rubra	-	-	-	-	+	+	+	+	+	+	±	-	-	-
C. neoformans	-	-	-	-	+	+	+	+	±	±	+	-	+	+
G. candidum	-	-	-	-	+	-	+	-	-	-	-	-	-	-

AG: Acid + gas production A: Acid A/-: Acid or no reaction for some strains
-/A: Negative or no reaction for some strains AG

Table 2. Identification of 700 strains of yeast-like organisms.

Origin	Total	C.albicans		C.tropi-calis		C.pseudo-tropicalis		C.krusei		C.parapsi-losis		C.stellato-idea		C.guiller-mondii		C.glabrata	
		No	%	No	%	No	%	No	%	No	%	No	%	No	%	No	%
Vaginitis	250	196	76.4	5	3.2	4	1.6	2	0.8	2	0.8	16	6.4	-	-	22	8.8
Oral mucosa	150	129	86.0	4	2.6	-	-	4	2.6	-	-	-	-	-	-	13	8.6
Skin + Nails	100	89	89	5	5	2	2	3	9	1	1	-	-	-	-	-	-
I.C.U.	50	45	90	2	4	1	2	-	-	-	-	-	-	-	-	2	4
Environment	250	98	39.2	36	144	20	8	39	15.6	34	13.6	7	2.8	6	2.4	10	4
Total	800																

Strains producing chlamydospores in PCB medium and forming germ tube were considered as C.albicans. In this case all the rest steps were omitted.

RESULTS AND DISCUSSION

The qualities of yeast-like organisms, useful for their identification, are presented in Table 1.

The results of the species where the 700 studied strains belong are shown in Table 2.

Obviously, the most frequently isolated species was C.albicans, isolated in a percentage of 76.4 % to 90 % from human excretions and of 39.2 % from the environment. This last finding proves the responsibility of man in spreading this pathogenic endogenous fungus in the environment.

A way to prevent environmental pollution is to make man understand that he must visit a specialist when he feels any disorder, because a prompt treatment helps himself and the environment as well.

REFERENCES

For references please contact O.Marcelou-Kinti, the first author of this paper.

IDENTIFICATION OF ANTIGENS SPECIFIC TO GERM-TUBES OF

CANDIDA ALBICANS BY MONOCLONAL ANTIBODIES

R.Robert[1]   J.Aubry[2]   C.Mahaza[1]   V.Annaix[1]
A.Leblond[1]   G.Tronchin[1]   J.M.Senet[1]

[1]Laboratoire d'Immunologie - Parasitologie - Mycologie,
Faculté de   Pharmacie, 49100 Angers, France
[2]U.211 INSERM, Faculté de Médecine - Pharmacie, 44035
Nantes, France

INTRODUCTION

Various ligands have been involved in the process of infection and tissue invasion of C. albicans[1]. It is adherence is enhanced when germ-tubes are emerging from blastospores. The importance of yeast germination raises the possibility of a molecular organization of cell walls different between blastospores and hyphae. Major biochemical differences have been described, but the functional significance of the molecules specific to germ-tubes remains to be determined. Further features should be given by the hybridoma technology to identify surface antigens. Paradoxically a limited number of informative monoclonal antibodies was obtained by immunization of mice with whole organisms. Generally mannoproteins of yeasts cell wall were recognized [2,5]. A different approach was attempted from fibrinogen binding factors released by the germ-tubes of C. albicans that we had reported previously [1]. We describe the reactivity and characteristics of a murine monoclonal antibody 3D9 directed to a component of germ-tube of C.albicans.

MATERIAL AND METHOD

1. Organisms and culture conditions

Several sources of C.albicans were used: C. albicans 1066 (serotype A), originally isolated from a case of Candida septicemia and several fresh clinical isolates from our medical center, maintained on Sabouraud broth (C. albicans, C. krusei, C. tropicalis, C. guillermondii). Organisms were grown either as blastospores in 199 medium at 22°C for 48 h or as germ - tubes and mycelia in the same medium at 37°C for 3 h and 48 h, respectively.

2. Immunization and generation of monoclonal antibodies

Adult male Balb/c mice received subcutaneous injections of fibrinogen binding factors as described elsewhere[1] containing 50 µg of

Candida and Candidamycosis, Edited by E. Tümbay *et al.*
Plenum Press, New York, 1991

protein in 250 µl of phosphate buffer saline emulsified with an equal volume of complete Freund adjuvant. Two other intraperitoneal injections at one month interval were given with incomplete Freund adjuvant.

Cell fusion and selection of hybrids were performed essentially as described by Dippold et al.[6]. Spleen cells of two immunized mice were fused with plasmocytoma X63/Ag.8.653 and distributed at the concentration of 10^5 cells by well prepared with a feeder layer of $2x10^4$ mouse irradiated peritoneal macrophages. Ten days after fusion the supernatants of growing hybridomas were checked both for their reactivity to the immunizing antigen by ELISA test and to C. albicans by indirect immunofluorescence with goat fluorescein conjugated antibodies anti-mouse IgM (µ chain specific) or anti mouse IgG (H+L).

3. Transmission electron microscopy

The protocol was carried out basically as described previously [7]. After absorption with germ-tubes of C. albicans, rabbit anti-mouse immunoglobulin M (IgM) was used to detect monoclonal antibodies and after several washings protein A/gold was added to the cells.

4. SDS-PAGE and immunoblotting techniques

Cell wall extracts of blastospores, blastospore bearing germ-tubes and mycelia of C. albicans were obtained by ultrasonication. Samples of these three homogenates were analyzed by a 6 % sodium dodecyl sulphate polyacrylamide gel electrophoresis essentially as described by Laemmli [8], then immunodetected after transfer to nitro-cellulose paper using the Mab 3D9 and an irrelevant antibody as negative control.

RESULTS

We have selected the reactivity of 45 antibody secreting hybridomas to fibrinogen binding factors by ELISA. The Mab 3D9 was isolated by double cloning as a murine IgM K, isotype determined by chain specific fluorescein-conjugated anti-immunoglobulin sera.

1. Distribution of the antigen

Indirect immunofluorescence with the monoclonal antibody 3D9 specifically revealed a labelling pattern restricted exclusively over the germ tube of C. albicans (Fig.1). This staining pattern was similarly observed with clinical isolates of C. albicans irrespective of serotype. Furthermore, no binding to the surface of the species C. tropicalis, C. krusei, C. pseudotropicalis and C. guillermondii could be seen either on blastospores or hyphae. Both observations are in favour of a new antigenic marker of C. albicans distributed along the germ tube detected by Mab 3D9.

2. Localization of the 3D9 epitope

By transmission electron microscopy with a rabbit immune serum anti-mouse IgM and protein A/colloidal gold, the outermost layer of the germ tube cell wall displayed a regular distribution of particles (Fig.1). By contrast no particle was seen associated with any layer of the cell wall of the blastospores. The specificity of the cytochemical reactions was demonstrated by the use of an irrelevant monoclonal antibody giving no labelling in control experiments.

224

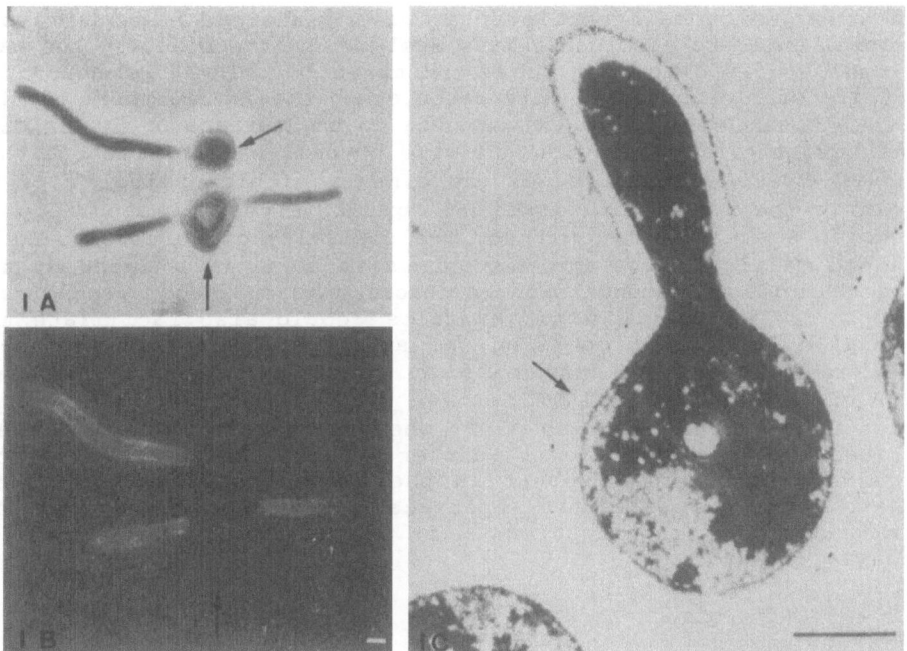

Fig. 1. Phase contrast (A) and immunofluorescence (B) photographs of
intact filamentous cells of C.albicans, stained with the Mab
3D9 raised against the fibrinogen binding factors specific to
the mycelial cell walls. Arrows point to blastospores that do
not stain with the Mab which, in contrast, strongly binds to
mature hyphae.

(C) Surface labelling observed by transmission electron
microscopy. The binding of Mab 3D9 revealed by a rabbit
immunserum anti-mouse IgM and colloidal gold labelled protein
A was distributed on the outermost layer of germ-tube cell
wall of C. albicans. No labelling was observed in blastospore.

3. Preliminary characterization of the antigen

The analysis of three cell wall extracts of C. albicans on a 6 %
SDS-PAGE then by immunoblotting with the Mab 3D9 revealed a high
molecular weight component with apparent molecular mass of 450 kilo-
daltons, expressed by mycelial cell walls and the germ-tubes of
blastospores contrasting with the absence of any material associated
to the blastospores.

DISCUSSION

The biochemical structure of adhesins of C. albicans remains
unclear. Antigenic differences were yet described between mannoproteins
of germ-tubes and blastospores of C. albicans, but none of the
prepared monoclonal antibodies were specific to germ tubes [4,5]. The
composition of immunizing material could explain numerous failures to
obtain such monoclonal antibodies [3,5]. Interestingly, a majority of
patients with disseminated candidiasis have high serum antibody

titers to germ tube specific antigens [9,10] whereas various rabbit polyclonal antisera adsorbed exhaustively with live blastospores have been made specific to germ tube structures [9,10]. The immunological repertoire of mice could be more restricted: the immunuzition of mice with C. albicans [3,4]. To detect antigens on the surface of C. albicans germ-tubes, the immunization of mice by cell wall extracts either produced by solubization of proteins by dithiothreitol [11] or by hydrolytic enzymes [12] or purified by affinity chromatography to fibrinogen seems a better method. The disparity of results could be explained mainly by the immunosuppressive activity induced by the polysaccharides as demonstrated in chronic mucocutaneous candidiasis patients [13]. The cell wall extracts could also facilitate the recognition and the processing of surface molecules masked by a preponderant amount of mannans. High molecular weight components (HMWC) of mycelial cell walls have been identified by our group specifically with the Mab 3D9 whereas similar molecules were described by others [11,12,14]. Further characterization of the antigenic determinants present in these HMWC is required to know whether germ-tube specific epitopes represent *de novo* protein determinants or modifications of preexisting structures by glycosylation.

REFERENCES

1. R.Robert, V.Tronchin, A.Bouali, J.M.Senet, in: "Proceedings X Congress ISHAM", J.M. Torres-Rodriguez, ed., pp. 185-191, Barcelona (1988).
2. T.Chardes, M.Piechaczyk, Y.Cavailles, S.L. Salhi, B. Pau, J.M. Bastide, Ann.Inst.Pasteur/Immunol., 137 C: 117 (1986).
3. V.Hopwood, D.Poulain, B.Fortier, G.Evans, A.Vernes, Infect. Immun., 54: 222 (1986).
4. W.L.Chaffin, J.Skudlarek, K.J.Morrow, Infect. Immun., 56: 302 (1988).
5. P.M.Sundstrom, M.R.Tam, E.J.Nichols, G.E.Kenny, Infect. Immun., 56: 601 (1988).
6. W.G.Dippold, K.O. Lloyd, L.T.C.Li, H.Ikeda, H.F.Oettgen, L.J.Old, Proc.Natl.Acad.Sci. USA, 77: 6114 (1980).
7. G.Tronchin, R.Robert, A.Bouali, J.M.Senet, Ann.Inst.Pasteur/ Microbiol., 138: 177 (1987).
8. U.K.Laemmli, (1970) Nature, 227: 680 (1970).
9. E.H.Smail, and J.M.Jones, Infect.Immun., 45: 74 (1984).
10. A.B.Mason, J.M.B.Smith, J.Infect.Dis., 153: 146 (1986).
11. J.Ponton, J.M.Jones, Infect.Immun., 53: 565 (1986).
12. M.Casanova, M.L.Gil, L.Cardenoso, J.P.Martinez, R.Sentandreu, Infect.Immun., 57: 262 (1989).
13. A.Durandy, A.Fischer, F.Le Deis, E.Drouhet, C.Griscelli, J.Clin. Immunol., 7: 400 (1987).
14. P.M.Sundstrom, E.Nichols, G.E.Kenny, Infect.Immun., 55: 616 (1987).

CANDIDA ANTIGEN DETECTION IN SERIOUS INFECTIONS: AN UPDATE

M. F. Price   L. O. Gentry

Infectious Disease Section, St. Luke's Episcopal
Hospital, 6720 Bertner, Houston, Texas 77030, U.S.A.

INTRODUCTION

Manifestations of Candida infections vary, so that disease may
involve only mucous membranes, and are considered "local" infections,
or may involve deep organs, in which case disseminated disease is
present. The ability to distinguish between the different types of
disease is important, since methods of diagnosis and therapy will
vary dependent upon the type of infection present.

Local mucous membrane infections include oral or vaginal thrush,
esophageal ulceration without dissemination, bladder infection and
cutaneous involvement. In contrast, disseminated disease, with deep
organ involvement, occurs when Candida reaches the bloodstream from
skin or gastrointestinal tract and is carried to other organs
including kidney, lung and, in leukemia patients, liver, to set up
life threatening disease.

With the continued incidence of serious Candida infections, in
1982 we set out to develop a simple and rapid test which would aid in
the diagnosis of disseminated Candida infection. In contrast to other
approaches which involved the search for defined antigenic
components, this test was developed to detect any serum antigen
present in patients with disseminated candidiasis: Latex was coated
with heterologous antibody to Candida and reacted with sera from
patients with documented infection. The latex preparation which gave
positive results in this patient population was used to test for
antigen in sera from a variety of patients. These included control
populations, such as healthy adults, colonized patients, patients
with other fungal or bacterial infections and patients in renal
failure without signs or symptoms of Candida infections and in whom
antigen should be absent. Two studies have also been carried out in
patients with high rheumatoid factor titers to determine the degree
of interference, or false positive results, obtained with this latex
test and this population, which has been reported to cause problems
with latex agglutination tests.

Patients with documented disseminated Candida, or those at high
risk for development of this disease, were also studied. The first
series involved 33 patients with positive blood cultures (not catheter
related) for Candida species, or autopsy proven infection. The second
study determined the practical application of the test for detection

of disease in high risk patients, and the third followed Candida antigen titers in selected patients with positive antigen titers undergoing treatment with amphotericin B. In all studies a control latex was also used to detect nonspecific agglutination.

CONTROL POPULATIONS

In 100 healthy young adults none had Candida antigen titer >1:2. In patients who were colonized with Candida species, some of whom had multiple sites positive for yeast, the antigen titer remained <1:4. Of 14 patients in renal failure, none showed accumulation of antigen, with all having titers <1:2. In a second study in this population, serum was collected from patients just prior to dialysis, and creatinine and Candida antigen titers tested. A total of 14 serum samples from 8 patients were evaluated in this way, to determine whether high creatinine levels caused "false positive" results. A total of nine sera were antigen negative. These had a mean creatinine value of 8.5 mg/dl, range 5.0-11.0. Four sera had an antigen titer =1:2. These had a mean creatinine of 8.3 mg/dl, range 6.4-12.6. Thus it appears that high creatinine levels do not cause "false positive" results in thist test.

Other control groups tested showed similar results, with the majority of those patients with bacteremia (n=20) and other fungal infections (n=25) having antigen titers <1:2. One patient in each group had an antigen titer >1:2, however in both cases other patients with the same pathogen had low antigen titers, thus indicating that false positive results were not consistantly seen in patients with these infections. The final control group tested consisted of patients with high rheumatoid factor titers. The initial study was carried out to determine the incidence of nonspecific agglutination in this population, the second was designed to correlate the RF titer with nonspecific agglutination at >1:2. The first study (n=25) showed that 24% of patients with high RF titers gave nonspecific agglutination >1:2 while the second study (n=100) confirmed that the incidence of nonspecific agglutination increased with increasing RF titer.

Thus in the control populations, including patients colonized with Candida or those in renal failure, Candida antigen is consistantly ≤1:2. False positive results may occur in patients with high RF titers.

PATIENTS WITH DISSEMINATED INFECTION

Our initial study in this population investigated Candida antigen titers in patients (n=33) with disseminated infection documented by a positive blood culture (if not catheter related) or autopsy. In contrast to control populations, the majority of patients in this group had antigen titers >1:2, while only two patients had titers <1:2. Nonspecific agglutination was found in only one patient. Eight patients had titers of 1:4, thirteen had titers of 1:8 and eight had titers of 1:16. In one patient the titer was 1:32. Infections were due to Candida albicans, C. parapsilosis and C. tropicalis. In all cases tested, antigen titers in patients who were treated with amphotericin B and survived returned to within normal levels.

Our second study looked at antigen titers in patients at high risk for development of disseminated Candida infection. One hundred and twenty-eight patients were evaluated in regard to their Candida antigen titer and signs and symptoms of disease. A total of sixty patients had titers ≤1:2. Only two of those patients were considered to have disseminated Candida infection and were considered false negatives (3%). Eight patients had local mucocutaneous infection and six had catheter induced Candida sepsis, which represented "false positive" blood cultures. The remaining fifty patients had no evidence of Candida infection. In contrast, a total of sixty-eight patients had Candida antigen titers ≥1:4. Eleven of these patients had disease documented by positive blood cultures (16%). Twenty-seven (40%) responded to amphotericin B therapy while thirteen (19%) did not require treatment. These patients had subsquent antigen titers <1:4. Seventeen patients (25%) expired soon after the antigen titer was obtained. No autopsy was performed and they were thus nonevaluable.

During these early studies, Candida antigen titers were followed sequentially whenever possible: Data on patients JS and MR were obtained during the first study, at which time the significance of a high antigen titer was still under investigation.

MR: This patient had a Candida antigen titer of 1:8 on days 1,6,9 and 13. Although the patient was febrile, no other evidence of disseminated Candida infection was preset until day 13, when Candida endophthalmitis was diagnosed and one blood culture grew C. albicans. The patient expired on the 14th day. At postmortem he was found to have disseminated Candida infection.

JS: The initial antigen titer in this patient was 1:8. Twenty-eight days later the patient had a titer of 1:32. On the 29th day the patient expired. On this day also multiple blood cultures drawn within the last few days grew C. albicans.

Thus, on completion of the early studies it became apparent that a high Candida antigen titer correlated with disseminated disease. It was also apparent that positive blood cultures were frequently unhelpfull for premortem diagnosis of disseminated Candida infection. Thus in the remaining studies, therapy was initiated on the basis of clinical indications and antigen titers. Confirmation of infection was considered to occur when the patient responded to antifungal therapy. Although this is less definitive than biopsy or autopsy proven infection, it remains the most realistic method of confirmation in a prospective study where blood cultures are negative.

AG: Had Candida antigen titers of 1:4 and 1:8 on the 1st and 3rd days, respectively. Although no blood cultures were positive for Candida species, C. tropicalis was grown from a pelvic abscess and from sputum. On the 4th day amphotericin B treatment was begun. The Candida antigen titer remained elevated for another ten days (1:8 on the 10th and 13th days) but fell to within normal limits (≥1:2) by the 26th day. This patient became afebrile on amphotericin B.

GA: This patient had low grade fever, multiple sputum cultures positive for C. tropicalis, a Candida antigen titer of 1:8, and had been treated with broad-spectrum antibacterial agents. Although all blood cultures were negative this patient was started on amphotericin B on the 3rd day. On the 4th and 5th days his Candida antigen titer was 1:8 and 1:16, respectively; the 7th day 1:4 and on the 11th, 12th, and 13th days had returned to negative. This patient also became afebrile on amphotericin B.

MCD: This patient had a similar clinical picture but multiple urine cultures were positive for C. parapsilosis. This patient had a Candida antigen titer of 1:4 on the 1st day and of 1:16 on the 7th day. Antifungal therapy was begun on the 7th day and the patient became afebrile within 48 hours. On the 14th day her antigen titer was 1:8 and remained at 1:4 on the 21st, 29th and 30th days but by the 42nd day had dropped to <1:2.

Thus, in the above cases, although positive blood cultures were not obtained, diagnosis of Candida infection was made. This was based on the clinical findings taken in conjunction with the Candida antigen titer. The response to amphotericin B, both clinically and in respect to the decrease in antigen titer, confirmed the original diagnosis.

For routine hospital care sequential antigen titers are now monitored in our institution in patients at high risk for development of disseminated Candida infections. These patients include those with organ transplants and represent one of the most difficult and complex populations to follow. A number of cases are now described.

RR: This patient was admitted for repair of an abdominal aortic aneurysm. On the day after he first became afebrile his Candida antigen was tested and found to be 1:8. Since his creatinine was high (2.1), the decision was made to hold of on treatment for a short time. Two days later however his antigen titer remained 1:8 and he still had a low grade fever. Next day his antigen was 1:16 and amphotericin B treatment was initiated. At this time, although sputum cultures had grown Candida species, blood cultures remained negative. At this time amphotericin B was discontinued. After 48 hours the patient remained afebrile and the antigen titer remained negative. A total of 260 mg amphotericin B had been administered to this patient.

DS: This patient had recently undergone a heart transplant. Initially he had done well, he was afebrile with negative cultures and Candida antigen. He had undergone a splenectomy sometime previously. He became febrile on "day 1". Next day his antigen titer was 1:8. Two days later however, he was afebrile. On day 9 he again developed a low grade fever. His Candida antigen titer was retested and found to be 1:16. Only sputum cultures were positive for C. albicans. Three days later his titer remained 1:16 and he was still febrile. On day 17, the site of Candida infection was found when his gallbladder was aspirated and cultured. The aspirate grew a pure culture of C. albicans. At this time Candida cholecystitis was diagnosed and the patient was started on amphotericin B. On day 21 the patient had a gastrointestinal bleed, possibly seeding Candida into the bloodstream once again. Although the amphotericin B was continued, on day 35 the bile remained positive for C. albicans and soon after liver involvement was suspected. The patient expired 41 days after becoming afebrile. At this time his antigen was still 1:16. He had received a total of 535 mg of amphotericin B. No autopsy was performed.

In summary, in this case of culture proven Candida cholecystitis the Candida antigen was high (1:8) within 24 hours of the development of fever. Unfortunately this patient failed to respond to amphotericin B therapy. The reasons for this are unclear, although it could have been related to his immunosuppressed state compounded by prior splenectomy, lack of penetration of amphotericin B to the site of infection or the presence of Candida infection in the liver, which is notoriously difficult to treat.

JM: This patient had a history of a duodenal ulcer and GI bleed. On the day that he received a heart transplant he also had catheter sepsis with Candida and a coagulase negative Staphylococcus. Next day he remained afebrile and his Candida antigen was also negative. Two days later his Candida antigen had risen to 1:16, however he was still afebrile. Twenty-four hour later he developed low grade fevers and amphotericin B was initiated. The patient responded to antifungal therapy and after 3 days he was once again afebrile, however his antigen titer remained elevated at 1:16. On day 20 amphotericin B therapy discontinued. At this time the patient was afebrile but the antigen was 1:16. On day 32 it had dropped to 1:4 and on day 45 was negative. On day 56, however, the antigen was once again elevated at 1:8 and the patient had tender hepatomegaly and an elevated alkaline phosphatase, suggesting possible liver involvement. At this time the patient developed pneumonia, due to a resistant Enterobacter, which failed to respond to treatment and the patient expired. No autopsy was performed.

In the early stages, this case is typical of the events seen in patients with catheter induced Candida sepsis: The period of positive blood cultures is the time when Candida is being seeded via the bloodstream to the deep organs, but prior to development of actual pathological changes. At this time the antigen remains negative. It appears that once the disease process begins, the antigen becomes elevated. In this case the period between sepsis and high antigen titer was 48 hours. The patient developed clinical evidence of infection, ie, fever 24 hours later. Although this patient became afebrile on amphotericin B therapy, the delayed decrease in Candida antigen titer suggested a source of infection recalcitrant to therapy. Although in part speculation, the sequence of events is consistant with an original catheter-induced sepsis resulting in disseminated disease, including hepatic involvement initiated at the time of original sepsis or secondary to a GI bleed. Since permission for autopsy was denied however, this could not be confirmed.

HK: This patient had a history of congestive myopathy and was awaiting heart transplantation. His initial antigen titer (day 1) was negative. On day 9 a Jarvik Mechanical heart was placed due to his deteriorating condition. By day 17 chest tube cultures were positive for C. albicans, and his Candida antigen was 1:2. Six days later the chest tube cultures were still positive and amphotericin B was stated for therapy of Candida wound infection. On the 31st day the Candida antigen was negative and on day 36 amphotericin B was discontinued. On day 40 he received a heart transplant and next day cultures of the Jarvik indicated that the region of the aortic graft was culture positive for C. albicans: amphotericin B was restarted. While the patient received antifungal therapy, he remained afebrile and his Candida antigen remained negative, however, he continued to have drainage from his sternal wound and had multiple positive blood cultures. After a long course of amphotericin B, on day 114 his creatinine levels became elevated. His Candida antigen was negative and antifungal treatment was temporarily discontinued. His Candida antigen titer was monitored for recurrence: Three days after the therapy was stopped, his antigen was 1:2. After another three days it had risen 1:4, however he remained afebrile for another two days, at which time low grade fever returned and amphotericin B was restarted. Ultimately this patient underwent surgery: Pathology confirmed the presence of Candida infection of the aortic graft and surrounding tissues. Consequently the patient received liposomal amphotericin B (5g) after which he remained afebrile with a negative antigen. The chest wound continued to have intermittent drainage with scant Candida present.

To summarize, this patient had a documented aortic graft infection. Since the Candida antigen titer remained low when the infection was localized within the graft, it was used to monitor the situation when suppressive antifungal therapy was temporarily discontinued due to toxic side effects.

The following case represents one of the few situations we have encountered, when a falsely low or negative Candida antigen titer is present in a case of disseminated disease. Antibody titers were determined using a latex agglutination test which detects antibody to cytoplasmic components of Candida.

ML: On day 1 this patient was found to have a Candida antigen titer of 1:2 and antibody titer of 1:4. The same results was obtained from serum obtained seven days later, however on day 8 an antigen titer of 1:4 and antibody 1:16 were reported. The following day the patient expired, and serum drawn earlier in the day had an antigen titer of 1:2 and antibody 1:128. At autopsy this patient was found to have disseminated Candida infection.

Thus in patients with high or increasing antibody titers to cytoplasmic components, antigen titers are lower than expected. This is seen infrequently however, and is likely to be a function of immune complex formation.

To summarize, the data suggest that a rising titer of antigen ≥1:4 is associated with serious Candida infection and the need for therapeutic intervention. Low or negative titers are associated with colonization or localized infections. Candida antigen titers are capable of changing within 24 to 48 hours. Early in catheter induced sepsis Candida antigen titers are negative and removal of the catheter has, in some patients, been sufficient to prevent disseminated infection. Successful treatment with amphotericin B is reflected in decreasing antigen titer.

The only population so far identified as giving false positive results are those with high rheumatoid factor titers, while patients with high antibody levels to cytoplasmic components of Candida have been associated with false negative results. Patients with C. krusei infection also appear to give false negative Candida antigen titers.

REFERENCES

1. M.F.Price and L.O.Gentry, Incidence and significance of Candida antigen in low-risk and high-risk patient populations, Eur.J.Clin.Microbiol., August: 416 (1986).
2. L.O.Gentry, I.D.Wilkinson, A.S.Lea, M.F.Price, Latex agglutination test for detection of Candida antigen in patients with disseminated disease, Eur.J.Clin.Microbiol., 2: 122 (1983).

PASTOREX CANDIDA, A NEW LATEX AGGLUTINATION TEST FOR MANNAN DETECTION IN SERUM OF PATIENTS WITH INVASIVE CANDIDOSIS

L.Meulemans[1]   A.Andremont[2]   F.Meunier[3]
A.M.Ceuppens[3]   M.-L.Garrigues[4]   D.Stynen[1]

[1]Eco-Bio NV, Woudstraat 25, B-3600 Genk, Belgium
[2]Institut Gustave Roussy, Rue C. Desmoulins
 F-94805 Villejuif Cedex, France
[3]Institut Bordet, Rue Heger Bordet, B-1000 Brussels
 Belgium
[4]Diagnostics Pasteur, Bd. R. Poincaré 3
 F-92430 Marnes-la-Coquette, France

INTRODUCTION

The increasing number of patients with virus and drug-induced immunodeficiencies has also lead to a higher incidence of invasive candidosis and other opportunistic infections. Because of the bad prognosis of systemic _Candida_ infections, early and accurate diagnosis is very important. Diagnosis is difficult, however, since traditional methods as blood culture and antibody determination are not very reliable. This has stimulated research on the presence of _Candida_ antigens circulating in the blood. Although several recent reports have focussed on a 47 kD cytoplasmic protein [1,2], the most extensively studied circulating antigen is mannan, a major cell wall polysaccharide [3,5]. Its detection in patient serum is a good indicator for deep-seated disease [6,7].

We have produced a rat monoclonal antibody (IgM) that recognizes mannan from a wide variety of _Candida_ species and coupled it covalently to latex beads. This resulted in a very sensitive reagent, called "Pastorex Candida", than can detect purified mannan spiked at concentrations as low as 2.5 ng/ml in human sera or immune rabbit sera. The first retrospective clinical evaluations, which we present in this paper, suggest that it is a valuable test for the screening of patients at high risk of invasive _Candida_ infections.

MATERIAL AND METHOD

Monoclonal antibodies against mannan were prepared by immunizing LOU/C rats with formalin fixed _Candida albicans_ yeast cells. Spleen cells were fused with IR 983 F production and affinity purification of monoclonal antibodies were carried out as described by Bazin [8]. Covalent coupling of monoclonal antibodies to 0.8 μm latex beads was performed essentially according to Faure et al [3].

The latex agglutination experiments were carried out according to the instructions in the kit. One hundred µl of treatment solution is added to 300 µl of serum sample in a microcentrifuge tube. The mixture ise heated at 100°C during 3 minutes and centrifuged in a microfuge for 10 minutes. Forty ml of the supernatant is transferred onto the disposable plastic cards, provided in the kit, and mixed with 10 µl of the sensitized latex beads. The plates are either manually rotated or rotated on an orbital shaker at 170 rpm. Results are scored after 10 minutes. A buffer control was included on every plate as a negative control. The performance of the latex reagent was controlled with a dilute mannan solution, included in the kit. The negative control latex consisted of latex beads coated with an irrelevant rat monoclonal IgM and was used in this study to estimate the number of false positive results.

The passive haemagglutination inhibition (PHA-I) experiments were performed as described [6].

RESULTS

The first evaluations of Pastorex Candida were performed at the Institut Gustave Roussy (Villejuif) and at the Institut Bordet (Brussels). At the Institut Gustave Roussy, serum samples from 6 patients with repetitive hemocultures positive for C. albicans, were tested. Pastorex Candida detected antigen in 3 samples (from 3 different patients). Two single serum samples from 2 patients with only one positive hemoculture were negative. An additional serum sample from a patient with fungemia due to Candida albicans was also negative.

Table 1 shows the results and patients' information from the study at the Institut Bordet. Most samples were tested with both Pastorex Candida and the passive haemagglutination-inhibition (PHA-I) assay described by Meunier-Carpentier and Armstrong [6]. From the group of 10 patients (a total of 18 serum samples) with autopsy proven systemic candidosis, 5 (6 samples) were positive with the latex reagent. Two (4 samples) of these were also positive with PHA-I. PHA-I, however, also detected antigen in a patient negative with Pastorex Candida, giving it an overall performance of 3 patients (6 samples) positive on 6 (12 samples) tested. In the group of suspected cases (7 patients, 10 serum samples), Pastorex Candida detected mannan in one serum sample. Three were positive with PHA-I.

In order to estimate the specificity of Pastorex Candida, serum samples from patients without evidence for invasive candidosis were treated according to the protocol and tested. These samples included 25 amples from a bloodbank, 59 samples from a clinical laboratory that receives its samples mainly from general practitioners, 14 samples from colonized patients, 5 from patients with proven invasive aspergillosis (3 of these were positive with the Pastorex Aspergillus prototype), and 5 from hospital patients without signs of fungal disease. No false positive reactions could be observed with any of these control sera. The serum samples tested at the Institut Bordet, were also incubated with a negative control latex. No agglutination was observed.

DISCUSSION

The detection of circulating mannan is an important diagnostic indicator for invasive candidosis [10]. The presence of mannan in the patients' serum is transient, however, and therefore patients at high

Table 1. Evaluation of Pastorex Candida on patients with invasive candidosis proven by autopsy (patients 1-10) and patients with candidemia (patients 11-17) at the Institut Bordet.

Patient# (Autopsy)	Species	Underlying disease*	Predisposing factors	Number of serum samples		
				Tested	PHA-I +	Pastorex+
1.(+)	C.albicans	AnLL	Neutropenia TPN, TBI...	3	2	2
2.(+)	C.albicans	CML	Neutropenia,TPN	2	2	1
3.(+)	C.tropicalis	Ovary	Neutropenia,TPN	3	0	1
4.(+)	C.albicans	Head and neck	TPN, upper respiratory tract fistula	1	0	1
5.(+)	C.albicans	Larynx	Cirrhosis	1	ND	1
6.(+)	C.albicans	Lung	Neutropenia	2	2	0
7.(+)	C.albicans	Uterus	Surgery	3	ND	0
8.(+)	Candida	Lymphoma	Prednisone	1	ND	0
9.(+)	Candida	Breast	Extensive cancer	1	ND	0
10.(+)	C.albicans	ALL	Neutropenia	1	0	0
11.(S)	C.albicans	Bladder	Multiple surgery	2	1	1
12.(S)	C.albicans	Cervix	Radiotherapy	1	1	0
13.(S)	C.albicans	Uterus	Pelvic surgery	1	0	0
14.(S)	C.albicans	Tongue	TPN	2	0	0
15.(-)	C.albicans	Head and neck	TPN	1	1	0
16.(-)	C.alb.+ C.glabrata	Head and neck	TPN	1	0	0
17.(ND)	C.tropicalis	AnLL	Neutropenia	2	0	0

* Type or location of neoplasm.
(+) Autopsy   (-) Autopsy negative   (S) Patient survived   ND: Not done
A(n)LL: Acute (non) lymphoblastic leukemia   CML: Chronic myelocytic leukemia   TPN: Total parenteral nutrition   TBI: Total body irradation
PHA-I: Passive haemagglutination inhibition

risk of invasive mycoses, should be screened regularly [4,11]. In this retrospective study, only small numbers of serum samples, usually one, were available from each patient. Despite this drawback, Pastorex Candida was positive with 10 out of 38 serum samples from 9 out of 26 patients with proven or suspected invasive candidosis. In the group of autopsy proven cases alone, 5 out of 10 patients were positive. The degree of invasion and dissemination in the tissues may be important to produce high amounts of circulating antigen. As an example, patient 9 had only evidence of pulmonary candidosis at autopsy and not of widespread infection. It is noteworthy that Pastorex Candida also detected mannan in a patient with an invasive C. tropicalis infection.

The comparison of Pastorex Candida, detecting mannan with a monoclonal antibody, and PHA-I, which uses polyclonal antibodies to detect several antigens, shows that there is no absolute correlation between the positive results of both reagents, which to some extent seem to be complementary. When both tests are combined, antigenemia was observed in 11 out of 23 serum samples from 9 out of 14 patients. Only one patient with autopsy proven systemic disease was negative with both tests.

No false positive results were observed with sera from control patients, including patients colonized by C. albicans, and patients with invasive aspergillosis. Therefore, Pastorex Candida seems to be highly specific. Furthermore, rheumatoid factors can not cause non-specific agglutination since the selected antibody is an IgM.

These initial evaluations are rather limited in terms of the number of samples tested. Taking into account, however, that only 1 to 3 samples per patient were available, the results are very promising. If serial samples are tested, the sensitivity is likely to increase significantly. Furthermore, no false positive reactions could be observed. More extensive prospective studies will now have to demonstrate how useful Pastorex Candida is as an easy screening test for patients at high risk of invasive candidosis.

ACKNOWLEDGMENT

The research project at Eco-Bio was partially funded by I.W.O.N.L.

REFERENCES

1. R.C.Matthews and J.P.Burnie, J.Clin.Microbiol., 26: 459 (1988).
2. N.A.Strockbine, M.T.Largen, S.M.Zweibe, H.R.Buckley, Infect. Immun., 43: 715 (1984).
3. L. De Repentigny, R.J.Kuykendall, F.W.Chandler, J.R.Broderson and E.Reiss, J.Clin.Microbiol., 19: 804 (1984).
4. M.A.Lew, G.R.Siber, D.M.Donahue, F.Maiorca, J.Infect.Dis., 45: 45 (1982).
5. E.Reiss, L.Stockmann, R.J.Kuykendall, S.J.Smith, Clin.Chem., 28: 306 (1981).
6. F.Meunier-Carpentier and D.Armstrong, J.Clin.Microbiol., 13: 10 (1981).
7. M.H.Weiner and M.Coats-Stephen, J.Infect.Dis., 140: 929 (1979).
8. H.Bazin, in: "Methods of Hybridoma Formation", A.H.Bartal and Y.Hirshout, eds., pp. 337-338, Humana Press, Clifton, N.J. (1987).
9. A.Faure, D.Bladier, F.Fabia, P.Cornillot, in: "Protides of the Biological Fluids, Proc. 20th Colloquim", H.Pecters, ed., Pergamon Press, Oxford (1973).
10. L.Kaufman and E.Reiss, in: "Manual of Clinical Laboratory Immunology", 3rd Edition, N.R.Rose, H.Friedman, J.L.Fahey, eds., pp. 446-466, American Society for Microbiology, Washington D.C. (1986).
11. S.Fujita, F.Matsubara, T.Matsuda, J.Clin.Microbiol., 23: 568 (1986).

PATHOLOGY OF SYSTEMIC CANDIDIASIS IN MICE INDUCED BY

CANDIDA PSEUDOTROPICALIS

R. Hazıroğlu[1]   K.S.Diker[2]   M. Arda[2]

Faculty of Veterinary Medicine, Ankara University
Ankara 06110, Turkey
[1] Department of Pathology
[2] Department of Microbiology

INTRODUCTION

Candida spp. are opportunistic fungal pathogens of immunosuppressed people, and Candida albicans is the most commonly isolated species of this genus. Candida pseudotropicalis, also known as the imperfect (asporogenous) form of Kluyveromyces fragilis, has been generally isolated from genital candidiasis in man [1,2].

Experimental infections with Candida spp. have been produced in normal and immunosuppressed mice by several workers [3-7]. In experimental mice, C.albicans and C.tropicalis exhibit progressive infection and cause death of animals[3,5,6,8], whereas infections associated with C.pseudotropicalis have not been documented. Therefore, this study was conducted to describe pathological lesions in experimental infection induced by C.pseudotropicalis in immunosuppressed mice.

MATERIAL AND METHOD

Organism. A C. pseudotropicalis strain isolated from the liver of a cat with systemic candidiasis was used for experimental infection. It was grown on Sabouroud dextrose agar (SDA) at 37° C for 24 hours.

Animals. Swiss albino mice weighing 20-25 g were used as experimental animals. All mice were immunosuppresed by injecting a single dose of cyclophosphamide (200 mg/kg of body weight) two days before the experimental infection.

Animal inoculation. Inoculum was prepared by suspending Candida colonies in sterile saline at a concentration of 10^8 yeast cells per ml. Cell suspension was injected intraperitoneally (0.2 ml, n:5) and intravenously (0.1 ml, n:5) into mice.Six control mice were inoculated with only sterile saline. The infected animals were observed for ten days, and surviving ones were killed on the tenth postinoculation day.

Mycological and pathological examinations. All died and sacrificed

Candida and Candidamycosis, Edited by E. Tümbay et al.
Plenum Press, New York, 1991

237

mice were necropsied. The livers and spleens were cultured for
C.pseudotropicalis on Sabouraud dextrose agar (SDA). Tissue samples
were fixed in 10 % neutral buffered formalin. Samples were processed
by conventional methods. Sections were stained with haematoxylin and
eosin (HE), Gridley and periodic acid-Schiff (PAS).

RESULTS

At necropsy, remarkable lesions were observed in the liver. The
liver of normal size was studded with numerous discrete, firm,
necrotic lesions that were 1 to 5 mm in diameter. The lesions were
white-yellow in colour on the surface and in the parenchyma. The size
of the lesions varied according to inoculation route. In mice inoculated intravenously, the liver lesions were more diffuse and smaller
than mice inoculated intraperitoneally. No gross lesions were found
in other organs.

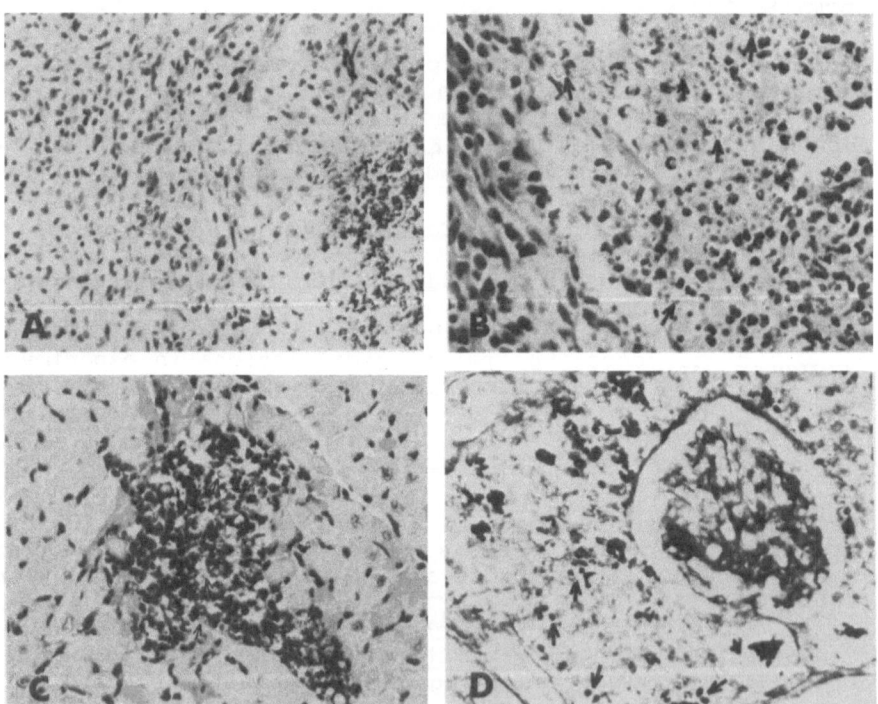

Fig. 1. A : Liver from affected mouse, with an area of necrosis
            and granulomatous response of neutrophils, lymphocytes,
            histiocytes and fibroblasts (IP inoculation), HEx240.
        B : Area of necrosis in the liver. Large number of yeasts
            (arrows) consistent with C.pseudotropicalis (IP
            inoculation), HEx380.
        C : Cellular reaction association with many yeasts (arrows)
            in the myocardium (IV inoculation), HEx240.
        D : Scattered yeasts (arrows) in the lesion of the kidney
            (IV inoculation), PASx380.

In mice inoculated intraperitoneally, the foci seen grossly in the liver were areas of necrosis with neutrophilic infiltration, sometimes involving only small areas and other times involving most or all of the lobules. There was no specific localisaton of the necrotic foci within the lobule. Most lesions were surrounded by mononuclear cells, multinucleated giant cells and fibroblast (Fig. 1A). Some lesions had only large necrotic area with neutrophilic infiltration and there was thin mononuclear cell zone around the lesion. Similar microscopic lesion were also present in the spleen, heart, mesenterial lymph nodes and serosa of the intestine.

In mice inoculated intravenously, however, mononuclear cell reaction with or without neutrophilic infiltration was predominant in the lesions of the liver. Hepatocytes around the lesions had vacuolar degeneration. Multiple focal areas of necrosis of varying size with neutrophilic infiltration were seen in the kidneys, heart (Fig. 1C), spleen, adrenals, mesenterial lymph nodes and intestine (Fig. 2A, B). In the cerebral cortex, medulla oblongata and thalamic regions, there were small granulomas, characterized by marked leucocytic infiltration with neutrophils predominating (Fig. 2C, D).

Yeast forms of C.pseudotropicalis were present in almost all lesions of affected organs (Fig. 1B, C and 2B, C). The same form were encountered within the blood vessels of the myocardium in both intravenous and intraperitoneal routes.Generally, fungal elements were clearly observed in the early lesions of affected organs, but not, or very few observed in the late lesions. In sections stained with PAS method, numerous yeasts were demonstrated in and about granulomas and also in areas of necrosis (Fig. 1D and 2D). C.pseudotropicalis was reisolated from the liver and spleen of all infected mice, but none of the controls.There were no pathological lesions in control animals.

DISCUSSION

Candida pseudotropicalis is a yeast infrequently isolated from systemic infection of humans and animals. Only a few reports of infection due to this organism could be found in the literature [2,8].

Mice have been used to evaluate the pathogenic potential of Candida spp. in several studies, and C.albicans is frequently seen mycelial form in affected tissues[3-7]. It has been reported that the pathogenic properties of C.albicans connected with the formation of pseudomycelia in vivo differ from other species of Candida genus that retain yeast-like morphology[3,6]. In our study, fungal cells observed in tissues were also in yeast form and were accompanied by a marked neutrophilic response. However, Holzschu et al[4] have not observed inflammatory responce in cortisone-treated mice infected with C.pseudotropicalis and Kluyveromyces fragilis from industrial origin. As comparing the pathogenicity of our strain and those of Holzschu et al.'s[4] it has been suggested that the source of organism might be effective in the development of the lesions,since our strain was from animal clinical source.

As conclusion, C.pseudotropicalis is able to produce marked lesions in immunosuppressed mice.

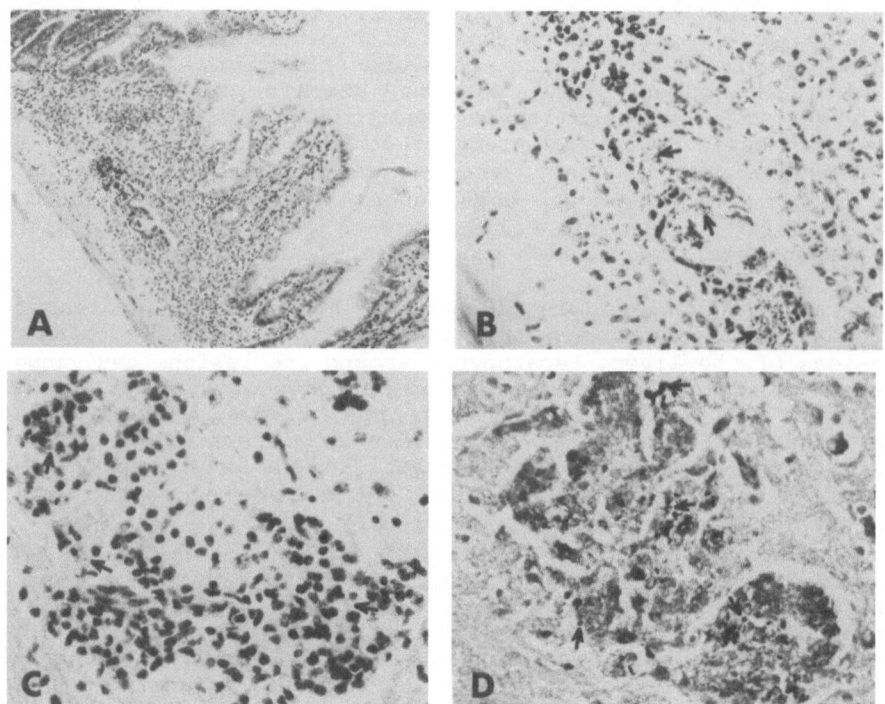

Fig. 2. A : Inflammatory foci involved almost all layers of the
          intestine (IV inoculation), HEx95.
       B : Higher magnification of A showing cellular infiltra-
          tion and yeasts (arrows) (IV inoculation), HEx240.
       C : Cellular response with yeasts (arrows) around a vessel
          of the brain (IV inoculation), HEx380.
       D : Same area of C showing large number of yeasts (arrows)
          (IV inoculation), PASx380.

REFERENCES

1. E. Tümbay (Ed), "Candida ve İnfeksiyonları", Türk Mikrobiyoloji
   Cemiyeti Yayınları, No : 6, Bilgehan Basımevi, İzmir (1986).
2. N.Van Uden and H.Buckley, Candida. Berkhout, In : "The Yeasts - A
   Taxonomic Study", J. Lodder, ed., pp. 893-1087, North-Holland
   Publishing Company, Amsterdam (1970).
3. D.W. Hill and L.P. Gebhardt,Morphological transformation of Candida
   albicans in tissues of mice, Proc.Soc.Exptl.Biol.Med.,92 : 640
   (1956).
4. D.L. Holzschu, F.W. Chandler, L. Ajello and D.G.Ahearn,Evaluation
   of industrial yeasts for pathogenicity, Sabouraudia, 17 : 71
   (1979) .
5. D.B. Louria, R.G. Brayton and G. Finkel, Studies on the pathogenesis of experimental Candida albicans infections in mice,
   Sabouraudia, 2 : 271 (1963).
6. N. Simonetti and V. Strippoli, Pathogenicity of the Y form as
   compared to M form in experimentally induced Candida albicans
   infections, Mycopath. Mycol. Appl., 51 : 19 (1973).
7. B. Winblad, Experimental renal candidiasis in mice and quinea
   pigs, Acta Pathol. Microbiol. Scand.Sect. A, 83 : 406 (1975).
8. L.I. Lutwick, H.J. Phaff and D.A. Stevens, Kluyveromyces fragilis
   as an opportunistic fungal pathogen in man, Sabouraudia, 18 : 69
   (1980).

IN VITRO ACTIVITIES OF NINE ANTIFUNGAL AGENTS ON CANDIDA SPECIES

ISOLATED AS CAUSATIVE AGENTS FROM CLINICAL MATERIALS

C.B.Johansson[1] S.Bilgin[2] J.Taşçıoğlu[2] G.Söyletir[1] B.Ener[1]

[1]Department of Microbiology, Faculty of Medicine, Marmara
University, Haydarpaşa, Istanbul, Turkey
[2]Microbiology Laboratory, Haydarpaşa Numune Hospital,
Haydarpaşa, Istanbul, Turkey

INTRODUCTION

The incidence of mycotic infection due to yeasts has increased
during the last few years because of the immunosuppressive therapies,
prolonged antibiotic therapy, organ transplantations, aggressive
instrumentations in medicine, AIDS, etc.

During the past several years many antifungal agents have been
produced and evaluated for use in the therapy of fungal infections.
Unfortunately resistant strains are reported by different workers
from time to time [1-4], and for this reason the clinical laboratory is
now gaining an important role in the selection and monitoring of
antifungal chemotherapy.

This paper presents the in vitro antifungal activities of nine
agents against 47 clinical isolates of Candida species.

MATERIAL AND METHOD

We studied the susceptibility of 47 strains of Candida spp.
isolated in Haydarpaşa Numune Hospital and Marmara University
Hospital from January 1 to April 1, 1989.

The yeasts were isolated from various clinical specimens such as
blood, gastric aspirate, bile, throat swabs, peritoneal dialysis
fluids, vaginal swabs, sputum and urine.

The identification of the isolates was based on classical
methods : The yeast's ability of germ-tube production; chlamydospore
production in bile agar; carbohydrate assimilation with glucose,
galactose, sucrose, xylose, maltose, trehalose, raffinose, cellobiose,
inositole; fermentation characteristics in glucose, maltose, saccharose
and lactose; urease activity and resistance to actidione.

To support good growth of the yeasts and not to interfere with
the action of the antifungal agent being tested and to obtain

Candida and Candidamycosis, Edited by E. Tümbay *et al.*
Plenum Press, New York, 1991

reproducible test data, we used buffered Yeast Nitrogen Base Broth supplemented with glucose and asparagine.

To prepare the inoculum, 24-48h yeast culture in Sabouraud dextrose agar was transferred into sterile distilled water. In a spectrophotometer set at 530nm the inoculum suspension was adjusted to 95-98% transmission. After proper dilution the final concentration of approximately $5X10^4$ CFU/ml was obtained.

The antimycotics were obtained from their manufacturers : The imidazole derivatives clotrimazole, isoconazole, ketoconazole, miconazole, oxiconazole, tioconazole; the polyene antibiotics amphotericin B, nystatin and pimaricin. Solutions of antimycotics were prepared in dimethylsulfoxide and a stock solution of 10000µg/ml was achieved. A working solution containing 1000µg/ml was prepared and stored in the refrigerator for approximately one week.

For MIC determinations macrodilution method in broth was used. Using the working solution twofold dilutions were made and final concentrations were ranging from 250µg/ml to 0.01µg/ml. Yeast solutions were added to the tubes. Controls were added to each series. Results were read after 24 and 48 hours of incubation at 37° C. For MIC determinations the last drug dilution showing no growth is the lowest concentration of the drug which inhibits the growth. The MIC was the last dilution which killed the yeast. The concentration at which no growth occured upon subculture was considered to be the MIC.

RESULTS

Table 1 lists the distribution of Candida strains among clinical specimens. From sputum C. albicans (25.5%); from urine samples C. albicans, C. tropicalis, C. guilliermondii and C. krusei (34.04%); from vaginal swabs C. albicans and C. tropicalis (19.1%); from peritoneal fluids C. albicans and C. tropicalis (6.4%); from throat swabs C. albicans (6.4%); from bile C. albicans (4.3%); from gastric aspirates and blood C. tropicalis (2.1% from each) were isolated.

Table 1. Distribution of yeasts isolated from clinical specimens.

Candida spp.	Number of Strains	Specimen	Total % of Isolates
Candida albicans	12	Sputum	25.5
Candida albicans	12	Urine	34.04
Candida tropicalis	2	"	
Candida guilliermondii	1	"	
Candida krusei	1	"	
Candida albicans	7	Vaginal swabs	19.1
Candida tropicalis	2	"	
Candida albicans	2	Peritoneal fluids	6.4
Candida tropicalis	1	"	
Candida albicans	3	Throat swabs	6.4
Candida albicans	2	Bile	4.3
Candida tropicalis	1	Gastric aspirates	2.1
Candida tropicalis	1	Blood	2.1

Table 2. *In vitro* antifungal activities of nine antifungal agents against 47 *Candida* species - MIC ranges.

Antifungal agent	C.albicans (38 strains)	Candida spp.* (9 strains)	Candida (Total) (47 strains)
Amphotericin B	<0.01-1.95	<0.01-1.95	<0.01-1.95
Clotrimazole	<0.01-31.25	0.12-15.6	<0.01-31.25
Isoconazole	0.24->100	0.48->100	0.24->100
Ketoconazole	0.12-62.5	0.12->100	0.12->100
Miconazole	0.02->100	<0.01->100	<0.01->100
Oxiconazole	0.12->100	0.12-31.25	0.12->100
Tioconazole	<0.01-31.25	0.12-7.8	<0.01-31.25
Nystatin	0.24-15.6	0.12-0.39	0.12-15.6
Pimaricin	0.97->100	0.24->100	0.24->100

* 7 *C.tropicalis*, 1 *C.guilliermondii*, 1 *C.krusei*,

The results of in vitro susceptibility of 47 *Candida* strains to nine antifungals are given in Table 2. The degree of activity varied over a wide range with most of the agents. For amphotericin B, clotrimazole and tioconazole the lowest MIC values were less than $0.01\mu g/ml$, and highest were more than $100\mu g/ml$ for pimaricin, isoconazole, miconazole and oxiconazole. Clotrimazole and tioconazole showed similar activities. MIC's on *C. albicans* with nystatin varied between $0.24\mu g/ml$ and $15.6\mu g/ml$.

Against other *Candida* species the best effect was observed with amphotericin B, nystatin, tioconazole and clotrimazole, Resistant strains were observed against isoconazole, ketoconazole, miconazole, oxiconazole, and pimaricin.

Table 3 shows the MIC_{50}, MIC_{90} and MLC_{50}, MLC_{90} values. Amphotericin B showed considerable activity on all *Candida* species, and it was followed by nystatin, clotrimazole, pimaricin, ketoconazole and others. Killing effects were best achieved with amphotericin B, nystatin and tioconazole. MLC_{50} values were almost similar with pimaricin, clotrimazole, ketoconazole, and miconazole. We read high MLC90's with most of the agents.

DISCUSSION

Candida spp. are normal inhabitants of respiratory, intestinal and female genital tracts and skin in man. *Candida*mycosis is usually caused by *C.albicans* and occurs predominantly in patients who have diabetes mellitus, leukemia, lymphoma or immunologic deficiency or who require treatment with antibacterial agents, corticosteroids or immunosuppressive drugs. Urinary catheters may also lead to candidal colonization; candiduria caused by *C.albicans* and *C.tropicalis* is common in hospitalized patients who have indwelling catheters and who are receiving antibacterial therapy [5,6]. In our study, *Candida* spp. were mostly isolated from sputum and urine in percentages 34.04 and 25.5, respectively (Table 1). The most frequently isolated species was *C. albicans* with a percentage of 80.9, *C.tropicalis* being the second species in rank (Table 1).

Table 3. MIC$_{(50-90)}$ and MLC$_{(50-90)}$ values of <u>Candida</u> strains against antifungal agents.

Antifungal agent	MIC$_{50}$	MIC$_{90}$	MLC$_{50}$	MLC$_{90}$
Amphotericin B				
C.albicans	0.24	0.97	0.78	7.8
Other Candida spp.*	0.48	1.95	1.56	31.25
Total	0.24	0.97	0.97	7.8
Clotrimazole				
C.albicans	1.95	31.25	31.25	>100
Other Candida spp.	3.9	15.6	62.5	>100
Total	1.95	31.25	31.25	>100
Isoconazole				
C.albicans	31.25	>100	>100	>100
Other Candida spp.	1.95	>100	>100	>100
Total	7.8	>100	>100	>100
Ketoconazole				
C.albicans	3.9	31.25	31.25	>100
Other Candida spp.	0.48	>100	1.95	>100
Total	3.9	31.25	31.25	>100
Miconazole				
C.albicans	3.9	31.25	15.6	>100
Other Candida spp.	7.8	>100	25	>100
Total	7.8	31.25	15.6	>100
Oxiconazole				
C.albicans	7.8	>100	>100	>100
Other Candida spp.	1.95	31.25	15.6	>100
Total	3.9	62.5	62.5	>100
Tioconazole				
C.albicans	3.9	15.6	15.6	31.25
Other Candida spp.	1.95	7.8	15.6	31.25
Total	3.9	12.5	15.6	31.25
Nystatin				
C.albicans	1.95	15.6	3.9	15.6
Other Candida spp.	0.97	3.9	3.9	15.6
Total	1.95	7.8	3.9	15.6
Pimaricin				
C.albicans	3.9	31.25	15.6	>100
Other Candida spp.	3.9	15.6	31.25	>100
Total	3.9	15.6	15.6	>100

* 7 <u>C.tropicalis</u>, 1 <u>C.guilliermondii</u>, 1 <u>C.krusei</u>

<u>Candida</u>mycosis was once rare, but now it is responsible for an increasing number of infections in patients where the normal microenvironment is changed or where the host's normal defence mechanisms are impaired. Under these conditions these usually harmless fungi can cause life-threatening systemic infections. The major problem in treating these infections is that both fungus and host are eukaryotic organisms, and the number of potential selective targets for antifungal drugs is limited. Therefore, <u>in vitro</u> susceptibility testing of antifungal agents is becoming increasingly important because of the introduction of new antifungal agents and the recovery of clinical

isolates that exhibit innate or have developed resistance during therapy [7,8]. However, the interpretation of _in vitro_ susceptibility data is difficult because of the lack of standardized testing methods [8,9]. As shown in Table 2, amphotericin B has been the most effective drug and pimaricin the least among polyene antifungals. In respect to azole compounds, clotrimazole, tioconazole had lower MIC ranges (<0.01-32.25 µg/ml) than others. MIC values ranged between 0.12->100 and 0.01->100 in the case of ketoconazole and miconazole, respectively. There are important differences among results reported by several investigators in the literature [8,10-12].

Although we cannot interprete that one drug is superior to another used in this study, we conclude that these findings may be useful data in determining susceptibility ranges of _Candida_ spp. to several antifungals.

REFERENCES

1. J.R. Graybill, J.N. Galgiani, J.M. Jorgensen and D.A.Strandberg, Ketoconazole therapy for fungal urinary tract infections, _J.Urol._, 129 : 68 (1983).
2. R. Guinet, J. Chanas, A. Coullier, G. Bonnefoy and P. Ambroise-Thomas, Fatal septisemia due to amphotericin B resistant _Candida lusitaniae_, _J. Clin. Microbiol._, 18 : 443 (1983).
3. R.J. Holt and A. Azmi, Miconazole resistant _Candida_, _Lancet_, i : 50 (1978).
4. E.M. Johnson, M.D. Richard and D.N. Warnock, In vitro resistance to imidazole antifungals in _Candida albicans_, _J. Antimicrob. Chemother._, 13 : 547 (1984).
5. R.F. Boyd and B.G. Hoerl, The pathogenic fungi, _in_ : "Basic Medical Microbiology", R.F. Boyd, ed., pp. 775-813; Little, Brown and Company; Boston (1986).
6. P.D. Hoeprich and M.G. Rinaldi, Candidosis, in "Infectious Diseases", P.D. Hoeprich, ed., pp 436-450, Harper and Row Publisher, Philadelphia (1983).
7. D. Kerridge and R.D. Nicholas, Drug resistance in the opportunistic pathogens _Candida albicans_ and _Candida glabrata_, _J.Antimicrob. Chemother._, Supp. B, 18 : 39 (1986).
8. M.R. Mc Ginnis and M.G. Rinaldi, Antifungal drugs : Mechanisms of action, drug resistance, susceptibility testing and assays of activity in biological fluids, _in_: "Antibiotics in Laboratory Medicine", Second Edition, V. Lorian, ed., Williams and Wilkins, Baltimore (1986)
9. J.N. Galgiani, J. Reiser et al., Comparison of relative susceptibilities of _Candida_ species to three antifungal agents as determined by unstandardized methods, _Antimicrob. Agents Chemother._, 31 : 1343 (1987).
10. S.P. Clissold and J.V. Cutsem, Overview of the antimicrobial activity of ketoconazole, _in_ : "Ketoconazole Today", H.E. Jones, ed., pp. 1-18, ADIS Press Limited, Manchester (1987).
11. D.J. Drutz, Newer antifungal agents and their use, including an update on amphotericin B and flucytosine, _in_ : "Current Clinical Topics in Infectious Disease", J.S. Remington and M.N. Swartz, eds., pp 97-135, Mc Graw-Hill Book Company, New York (1982).
12. S. Shadomy and A. Espinel-Ingroff and R.Y. Cartwright, Laboratory studies with antifungal agents : Susceptibility test and bioassays, _in_ : "Manual of Clinical Microbiology", E.H. Lennette, ed., pp. 991-999, American Society for Microbiology, Washington D.C. (1985).

THE EFFECTS OF VARIOUS DISINFECTANTS AND ANTISEPTICS

ON CANDIDA ALBICANS STRAINS

Ayşe YÜCE

Department of Microbiology, Faculty of Medicine
Dokuz Eylül University, Inciraltı, Izmir, Turkey

INTRODUCTION

Due to the extensive administration of antibiotics, corticosteroids, various immunosuppressive and antitumoral agents, the normal body flora is altered and thus susceptibility to infections is enhanced. Many bacteria and particularly opportunistic fungi are the major causative agents in such infections. Among fungi; yeasts such as Candida albicans and other Candida species, Candida (Torulopsis) glabrata, Cryptococcus neoformans, Geotrichum and Rhodotorula and molds such as Aspergillus and Mucor species are the most often detected agents [1,5].

The aim of this study is to evaluate the susceptibility of Candida albicans - the most frequently recovered fungus in nosocomial infections - to various disinfectants and antiseptics.

MATERIAL AND METHOD

Test Strains

1) A local C.albicans strain isolated from an intensive care unit patient
2) C.albicans ATCC 10231

Disinfectants

	Trade Name	Abbreviation
1) 6 % formaldehyde + 1.8 % glutaraldehyde	(Lisoformin)[R]	For.Gl
2) Sodium dicholoroisocyanurate	(Presept)[R]	NaDCC
3) 2 % glutaraldehyde	(Cidex)[R]	G l
4) 15 % cetrimide + 1.5 % chlorhexidine	(Savlon)[R]	Cet.Chlx
5) Benzalkonium chloride	(Zefir-Sol)[R]	Benz.Ch
6) 4 % chlorhexidine	(Hibiscrub)[R]	Chlx
7) 10 % povidone - iodine	(Poviod)[R]	Pov
8) 70 % alcohol	(home made)	
9) 2 % phenol in water	(home made)	

Candida and Candidamycosis, Edited by E. Tümbay *et al.*
Plenum Press, New York, 1991

The disinfectants and antiseptics used were diluted to desired working concentrations with water of standard hardness (WSH), and susceptibility of C.albicans strains to them were evaluated by the "cotton germ carrier" method proposed by German Society for Hygiene and Microbiology [6].

Preparation of C.albicans suspension

C.albicans was grown on Saburaud dextrose agar (SDA) slants at 37° C for 72 hours and harvested with Sabouraud broth (SB). The suspension was adjusted to $10^7 - 10^8$ cells/ml. Thirty ml of this suspension were poured on previously sterilized cotton fabric squares of approximately $1cm^2$ and incubated at $37^{O}C$ for 15 minutes. A number of contaminated germ carriers which were wet and corresponded to the scheduled testing periods -15 , 60, 120 minutes- were placed in a petri dish and 10 ml of the disinfectant solution (made with WSH) to be tested poured on. Germ carriers were taken out at the end of the exposure times and rinsed twice in 10 ml of SB containing Tween 80 + 3 % saponin + 0.1 % histidin + 0.1 % cystein as neutralising agents. Germ carriers were then placed on SDA slants, smeared briefly by moving cotton squares to and for with a loop, and finally one of the squares was placed on the surface of the culture medium, just above the level of the condensed water and the other one in the middle of the agar surface.

The same procedure was applied to the control test where squares of cotton cloth were kept in WSH for 120 minutes instead of in a disinfectant solution.

The cultures were incubated at $37^{O}C$ for 7 days. The results were recorded as follows:

(-)   : no growth
(+)   : growth of isolated colonies
(++) : strong growth (more than 10 colonies)

RESULTS

None of the disinfectants and antiseptics used in this study had an effect on C.albicans in 15 and 60 minutes. However; Savlon, Hibiscrub, Zefir-Sol, Lisoformin and Cidex definitely inhibited the growth in 120 minutes (Table 1).

DISCUSSION

The fungi as opportunistic microorganisms cause serious nosocomial infections either by interfering with the host's defense mechanisms or as a result of theurapeutic and diagnostic invasive attempts[12,13,15,21]. Fungal infections are increasingly being observed in AIDS patients and in those fed parenterally for a long time [7,14]. Obviously, sterilization and disinfection play an important role in the prevention of nosocomial infections.

Among the widely used antiseptics; cetrimide, chlorhexidine, and benzalkonium chloride are known to be effective on various bacteria, fungi and some viruses [15,17]. In our study, Savlon (Cet.Chlx), Hibiscrub (Chlx) and Zefir-Sol (Benz.Ch) were found to be definitely effective at 120 minutes on both C.albicans strains, whereas Poviod (Pov), 70 % alcohol and 2 % phenol were ineffective.

Table 1. The effects of disinfectants and antiseptics on
C.albicans strains.

Products	Candida Strains					
	Local strains			ATCC 10231		
	15min	60min	120min	15min	60min	120min
Antiseptics						
Cet Chlx 1/100 (Savlon)	++	++	−	++	++	−
Benz Ch 1/100 (Zefir Sol)	++	++	−	++	++	−
Chlx pure (Hibiscrub)	++	++	−	++	+	−
Pov pure (Poviod)	++	++	+	++	++	+
Alcohol 70% (home made)	++	++	++	++	++	++
Phenol in water 2% (home made)	++	++	++	++	++	++
Disinfectants						
For GL 1/100 (Lisoformin)	++	+	−	++	+	−
GL 2/100 (Cidex)	++	++	−	++	++	++
NaDCC (Presept)	++	++	+	++	++	+

(−) : No growth, (+) : Growth of isolated colonies,
(++) : Strong growth (more than 10 colonies)

The disinfectants of aldehyde group are known to be effective
not only on bacteria, spores and viruses but also on mycobacteria and
fungi. They do not interfere with the organic substances and have a
long lasting action due to their high diffusing capacity [5,15,16,18]. Of
the tested disinfectants, Lisoformin (For-GL) and Cidex (Gl) were
found to be definitely effective on both C.albicans strains in 120
minutes, but, however, Presept (NaDCC) was only partially effective.
As it is known, hypochlorides are effective to some extent on
bacteria and fungi, but they lose most of their activity in environ-
ments rich in alkaline and organic substances [16,18].

Carrera et al.[19] proposed Savlon (Cet-Chlx), Hibiscrub (Chlx) and
Armil (10 % Benz.Ch) as antiseptics and formalin and Cidex (Gl) as
disinfectats of first choice against Candida species.

Unat et al.[20] reported that Zefiran (Benz.Ch), Savlon (Cet.Chlx),
Batticon (10 % Pov. iodine) and 95 % alcohol were effective in 1-5
minutes and 1 % phenol at 60 minutes against Candida species. These
results are not in accordance with our findings in which a growth at
15 and 60 minutes was observed with all the disinfectants and
antiseptics tested. This may be due to the testing of different
strains.

Consequently, an effective disinfection procedure and the choice
of appropriate preparation(s), which are particularly important for

the prevention of nosocomial infections, should depend on the tests to be performed periodically with numerous local strains to reveal the local susceptibility patterns. Regarding this point of view, it should be stressed that each hospital has to have its own disinfection policy.

REFERENCES

1. J.S.Hayes, B.M. Soule, and M.T. La Rocco, Nosocomial infections: An overview,in:"Clinical and Pathogenic Microbiology", D. Carson and S.Bircher,eds.,p.67, The C.V. Mosby Company, St.Louis (1987).
2. K.Kılıçturgay, O.Töre, Hastane infeksiyonları ve opportünist mikroorganizmalar, J. KÜKEM 8: 9 (1985).
3. S.Koşay, ed.,: "Hastane İnfeksiyonları", Bilgehan Matbaacılık, Izmir (1981).
4. K.Töreci, N.Gürler,: Hastane infeksiyonlarından izole edilen mikroorgangzmalar, J.KÜKEM, 8: 44 (1985).
5. R.P.Wenzel, Nosocomial infections, in: "Principles and Practice of Infectious Diseases", G.L.Mandell, R.G.Douglas and J.E.Bennett, eds., p. 2213, John Wiley and Sons, New York (1979).
6. German Society for Hygiene and Microbiology, "Directives for the Testing and Evaluation of Chemical Disinfecting Procedures, Part I (State of 1.1.1981).
7. K.Akiyama, H.Takizawa, M.Suzuki, S.Miyachi, M.Ichinohe and Y.Yanagihara, Allergic bronchopulmonary aspergillosis due to Aspergillus oryzae, Chest, 91: 285 (1987).
8. E.W.Benbow, R.E.Banshek and R.W.Stoddart, Endobronchial zygomycosis, Thorax, 42: 553 (1987).
9. R.A.Calderone, R.L.Cihlar, D.Lee, D.S.K.Hoberg and W.M.Scheld,: Yeast adhesion in the pathogenesis of endocarditis due to Candida albicans studies with adherence-negative mutants, J.Infect.Dis., 152: 710 (1985).
10. R.M.Hoffman, Chronic endobronchial mucormycosis, Chest, 91: 469 (1987).
11. M.C.Kearon, J.T.Power, A.E.Wood and L.J.Clancy, Pleural aspergillosis in a 14 year old boy, Thorax, 42: 477 (1987).
12. B.R.C.O'Driscoll, R.D.P. Cooke, H.Mamtora, M.H.Irving and A.Bernstein, Candida lung abscesses complicating parenteral nutrition, Thorax, 43: 418 (1988).
13. R.M.Petrak, J.L.Pottage and S.Levin, Invasive external otitis caused by Aspergillus fumigatus in an immunocompromised patient, J.Infect.Dis., 151: 196 (1985).
14. L.Wasser and V.W.Kala, Pulmonary cryptococcosis in AIDS, Chest, 92: 692 (1987).
15. J.E.Clarridge, Disinfection and sterilization, in: "Clinical and Pathogenic Microbiology", D.Carson and S. Bircher, eds., p.83, The C.V.Mosby Company, St.Louis (1987).
16. D.A. Coates, The use of disinfectants in hospitals, in: "Mediterranean Conference on General Surgery, 18 June 1987, Istanbul (Monograph).
17. J.E.Sebben, Avoiding infection in office surgery, J.Dermatol. Surg. Oncol. (USA), 816: 455 (1982).
18. İ.Erefe, "Hastane İnfeksiyonlarıyla Savaş İlkeleri ve Hemşirelik Uygulamaları", Ege Univ. Yük.Hem.Okulu Yayın No: 1, Ege Üniversitesi Matbaası, Izmir (1983).
19. R.H.Carrera et al, Comparative study of the efficacy of a number of antiseptics and disinfectants, Rev.Diag.Biol., 29: 334 (1980).
20. E.K.Unat, A.Yücel, M.Mamal, B.Çokneşeli, Ç.Çetinkale, N.Akgül, Ticaretle bulunan bazı dezenfeksiyon maddeleri üzerine bir araştırma, Türk Mikrobiol. Cem.Derg., 15: 88 (1985).

THE SENSITIVITY OF CANDIDA SPECIES ISOLATED FROM CHILDREN

WITH DIARRHEA TO ANTIFUNGAL COMPOUNDS

G. Hasçelik   Y. Akyön   N. Yuluğ

Department of Microbiology, Faculty of Medicine
Hacettepe University, Ankara, Turkey

INTRODUCTION

Candida species are opportunistic microorganisms which are members of the normal flora of the mucous membranes in the respiratory, gastrointestinal and female genital tracts. In the presence of suitable conditions, they may gain dominance and be associated with pathologic conditions [1]. In the past 30 years, there has been a significant increase in the number of infections caused by Candida species [2].

In this study, the causative Candida species in childhood diarrhea and their in vitro susceptibility to the antimycotic agents were investigated.

MATERIAL AND METHOD

Specimen were obtained from 150 children admitted to the Hacettepe University Children's Hospital between 29.9.1988-1.12.1988. All were outpatients of age 13 days - 13 years and had diarrhea for one week. Specimens were rectal swabs. The rectal swabs were cultured on Sabouraud dextrose agar (Difco) and also on EMB agar (Difco), Butzler's selective agar (Oxoid) and SS agar (Difco) for isolation of Candida, enteric bacteria like Salmonella, Shigella and Campylobacter. Only pure Candida cultures were included in the study. Candida species were identified by germ tube formation, sugar assimilation and fermentation tests and by chlamydospore formation on corn meal agar (Oxoid) [3,4].

Nystatin (Squibb), 5-fluorocytosine (Roche), amphotericin B (Sigma-Chemical Co) and ketoconazole (Janssen Pharmaceutica) were used in the antimycotic resistance tests.

For 5-fluorocytosine, nystatin, ketoconazole the tube dilution technique and for amphotericin B microdilution in microtiter plates were used. All tests were performed in Saboraud dextrose broth, pH 5.6 [2,4].

Dilutions were made as follows: for nystatin (1μgr was accepted equivalent to 3 units) 100, 50, 30, 20, 10, 1, 0.1 μgr/ml; for 5-fluorocytosine and ketoconazole 1000, 100, 10, 1, 0.1, 0.01, 0.001 μgr/ml; and for amphotericin B 20, 10, 5, 2.5, 1.25 μgr/ml.

Candida suspensions were prepared from overnight cultures which were incubated at 37°C. 10^5 cfu/ml was inoculated in Sabouraud dextrose broth which contained the antifungal agents. All isolates were tested in dublicate. The results were read after incubation at 30°C at the end of 24 hours [2,5].

RESULTS

In the 150 cultures examined, 27 (18 %) were pure Candida. Of these, 12 were C. albicans and 15 were C. tropicalis. The rate of isolation was highest under 2 years of age.

The MIC 50 and MIC 90 values for C.albicans and C. tropicalis are shown in Tables 1 and 2.

DISCUSSION

It is advised to treat candidosis according to drug suscepti-bility tests [4]. In this study, we tried to evaluate the in vitro effect of antifungals on Candida.

We have not encountered natural resistance to nystatin, amphotericin B, ketoconazole for both C. albicans and C. tropicalis. We have found primary resistance only against 5-fluorocytosine in two C. albicans strains (7.4 %). This result is similar to those found in some European studies [6,7]. The strains found resistant against 5-fluorocytosine were isolated from children who had never received any antifungal agent.

Table 1. The MIC values for C. albicans.

Compound	Range	MIC 50 (μgr/ml)	MIC 90	Resistance %
Nystatin	1-10	1	1	—
5-Fluorocytosine	0.001-1000	1	1000	7.4
Ketoconazole	0.001-1	0.1	1	—
Amphotericin B	0.62-1.25	0.625	1.25	—

Table 2. The MIC values for C. tropicalis.

Compound	Range	MIC 50 (μgr/ml)	MIC 90	Resistance %
Nystatin	1-10	1	10	—
5-Fluorocytosine	0.01-1	0.1	1	—
Ketoconazole	0.1-1	0.1	1	—
Amphotericin B	0.62-0.625	0.625	0.625	—

It is certain that sensitivity tests should be performed for candidal infections just like for bacterial infections, but MIC values vary greatly as the conditions are experimental. The type of medium, temperature, pH, duration of incubation, inoculum size, phase of fungal growth and presence or absence of serum may influence the results [8].

In recent years studies are being performed to standardize antimycotic sensitivity tests [9]. We hope that in a few years' time standard techniques will be developed for *in vitro* antimycotic sensitivity determinations.

REFERENCES

1. E. Jawetz, J.L. Melnick and E.A. Adelberg, Medical mycology, *in*: "Review of Medical Microbiology", 17th ed., pp. 318- 337, Lange Medical Publications, Los Altos (1987).
2. E.D. Spitzer, S.J. Travis and G.S. Kobayashi, Comparative in vitro activity of LY121019 and amphotericin B against clinical isolates of *Candida* species, *Eur. J. Clin. Microbiol. Infect. Dis.*, 7: 80 (1988).
3. K.N. Nugent and K.R. Couchet, Effects of sublethal concentrations of amphotericin B on *Candida albicans*, *J. Infect. Dis.*, 154: 665 (1986).
4. Y. Akgün and F. Akşit, Klinik olgulardan izole edilen Kandida'ların antimikotiklere duyarlılıkları, *Mikrobiyol. Bült.*, 15: 105 (1981).
5. M.R. Mc Ginnis and M.G. Rinaldi, Antifungal agents, *in*: "Antibiotics in Laboratory Medicine", 2nd ed., pp. 223-281 V.Lorian, ed., Williams and Wilkins, Baltimore (1986).
6. K.S. Defever, W.L. Whelan, A.L. Rogers, E.S. Beneke, J.M. Veselenak and D.R. Soll, *Candida albicans* resistance to 5-fluorocytosine frequency of partially resistant strains among clinical isolates, *Antimicrob.Agents Chemother.*, 22: 810 (1982).
7. R.J. Holt,Clinical problems with 5-fluorocytosine, *Mykosen*, 21: 363 (1978).
8. S.P. Clissold and J. Van Cutsem, Overview of the antimicrobial activity of ketoconazole, *in*: "Ketoconazole Today:A Review of Clinical Experience", H.E. Jones, ed., pp.5-12, ADIS Press Limited (1987).
9. J.P. Galgani et al., Comparison of relative susceptibilities of three antifungal agents as determined by unstandardized methods, *Antimicrob. Agents Chemother.*, 31: 1343 (1987).

EVALUATION OF HALOGENOPHORE EFFECTIVENESS ON STRAINS OF CANDIDA ALBICANS ISOLATED FROM MARE'S REPRODUCTIVE ORGANS

K. Bukowski[1]   A. K. Kłopotek[2]   G. Działa[2]

[1] Department of Microbiology, Institute of Applied Biology, Agricultural and Pedagogical University 08,110 Siedlice, Prusa 12, Poland
[2] Industrial Chemistry Research Institute, 01-793 Warsaw, Rydygiera 8, Poland

INTRODUCTION

Preparations containing active halogens make a group of agents applied for preserving hygiene and for overcoming infectious sources, considering their high microbiocidal activity.

The active halogens connect with constituents of living cell through coordinate bonds. Reactions of active halogens with nutritive elements and enzymes lead to repression processes of protein synthesis in microorganisms and damage of cytoplasmatic membrane [1,2].

The subject of our investigations were complex compounds of halogens with surface active agents. Complex compounds of halogens with surface active agents are obtained by treating saturated aqueous solutions of potassium trihalides with surface active agents, especially nonionic, in optimal temperatures $20°-40°C$ and atmospheric pressure, with molar ratio of substrates not exceeding 1:2. These compounds have the general formula

$$R-O-H \ldots X - 0 \ldots H-O-R,$$

where halogen X (iodine, bromine, or chlorine) is linked to the hydrogen of the hydroxyl group of surface active agent through a coordinate bond (hydrogen bridge type).

Complex compounds of halogens with surface active agents are the main components of halogenophore wash-disinfecting preparations. These preparations also contain aid agents like improving wash properties, pH correctors, corrosion inhibitors and foam regulators.

These compounds have good washing properties and strong fungicidal, bactericidal and virustatic activites [3,4].

The purpose of this work was evaluation of halogenophore effectiveness against strains of Candida albicans isolated from mare's reproductive organs.

Candida and Candidamycosis, Edited by E. Tümbay *et al.*
Plenum Press, New York, 1991

MATERIAL AND METHOD

Studies were carried out on 30 strains of <u>Candida albicans</u> isolated from infections in mare's reproductive organs. For investigation, the following breeding-grounds and liquids were used: physiological salt solution, broth bouillon, Sabouraud agar and fungiphil breeding ground.

For halogenophore inactivation, sodium thiosulphate in were a concentration of 0.5 % was used. Four types of preparations studied: Pollena Jod K, Pollena Jod Z, iodine-chlorine-Polchlor, Polchlor K, iodine-bromine-Bromeks SK, Bromeks A, bromine-Pollena Brom K, Pollena Brom Z.

The activity of above mentioned preparations against 18-hour culture strains of <u>Candida albicans</u> was estimated.

Details of metabolic descriptions are described in papers by Krzywicka and et al [5]. and Bukowski et al [3,6].

RESULTS AND DISCUSSION

Sick animals are a source of infection for the environment and for medical-veterinary and technical service. Halogenophores are assigned for disinfection of equipment and apparatus, rooms, and for washing and disinfection of hands, animals skin and their external organs. These preparations can be applied near medicines; and considering their low toxicity, lack of acridity and sensintivity they can be used in the environment of man and animals.

During microbiological investigations the influence of individual halogens (iodine, bromine) and synergistic system of halogens (chlorine+iodine, bromine+iodine) should be considered. This last system increases antimicrobial activity and at the same time lowers the concentrations of these elements in the environment.

Chemical characteristics of these preparations ara shown in Table 1.

Table 1. Chemical characteristics of halogenophore preparations.

Preparation	Halogen	Content of active halogen, %	pH of 1% aq.sol.
Pollena Jod K	iodine	1.9-2.2	1.9
Pollena Jod Z	iodine	2.0-2.4	8.4
Polchlor	chlorine+iodine	1.5-1.8	2.1
Polchlor K	chlorine+iodine	1.7-2.0	1.8
Bromeks SK	bromine+iodine	1.4-1.7	1.9
Bromeks A	bromine+iodine	0.5-0.8	8.5
Pollena Brom K	bromine	1.8-2.2	2.3
Pollena Brom Z	bromine	0.4-0.8	8.5

The results of halogenophore effectiveness on strains of <u>Candida albicans</u> are presented in Table 2.

Table 2. Action of halogenophore preparations on strains of <u>Candida albicans</u>.

Preparation	Fungicidal activity of preparations against evaluated strains, wt%
Pollena Jod K	0.9-1.1
Pollena Jod Z	2.0-3.0
Polchlor	0.65-0.75
Polchlor K	0.2-0.5
Bromeks SK	0.1-0.44
Bromeks A	0.3-0.63
Pollena Brom K	0.1-1.5
Pollena Brom Z	2.0-4.0

The obtained results of activity of halogenophore preparations, which were used for evaluation, indicate their different actions. <u>Candida albicans</u> strains show different susceptibility to the tested halogenophores. The most active are: Polchlor K, which contains chlorine+iodine; Bromeks SK, whick contains bromine+iodine; and Pollena Jod K, which contains iodine. All of them are strong acid liquid preparations.

REFERENCES

1. W. Kedzia, "Dezynfekcja w medycynie i farmacji", PZWL, Warszawa (1981).
2. S. S. Block, "Disinfection, Sterilisation and Preservation", Lea and Febiger, Philadelphia (1977).
3. K. Bukowski, G. Dziala, A. Klopotek, <u>Medycyna Wet</u>., 41:613, (1985).
4. A.Klopotek, Zesz. Nauk. SGGW-AR Warszawa, Rozprawy Naukowe (1978).
5. H. Krzywicka, A. Bielicka, J. Janowska, E. Jaszszczuk, B. Tadeusiak, Metody badania aktywnosci bakteriobojczej preparatow dezynfekcyjnych, <u>Wyd. Met. PZH</u>, Warszawa 33 (1981).
6. K. Bukowski, A. Konarzewski, <u>Medycyna Wet</u>., 36:61 (1980).

THE INFLUENCE OF NEWLY INTRODUCED COMPOUNDS

AND PREPARATIONS ON <u>CANDIDA ALBICANS</u>

K. Bukowski[1]   A. Kłopotek[2]   G. Dziafa[2]

[1] Department of Microbiology, Institute of Applied Biology, Agricultural and Pedagogical University, 08-110 Siedlce, Prusa 12, Poland
[2] Industrial Chemical Research Institute, 01-793 Warsaw, Rydygiera 8, Poland

INTRODUCTION

The process of antimicrobial action depends on the following factors: kind and properties of chemical substances, their concentrations in the medium, temperature, biological and mineral factors, and mutable sensibility of microorganisms on which the antimicrobial agent acts [1,8].

Continuous searh for new compounds and preparations is dictated by the need of preparations which are more and more perfect in their antimicrobial activity, but harmless for man and animals.

The purpose of this work is the evaluation of the effect of new compounds and preparations on <u>Candida albicans</u> strains, standard and isolated from man and animals infected.

MATERIAL AND METHOD

The study was carried out on strains of <u>Candida albicans</u> (standard strain <u>C. albicans</u> 887/77 and 58 strains isolated from cases of human or animal mycoses). Breeding-ground, Sabouraud bouillon and agar, physiological salt solution and Fungiphil breeding-ground (Biomerieux) were used as media.

The antifungal activities of various disinfecting compounds like : halogen and oxygen compounds, cationic, ampholytic and anionic surface active compounds, aldehydes, alcohols, esters and preparations with these compounds were studied.

The effect of above mentioned compounds and peraparations on 18-hour cultures of <u>Candida albicans</u> strains were investigated. A formerly described method was used.

Susceptibility of <u>C. albicans</u> strains to the compounds and preparations was defined according to the following classification:

Candida and Candidamycosis, Edited by E. Tümbay *et al.*
Plenum Press, New York, 1991

Mean fungicidal concentration	Group of activity
0.0-1.0%	active
1.0-2.5%	medium-active
>2.5%	weakly-active

RESULTS AND DISCUSSION

The results of antifungal activity of the examined compounds and preparations are shown in Table 1.

The obtained results of activity of compounds and preparations show different fungicidal action. The susceptibility degrees of the strains are various. Differences in susceptibility depend upon the chemical character of each substance tested (Table 2).

Table 1. Action of newly indroduced compounds and preparations on strains of <u>Candida albicans</u>.

Chemical substance	Fungicidal concentration of substance, wt %		Mean fungicidal concentration, wt %
	Standard strain 887/77	Strains isolated from man and animals	
Liquid chlorinated phosphate (2.28% Cl_a)	2.0	1.6-2.75	2.12
KJ_3	0.005	0.001-0.008	0.005
Sansed I	3.0	1.5-4.8	3.10
Sansed II	3.0	1.1-3.75	2.62
Urea peroxide	3.75	1.8-3.85	3.13
Preparation HN	3.25	3.75-5.0	4.00
Kamisol RC	0.1	0.008-2.75	1.01
Kaminox R2RM	0.1	0.3-3.25	1.22
Rokamin S8	0.15	0.4-1.75	0.77
Condensate of glycerin chloroacetate with tertiary amine	0.1	1.8-4.2	2.03
Liquid with antistatic and disinfecting properties	0.5	0.2-1.2	0.63
Wash-disinfecting preparation with acetic betaine (A)	0.01	0.04-0.75	0.27
Wash-disinfecting preparation with acetic betaine (B)	0.01	0.08-0.95	0.35
Malester X-12	0.1	0.3-2.8	1.07
Wash-disinfecting preparation with α-olefines	0.01	0.02-0.8	0.28
Glutaraldehyde	0.5	0.75-2.1	1.17
Benzyl alcohol	0.92	0.75-1.25	0.97
Nipagina M + Nipagina P	0.02	0.01-1.75	0.59

Table 2. Antifungal activity of newly introduced compounds and preparations depending on their chemical character.

Chemical character of disinfectant	Group of activity
Halogen (iodine)	active
Halogen (chlorine)	medium-active
Oxygen	weakly-active
Cationic surface active compound	medium-active
Amphoteric surface active compound	medium-active
Anionic surface active compound	medium-active
Aldehyde	medium-active
Aromatic alcohol	active
Aromatic esters	active
Mixtures of various disinfectants	active

Among examined compounds and preparations the most active are iodine (KJ_3), benzyl alcohol, aromatic esters (Nipagina M+Nipagina P) and mixture of various disinfectants (preparations with acetic betaine and $\propto$-olefines). The medium-active are chlorine and glutaraldehyde. Compounds with active oxygen have the weakest antifungal activity.

REFERENCES

1. K. Bukowski, A. Kłopotek, G. Działa, Evaluation of halogenophore effectiveness on strains of Candida albicans isolated from mare's reproductive organs, in: "FEMS Symposium on Candida and Candidamycosis", 24-28.04.1989, Alanya, Turkey-Abstracts", Bilgehan Publishing House, Izmir (1989).
2. K. Bukowski, G. Działa, A. Kłopotek, Medycyna Wet., 41:613 (1985).
3. K. Bukowski, A.Klopotek, G.Dizaka, in:"Symposium on Resistance of Gram-Negative Rods", p.173, Polish Academy of Sciences, Poznan (1980).
4. A. Kłopotek, Zesz, Nauk, SGGW-AR Warszawa, Rozprawy Naukowe 100, (1978).
5. W.Kedzia,Dezynfekcja w medycynie i farmacji,PZWL, Warszawa (1981).
6. S.S.Block,"Disinfection,Sterilisation and Preservation", Lea and Febiger, Philadelphia (1977).
7. B. Zyska, Mikrobiologiczna korozja materialow, WNt, Warszawa (1977).
8. K.Bukowski,A.Kłopotek,G.Dizała,"Materialy VI Zjazdu PTNW", pp.197-198, Wroclaw (1978)

IN VITRO INHIBITORY EFFECT OF CHLORPROMAZINE ON THE GROWTH

OF CANDIDA SPECIES: A PRELIMINARY REPORT

K.Kurtar[1]   İ.Çifter[2]   F.Ulutan[1]   N.Sultan[3]   N.Yüksel[2]

Faculty of Medicine, Gazi University, Ankara, Turkey
[1] Department of Infectious Diseases
[2] Department of Psychiatry
[3] Department of Microbiology

INTRODUCTION

Phenothiazines belong to the group of antipsychotic drugs which are frequently used in the treatment of psychotic patients. Chlorpromazine having aliphatic side chains is a prototype for phenothiazines, with significant sedative and antipsychotic effects. This drug has been widely used in the treatment of psychotic patients since 1950's. Recently, attention was drawn to beneficiary effects on the alleviation of gastrointestinal symptoms in cholera, as the drug inhibits the enzyme adenyl cyclase, consequently decreasing the intestinal secretions, which leads to the reduction of the loss of electrolytes [1,2]. It was also demonstrated that the drug had an in vitro antibacterial effect, acting as a membrane stabilizer on the strains of Staphylococcus aureus and Vibrio cholerae [3].In a previous study, we observed its outcome on a wide range of bacteria and noted the bacteriostatic results especially on Gram-positive bacteria [4]. This prompted us to examine whether this drug had similar consequences on different microorganisms, and an investigation was carried out on the Candida species. According to the best of our knowledge, no such study exists in the literature.

MATERIAL AND METHOD

Twenty-two strains of four different Candida species (C. albicans, C. tropicalis, C. stellatoidea, C. guillermondii) identified by fermentation and assimilation tests were obtained from various clinical specimens in the Department of Microbiology. Chlorpromazine powder was obtained from the Eczacıbaşı Pharmaceuticals. Kirby-Bauer disc diffusion method [5] was used to investigate the antibacterial effect of chlorpromazine. The disc sensitivity test agar (Oxoid) was used as culture media. One gram of chlorpromazine was dissolved in 1 ml of distilled water and further serial dilutions were prepared [6]. Consequently, a range of discs in 2000, 200, 20, 2, 0.2, 0.02 mg potency was made available, each capable of holding 20 microliters of water. The growth of all Candida species were inhibited by 20 mg disc whereas 2 mg discs were proven to be ineffective. The same procedure

Candida and Candidamycosis, Edited by E. Tümbay et al.
Plenum Press, New York, 1991

was repeated with freshly prepared discs having a potency of 10, 5, 4, 3 mg, respectively.The zone of inhibition around each disc was measured in millimeters at the end of the test.

In order to find out whether the effect was reversible or not, smears were prepared from the area of no grawth around discs, and passages were made which were examined after 48 hours to detect the growth of Candida species.

RESULTS

The mean diameters of inhibition zones for different Candida species are shown in Table 1.

The cultures prepared from the no-growth area around the discs displayed adequate growth of all Candida species. It was concluded that the inhibition of growth was reversible, the main effect being mycostatic.

DISCUSSION

Chlorpromazine belongs to a group of agents called membrane stabilizers. Such agents have stabilizing consequences on cell membranes and comprise different chemical compounds used in clinical practice (e.g.chlorpromazine, acetyl salicylic acid, chloroquine, barbiturates). Chlorpromazine, in close resemblance to other membrane stabilizers, decreases the permeability of cell membrane for water molecules as well as sodium and potassium ions, while transportation of glucose through the membrane is also altered which explain its antimicrobial effect [3,7].

As shown in Table 1, regardless of the Candida species, 5-10 mg/disc dosage was found to be effective on the inhibition of growth.

In our study, lower doses of chlorpromazine were found to be effective in vitro, compared to average daily antipsychotic doses (75-1200 mg/day). The drug might have an antimycotic protection for patients under such traetment.

We present this article as a preliminary report on the inhibitory effect of chlorpromazine on the growth of Candida species. Further studies need to be conducted to affirm its antimycotic properties and utility in the clinical context.

Table 1. The averange diameters (mm) of the no growth zone in different concentrations of chlorpromazine.

Candida species	Number of strains tested	Concentration of the drug per disc (microgram)									
		2000	200	20	10	5	4	3	2	0.2	0.02
C. albicans	6	37	28	16	10	7	-	-	-	-	-
C. tropicalis	6	38	30	16	11	8	7	-	-	-	-
C. stellatoidea	5	42	30	22	12	8	7	-	-	-	-
C. guillermondii	5	40	26	17	9	7	-	-	-	-	-
Total	22										

REFERENCES

1. G.H. Rabbani, J. Holmgren, W.B. Greenough III and I. Lönnroth, Chlorpromazine reduces fluid loss in cholera, Lancet, 1: 410 (1979).
2. G.H. Rabbani, W.B. Greenough III, J. Holmgren and B. Kirkwood, Controlled trial of chlorpromazine as antisecretory agent in patients with cholera hydrated intravenously, Br. Med. J., 284: 1361 (1982).
3. J.E.H. Kristiansen and K. Gaarslev, The antibacterial effects of selected neroleptics on Vibrio cholerae, Acta Pathol. Microbiol. Immunol. Scand. B, 93: 49 (1985).
4. İ. Çifter, K. Kurtar, N. Sultan, N. Yüksel and F. Ulutan, A preliminary study: In vitro antibacterial activity of chlorpromazine. Gazi Üniv. Tıp Fak. Derg., 4: 505 (1988).
5. A.W. Bauer, W.M.M. Kirby, J.C. Sherris and M. Turck, Antibiotic susceptibility testing by a standardised single disc method, Am. J. Clin. Pathol., 45: 493 (1966).
6. A.E. Swingyard, Psychopharmacologic agents, in: "Remington's Pharmaceutical Sciences", 16th edition, pp. 1029-1042, A. Osol, ed., Mack Publishing Company, (1980).

ANTIMYCOTIC   SUSCEPTIBILITY TEST OF GERM-TUBE FORMING CELLS OF

CANDIDA ALBICANS BY ROSCO NEO-SENSITABS METHOD

T.-M.Jen [1,2,5]   S.-J.Kuo [1]   H.-D.Shen [3,5]
J.-C. Wei [4]   S.-F.Cheng [6]   S.-P. Chen [1]

Mycosis Research Laboratory, Department of Pathology;
[1] Mycosis Clinic, Department of Dermatology;
[2] Department of Medical Research [3] and Department of
Pediatrics, [4] Veterans General Hospital, VACRS, Taipei.
Department of Microbiology and Immunology, National
Defense Medical Center, [5] Taipei. Department of
Dermatology, China Medical College Hospital,
[6] Taichung., Taiwan, Republic of China.

INTRODUCTION

   The antifungal susceptibility tests used in clinical mycology
have been adopted from bacteriology for more than 30 years with only
slight modifications[1]. Some investigators [1,3] have found an increasing
number of weakness in corresponding tests on some diphasic fungi such
as C.albicans with morphological conversion from the budding cell form
into the parasitic hyphal form. In Taipei, we have found discrepancies between in vitro and in vivo drug susceptibility tests since
1986 [2]. The prevalence of clusters of miconazole, 5-fluorocytosine
and amphotericin resistant strains of C.albicans was once considered
to be too high; over 10 %, in patients who have never ever received
any antimycotic therapeutics [2]. The main purpose of this report is to
share our experience using a simple and new method by means of ROSCO-
SENSITABS agar plate technique [2-4] on GTF cells of C. albicans [5,6] with
the profession.

MATERIAL and METHOD

Organisms

   Totaly 23 strains of C. albicans (CA) including 2 strains of CA
of ATCC 14053 and ATCC 18804 and 21 strains isolated from the
clinical specimens of the suspicious cases of candidiasis of this
hospital, were used in this study. Only 6 strains of profuse GTF CA;
i.e. the number of GTF cells was counted to be constantly over 90 %
of a total cell population, were selected for the drug susceptibility
test. All 24 hrs' strains of CA were identified carefully [2,7-12] and
used for study.

Preparation of inocula and performance of test [2,4,13]

   Instead of human serum, 2.5 % Difco-bovine albumin (0668-62) in
distiled water was used to produce GTF cells for this study [2,14]. The
inocula were prepared by distributing the same amount of 2 ml yeast
cell-2.5 % bovine albumin suspensions into 2 sterilized test tubes.

Candida and Candidamycosis, Edited by E. Tümbay et al.
Plenum Press, New York, 1991

The tube for GTF cell production was incubated at 37°C for 2-4 hours. Another tube for preparation of non-GTF cells was kept at 4-8°C and used for comparison. Before inoculation for susceptibility test, the concentration of the inoculum was adjusted to about 10^6 CFU/ml. The final concentration of fungal cells in each culture plate for test should be around 10^3 cells/ml. Other procedures for drug test were done according to instructions [4]. The first series of tests with 2 strains of CA in GTF and non-GTF cell forms, respectively, by ROSCO NEO-SENSITABS (RNS) method using 9 cm sized, 20 ml amt Bacto-Sabouraud dextrose agar pour plate was done by one of us (TMJ) in 1986[2]. The medicated tablets consisted of amphotericin B, clotrimazole, econazole, 5-fluorocytosine, miconazole and nystatin. The second series of tests by the same method using 14.5 cm sized, 45 ml amt modified Shadomy agar plates was done in 1987 [14]. The 12 medicated tablets were amphotericin B, ciclopirox, clotrimazole, econazole, 5-fluorocytosine, griseofulvin, isoconazole, ketoconazole, miconazole, natamycin, tioconazole and nystatin. The third series of tests with 2 strains of CA using only a single RNS medicated tablet in each 5 cm sized, 10 ml amt modified Shadomy agar pour plate has been carried out recently. The medicated tablets consisted of amphotericin B, clotrimazole, econazole, 5-fluorocytosine, ketoconazole, miconazole, nystatin and tioconazole. In the third series of tests, we detected, from 24 to 48 hours after incubation, the concentration of fungus cells in each agar plate which contained the GTF and non-GTF cells, respectively, by means of direct colony count per cm^2 with aid of a hand magnifier. Records were set down in written form daily and kept for references.

RESULTS

We proved for the first time in this hospital that Difco Bacto-bovine albumin can be used as a substitute for human serum to produce larger amount of GTF cells of CA (Fig.1). The efficacy of it is presented in Table 1. A method to produce rapid, mycelial and chlamydospore formations for 4 fresh clinical isolates of CA in yeast extract-wool fiber medium was satisfactory (Fig.2). On the basis of duplicate tests with 6 strains of profuse GTF cells of CA by RNS method, we demonstrated on 3 occasions that some imidazoles such as clotrimazole, econazole, ketoconazole (demonstrated in 2/2 times), miconazole and tioconazole (demonstrated in 2/2 times) were more fungistatic to GTF cells of CA than the non-GTF cells of the same fungus. Results of 2 series of tests using modified Shadomy medium are listed in Tables 2 and 3. Additionally, a phenomenon of "substandard" or "extra-inhibition zone" fungistatic effect of all 12 antimycotics except griseofulvin on GTF cells of CA was, for the first time remarkably demonstrated in this study (Fig. 3-5).

DISCUSSION

This is the first as well as the elementary experiment to prove that the GTF cells; i.e. a parasitic form of CA is more susceptible to imidazoles than its yeast form cells which are now used in most clinical microbiology laboratories for routine antimycotic susceptibility test. As MICs (Minimal Inhibition Concentrations) for some new imidazoles are subject to considerable variation [3], we have used a diffusion disk method 3 in this study. We hope that this report may offer our medical profession a practical method which may be helpful to resolve problems concerning discrepancies between _in vitro_ and _in vivo_ drug susceptibility tests for candidiasis.

Table 1. Efficacy of Difco Bacto-bovine albumin to induce germ tube forming (GTF) cells of <u>C.albicans</u>. Analysis of 23 strains isolated from clinical specimens.

	Positive rate of GTF cells examined by								
	High power lens tests			Showing frequencies & % on average of duplicate					
	1 hour			2 hours			4 hours		
	BA 2.5	BA 1.25	HS	BA 2.5	BA 1.25	HS	BA 2.5	BA 1.25	HS
C.albicans	20/21	19/21	4/21	21/21	21/21	21/21	21/21	21/21	21/21
API-20C-2021 file 2576170,Type A 14 strains	4.00	3.09	0.35	36.04	40.57	1870	88.52	83.85	76.66
API-20C-2021 file 2576170,Type B 3 strains									
API-20C-2021 file 2566170,Type A 2 strains									
API-20C-2021 file 2572170,Type A 1 strain									
API-20C-2021 file 6576170,Type A 1 strain									
C.albicans used as controls									
API-20C-2021 file 2576170,Type A 2 strains	10/10	10/10	2/10	10/10	10/10	10/10	10/10	10/10	10/1
	10/10	10/10	4/10	10/10	10/10	10/10	10/10	10/10	10/1
	10.96	11.60	0.35	60.85	68.35	27.63	89.95	88.65	87.0
C.tropicalis used as controls									
API-20C-2021 file 2556371,9 strains	1/9	1/9	!/9	!/9	!/9	!/9	!/9	!/9	!/9
	0.1	0.1	0.1	0.3	0.1	0.1	0.66	0.3	0.1

BA 2.5:2.5, Bacto-bovine albumin. BA 1.25:1.25, Bacto-bovine albumin. HS: Sterile fresh human sera obtained from Blood Bank, Veterans General Hospital, VACRS, Taipei, Taiwan, ROC.

Table 2. Susceptibility test of germ tube forming cells of <u>C. albicans</u> to 12 antibiotics by Rosco Neo-sensitabs method with modified Shadomy agar medium.

Antimycotics (diffusible amt of antifungal/interpretative standards in mm; resistant, intermediate and sensitive)	Strain numbers and inhibition zone* size in mm. Average of duplicate tests			
	VGH-76-2-3 non-germ tube forming cells	germ tube forming cells	non-germ tube forming cells	germ tube forming cells
Amphotericin B (20 I.U.,9,10-14,15.)	14	<u>15</u>	14	14
Ciclopirox (50 mcg, 11,12-19,20.)	<u>29</u>	26	25	25
Clotrimazole (10 mcg, 11,12-19,20.)	29	<u>30</u>	24	<u>27</u>
Econazole (10 mcg, 11,12-19,20.)	18	<u>40</u>	22	<u>40</u>
5-Fluorocytosine (10 mcg, 22,23-29,30.)	40	40	40	40
Griseofulvin (25 mcg, 9,9.5,10.)	0	0	0	0
Isoconazole (10 mcg, 11,12-19,20.)	<u>25</u>	23	0	<u>20</u>
Ketoconazole (15 mcg, 11,12-19,20.)	28	<u>40</u>	12	<u>40</u>
Miconazole (10 mcg, 11,12-19,20.)	17	<u>40</u>	12	<u>32</u>
Natamycin (50 mcg, 9,10-14,15.)	16	<u>22</u>	18	<u>20</u>
Nystatin (50 mcg, 9,10-14,15.)	<u>24</u>	23	18	<u>26</u>
Tioconazole (10 mcg, 11,12-9,20.)	34	<u>43</u>	23	<u>25</u>

* The more susceptible one which each of the underlined numbers belongs to.

Table 3. Susceptibility test of germ tube forming cells of
C. albicans to 8 antimycotics by Rosco-neosen-
sitabs method with modified Shadomy agar medium
and single medicated tablet technique.

Antimycotics (diffusible amt of anti-fungal/interpretative standards in mm; resistant, intermediate and sensitive)	Strain numbers and inhibition zone* size in mm. Average of duplicate tests			
	VGH-77-1-A		VGH-77-1-B	
	non-germ tube forming cells	germ tube forming cells	non-germ tube forming cells	germ tube forming cells
Amphotericin B (20 I.U.,9,10-14,15.)	20.00	20.00	20.00	20.00
Clotrimazole (10 mcg, 11,12-19,20.)	21.00	<u>38.50</u>	37.25	<u>38.50</u>
Econazole (10 mcg, 11,12-19,20.)	19.00	<u>34.75</u>	32.25	<u>34.75</u>
5-Fluorocytosine (10 mcg, 22,23-29,30.)	14.5	<u>40.5</u>	39.75	<u>40.50</u>
Ketoconazole (15 mcg, 11,12-19,20.)	15	<u>29.75</u>	26.50	<u>29.75</u>
Miconazole (10 mcg, 11,12-19,20.)	24.00	24.00	23.62	<u>24.00</u>
Nystatin (50 mcg, 9,10-14,15.)	20.00	<u>26.00</u>	25.75	<u>26</u>.00
Tioconazole (10 mcg, 11,12-19,20.)	17.50	<u>45.25</u>	41.75	<u>45.25</u>

* The more susceptible one which each of the underlined numbers belongs to.

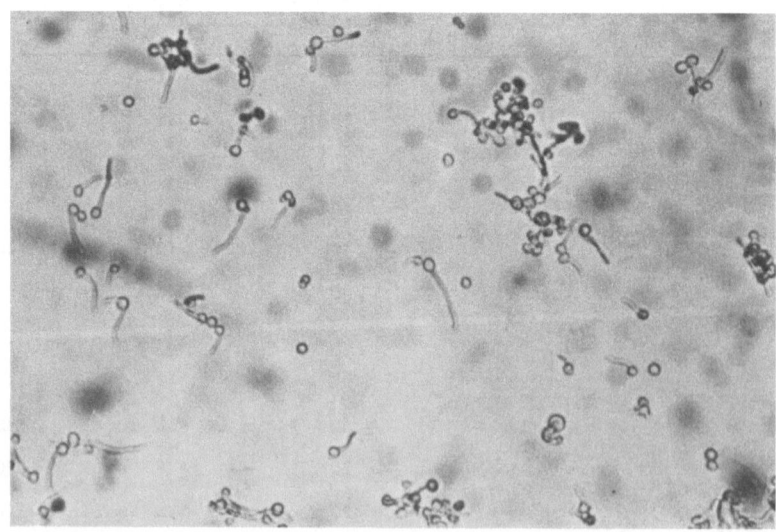

Fig.1. Germ-tube forming cells of C. albicans produced in
2.5 % Difco Bacto-bovine albumin. 24 hrs' culture on
Difco modified SDA slant at 27°C. Examined 2 hours
after incubation. 100 X.

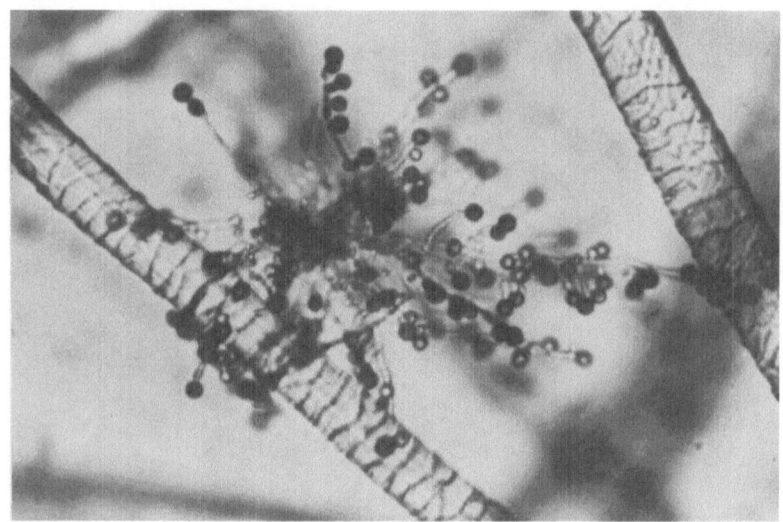

Fig.2. Mycelia and chlamydospores of <u>C. albicans</u> in yeast
extract-wool fiber medium [9].

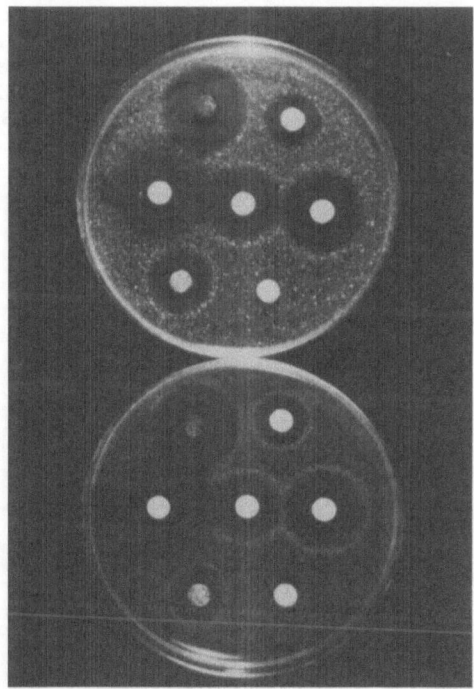

Fig. 3. Antimycotic susceptibility test of GTF cells and non-
GTF cells of <u>C. albicans</u> by Rosco Neo-sensitabs
method with Shadomy modified agar. Upper plate: GTF
cells. Lower plate; non-GTF cells. Seven antimycotics from left to right: amphotericin B, ciclopirox,
clotrimazole, econazole, 5,fluorocytosine, griseofulvin and isoconazole. Difference: 0.8-1.5 mm. An
occasional instance showing no remarkable
difference.

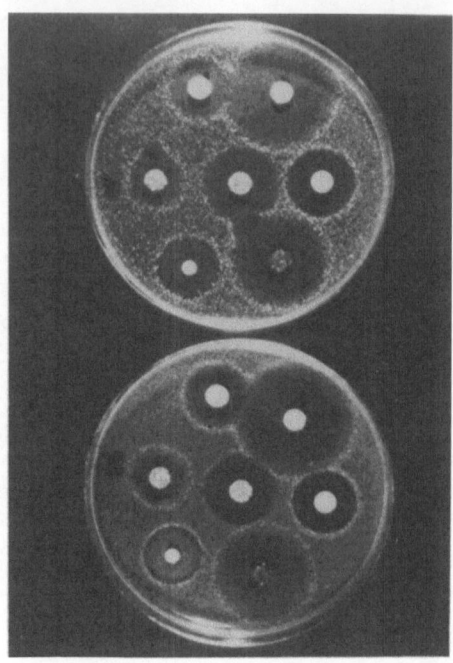

Fig. 4. Same test as described in Fig. 3. Upper plate; GTF cells. Lower plate; non-GTF cells. Seven antimycotics from left to right: Ketoconazole, miconazole, natamycin, nystatin, oxiconazole (ROSCO 3621271), tioconazole and gamastan [15]. Difference: 0.8-1.5 mm. An occasional instance showing no remarkable difference.

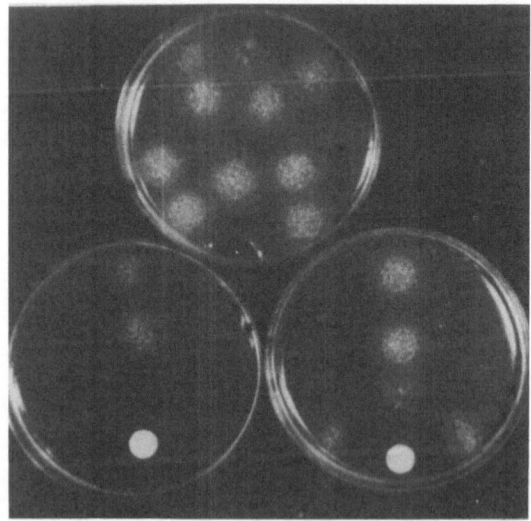

Fig. 5. Substandard or "extra-inhibition zone" fungistatic effects of ketoconazole on GTF and non-GTF cells of C. albicans ATCC 14053 by Oxford P-7000 (50) sampler technique. Upper plate (above a line); 5 drops of GTF cells. Upper plate (below a line); 5 drops of non-GTF cells. Lower left plate; 5 drops of non-GTF cells against a ROSCO ketoconazole disc. Lower right plate; 5 drops of GTF cells against a ROSCO ketoconazole disc.

REFERENCES

1. M.Plempel, D.Berg, K.H.Büchel and D.Abbink, Test method for anti-fungal agents; a critical review, _Mykosen_, 30: 28 (1987).
2. T.M. Jen and S.J.Kuo, Influence of germ tube forming cells of _Candida albicans_ to in vitro antimycotic susceptibility test. _Chinese Journal of Microbiology and Immunology_ (Taipei), 19: 120 (1986).
3. S.Shadomy, A.Espinal-Ingroff and R.Y.Cartwright, Laboratory studies with antifungal agents; susceptibility tests and bioassays, _in_: "Manual of Clinical Microbiology", 4th Ed., E.H.Lennette, A. Balows, W.J.Hausler Jr., H.J.Shadomy, eds., pp. 991-999, American Society for Microbiology, Washington D.C. (1985).
4. J.B.Casals and N.Pringler, "Antimicrobial Sensitivity Testing Using Neo-Sensitabs", 8th Ed., pp. 42-43, Rosco Diagnostica, Taastrup, Denmark (1985).
5. M.Bogers, M.D.Brabander, H.V.D.Bossche and J.V.Cutsem, Promotion of pseudomycelium formation of _Candida albicans_ in culture; a morphological study of the effects of miconazole and ketoco-nazole, _Postgrad. Med.J._, 55: 687 (1979).
6. M.Nimi, A.Kamiyama, M.Tokunaga, J.Tokunaga and H.Nakayama, Germ tube-forming cells of _Candida albicans_ are more susceptible to clotrimazole-induced killing than yeast cells, _J.Med.Vet.Mycol._, 23: 63 (1985).
7. J.A.Barnett, R.W.Payne and D.Yarrow, Summary of specific charac-teristics of yeasts, _in_: "Yeasts; Characteristics and Identification", Ist Ed., J.A.Barnett, R.W.Yarrow and L.Barnett, eds., pp. 39-59, Cambridge University Press, Cambridge (1985).
8. B.H.Cooper, and M.Silva-Hutner, Yeasts of medical importance _in_: "Manual of Clinical Microbiology", 4th Ed., pp. 526-541, E.H.Lennette, A.Balows, W.J.Hausler Jr. and H.J.Shadomy, eds., American Society for Microbiology, Washington D.C. (1985).
9. T.M.Jen, A modified method to detect dermatophytes in the diag-nosis of tinea unguium, _Journal of Formosan Medical Association_, 86: 1021 (1987).
10. T.M.Jen, S.J.Kuo and S.P.Chen, Rapid sucrose fermenters of _Candida albicans_, _Chinese Journal of Microbiology and Immunology (Taipei)_, 20: M-31 (1987).
11. T.M.Jen, S.L.Shern and S.J.Kuo, Studies on _Candida_ isolated from sputa of chronic cases of pulmonary disease in China, _Journal of Formosan Medical Association_, 66: 187 (1967).
12. H.D.Shen, K.B.Choo, W.C.Tsai, T.M.Jen, J.Y.Yeh and S.H.Han, Differential identification of _Candida_ species and other yeasts by analysis of (^{35}S) methionine-labeled polypeptide profiles, _Analytical Biochemistry_, 175: 548 (1988).
13. J.B.Casals and O.G.Pedersen, "Antimicrobial Sensitivity Testing Using Neo-Sensitabs A/S Rosco", 5th Ed., pp. 26-27, Denmark (1980).
14. T.M.Jen and S.J.Kuo, Germ tube test negative strains of _Candida albicans_ isolated in Taipei, ROC, _Chinese Journal of Microbiology and Immunology (Taipei)_, 20: 1 (1987).
15. T.M.Jen and S.J.Kuo, The fungistatic effect of a human immuno-globulin preparation, Gamastan, on _Phialophora dermatitidis_, _Mycologia_, LXX: 802 (1978).

SENSITIVITY OF CANDIDA SPECIES ISOLATED FROM TINEA PEDIS

OF SOLDIERS TO IMIDAZOLES

N.Yuluğ[1] Ö.Öztunalı[2] Y.Akyön[1] N.Çerikçioğlu[1] M.Özalp[1]

[1] Department of Microbiology, Faculty of Medicine
    Hacettepe University, Ankara, Turkey
[2] Department of Microbiology, Faculty of Medicine
    Cumhuriyet University, Sivas, Turkey

INTRODUCTION

Mass screening programs for superficial mycoses are rare in Turkey. The only study on soldiers in Turkey is a preliminary study [1]. Hundreds of new soldiers are trained in Sivas which is a military area. In our study, we determined the incidence of Candida species among newly recruited soldiers in Sivas and their sensitivity to antifungal agents.

MATERIAL AND METHOD

Three thousand newly recruited soldiers were screened in our study. Specimens were taken by scraping from 350 soldiers with superficial mycoses. The specimens were inoculated onto Sabouraud dextrose agar (Difco) with or without cycloheximide. The Candida species were identified with germ tube formation, chlamydospore formation on corn meal agar (Oxoid) and fermentation tests and assimilation reactions in yeast-nitrogen base broth (Difco) [2,3,4,5].

The tube dilution technique was used to determine the sensitivity to antifungal agents. A hundred percent dimethylsulphoxide (DMSO) was used to dissolve the Bifonazole, Econazole, Oxiconazole. Concentrations of 40, 20, 10 ... 0.04 mg/ml were prepared with serial dilutions using Sabouraud dextrose broth (Oxoid).

Microorganism suspensions prepared from cultures were kept for 24 hours at $37^{\circ}C$ were added to the serial test tubes. The final concentrations of microorganisms in the test tubes were 10^5 c.f.u./ml. The test tubes were incubated at $37^{\circ}C$ for 24 hours, and the MIC values determined. All tests were performed in duplicate.

RESULTS

Of the 3000 soldiers screened, 350 (11.6 %) had superficial mycosis, and 160 of the 350 cultured specimens (45.7 %) showed

Candida and Candidamycosis, Edited by E. Tümbay *et al.*
Plenum Press, New York, 1991

Table 1. <u>Candida</u> species isolated from tinea pedis in soldiers.

	Number	%
Candida albicans	25	62.5
Candida guilliermondii	8	20.0
Candida parapsilosis	5	12.5
Candida tropicalis	2	5.0
Total	40	100.0

Table 2. The MIC values of the anti-fungals used.

Candida spp.	Anti-fungal	Concentrations (μg/ml)									
		≤0.02	0.04	0.08	0.16	0.31	0.62	1.25	2.5	5	10
C.albicans (25)	B			1		3	1	3	2	3	12
	E	2		1	2	4	4	7	5		
	O		2	4	1	6	6	8	3		
C.guilliermondii (8)	B							1	2		5
	E					1	1	2	1	3	
	O					1	2	2	3		
C.parapsilosis (5)	B								2	2	1
	E		1				2	1	1		
	O		2			1			2		
C.tropicalis (2)	B										2
	E						1		1		
	O					1					1

B : Bifonazole   E : Econazole   O : Oxiconazole

Table 3. The cumulative percentages of the inhibited <u>Candida</u> strains with the concentrations used.

Candida spp.	Anti-fungal	Concentrations (μg/ml)									
		≤0.02	0.04	0.08	0.16	0.31	0.62	1.25	2.5	5	10
C.albicans (25)	B			4		16	20	32	40	52	100
	E	8		12	20	36	52	80	100		
	O		8	24	28	52	76	88	100		
C.guilliermondii (8)	B							12.5	37.5		100
	E				12.5	25	50		62.5	100	
	O				12.5	37.5	62.5	100			
C.parapsilosis (5)	B								40	80	100
	E		20				60	80	100		
	O		40			60			100		
C.tropicalis (2)	B										100
	E						50		100		
	O					50					100

B : Bifonazole   E : Econazole   O : Oxiconazole

growth. Of these, 120 (34.27 %) yielded dermatophytes and 40 (11.42 %) _Candida_ species. All _Candida_ species were isolated from specimens from lesions of Tinea pedis cases. The isolated _Candida_ species are shown in Table 1.

The MIC values for the imidazole derivatives used in our study are shown in Table 2 and the cumulative percentages of the inhibited strains with the concentrations used are shown in Table 3.

DISCUSSION

The most widely used imidazole derivatives in recent years are Bifonazole, Econazole and Oxiconazole [6,10]. The MIC values of bifonazole for isolated _C.albicans_ were 0.08-10 mg/ml, for _C.guilliermondii_ 1.25-10 mg/ml, for _C.parapsilosis_ 2.5-10 mg/ml and for _C.tropicalis_ 10 mg/ml. In another study [7], the MIC values of Bifonazole for _C.albicans_, _C.tropicalis_ and _C.parapsilosis_ were ≤ 1-16 mg/ml.

The MIC values of Econazole for _C.albicans_ were ≤ 0.02-5 mg/ml, for _C.parapsilosis_ 0.04-2.5 mg/ml and for _C.tropicalis_ 0.62-2.5 mg/ml. Another study [9] had determined the MIC value of Econazole for _C.albicans_ and _C.tropicalis_ as 100 mg/ml and for _C.parapsilosis_ as 10 mg/ml.

The MIC values of Oxiconazole for _C.albicans_ and _C.parapsilosis_ were 0.04-2.5 mg/ml, for _C.tropicalis_ 0.31-10 mg/ml and for _C.guilliermondii_ 0.16-1.25 mg/ml. Another study [10] has determined the MIC values of the same agent for _C.albicans_ as 0.1-100, for _C.tropicalis_ as 100, and for _C.parapsilosis_ as 0.03-10 mg/ml.

The different results in our study can be the result of differences in test conditions and a wide variation among the same species for MIC values [11].

The differences in MIC values of imidazole derivatives can be abolished by an international standardization of antimycotic sensitivity tests.

REFERENCES

1. A.Karaman, E.Tümbay and O.Demir, İzmir'de askerlerde görülen dermatomikoz insidansı ve etkenleri (Dermatomycoses in soldiers in Izmir), _Lepra Mec._, 12: 136 (1981).
2. The yeast like fungi, _in_: "Laboratory Manual for Medical Mycology CDC.", pp. E1-E28, U.S. Department of Health, Education and Welfare, U.S. Government Printing Office (1963).
3. E.W.Koneman, G.D.Roberts and S.E.Wright, eds., The yeast, _in_: "Practical Mycology", 2nd ed., pp. 103-117, The Williams and Wilkins Company, Baltimore (1979).
4. A.Stenderup, Oppurtinistic yeast-pathogens in mycoses. Isolation and identification, _Scand.J.Infect. Dis._ Suppl ., 16: 23 (1978).
5. E.Jawetz, J.L. Melnick, and E.A. Adelberg, _in_: "Review of Medical Microbiology", pp. 330-332, Appleton and Lange, Middle East Edition (1987).
6. G.Cauwenbergh, New and prospective developments in anti-fungal drugs, _Acta Dermatol. Venereol._ (Stockh) Suppl., 121: 147 (1986).
7. M.Plempel, E.Regel and K.H.Büchel, Antimycotic efficacy of Bifo-

nazole in vitro and in vivo, <u>Arzneim.-Forsch/Drug Res</u>., 33: 517 (1983).

8. New topical antifungal drugs, <u>Medical Letter on Drugs and Therapeutics</u>, 647: 98 (1983).

9. D.Thienpont, J.Van Cutsem J.M. Van Nueten, C.J.E. Niemegeers and R. Marsboom, Biological and toxicological properties of econazole, a broad spectrum antimycotic, <u>Arzneim.-Forsch/Drug Res</u>., 25: 224 (1975).

10. A.Polak, Oxiconazole, a new imidazole derivative. Evaluation of antifungal activity in vitro and in vivo, <u>Arzneim.-Forsch/Drug Res</u>., 32: 17 (1982).

11. S.P. Clissold and J. Van Cutsem, Overview of the antimicrobial activity of ketoconazole, <u>in</u>: "Ketoconazole Today: A Review of Clinical Experience", H.E.Jones, ed., pp. 5-22, ADIS Press Limited (1987).

IMPROVEMENT OF PSORIASIS WITH ORAL NYSTATIN

Ö. Aşçıoğlu   Ü. Soyuer   E. Aktaş

Department of Dermatology, Faculty of Medicine
Erciyes University, Kayseri, Turkey

INTRODUCTION

Psoriasis may result from the interaction of yeasts and other microorganisms with an abnormally responsive alternative complement pathway [1-3]. Meanwhile, we observed a significantly increased yeast colonization of gut of psoriatic patients [4].

This suggested the possibility of improving psoriasis by reducing Candida in gastro-intestinal tract of patients.

MATERIAL AND METHOD

Fourty eight psoriasis patients (20 males and 28 females) aged 14-49 were studied. In all patients topical and systemic treatments were discontinued for two weeks prior to therapy. Patients were selected randomly for double blind placebo controlled cross-over study.

Twenty four patients were given 2 million units of oral nystatin in four divided doses per day for three months.

Twenty four patients were given placebo tablets four times a day. Assessment of psoriasis was made by the PASI score (Psoriasis area and severity index)[5].

Patients were seen at weekly intervals for the first month and then fortnightly.

RESULTS

The follow-up period was three months. Eight patients treated with placebo were excluded from the study because they had not come for re-examination.

The mean PASI score was similar in two groups before the treatment, and there was no significant change at the end of the first month. The mean PASI score fell significantly after two months and remained unchanged at the end of the third month in the nystatin group, and there was a considerable reduction in PASI score (Fig. 1).

Candida and Candidamycosis, Edited by E. Tümbay *et al.*
Plenum Press, New York, 1991

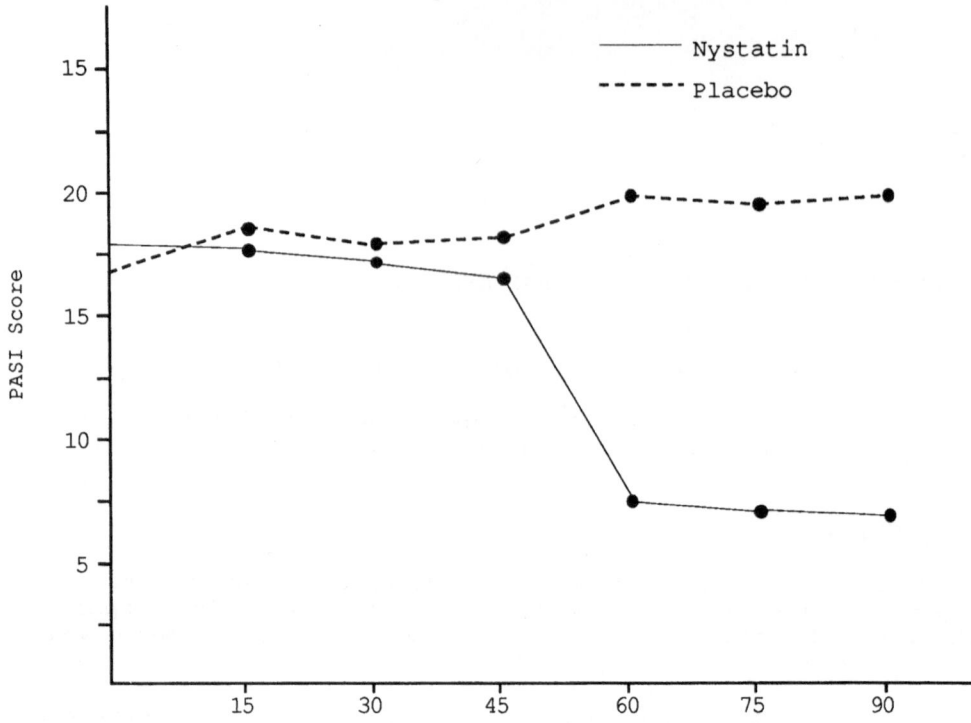

Fig. 1. The course of PASI score in nystatin and placebo groups.

PASI scores of placebo group did not show a significant change. In nystatin group; 10 cases showed excellent improvement, 10 cases showed moderate improvement, and one case deteriorated. In placebo group 6 cases showed minimal improvement and others showed no changes.

DISCUSSION

There are many topical and systemic treatment modalities for psoriasis vulgaris. Recently, it was reported that systemic antifungals could be effective in psoriasis. Some authors suggested that Malassezia ovalis (Pityrosporum ovale) residing in the hair roots of so-called seborrheic areas may play a role in the pathogenesis of psoriasis. However, clearing of psoriatic patches in non-seborrheic areas with antifungal treatment suggested the possibility of the systemic effects of these drugs on the yeasts residing in the gut [1-3].

Oral nystatin is poorly absorbed form the intestinal mucosa, and its effect is considered to be almost solely on yeasts in the gut microflora.

Nystatin is non-toxic and has no serious side effects. Although it is not a routine treatment for psoriasis, it may be an alternative therapy method especially for childhood psoriasis. We believe that oral nystatin may be a safe and effective drug for psoriasis vulgaris.

REFERENCES

1. E.W. Rosenberg and P.W. Belew, Microbial factors in psoriasis, <u>Arch. Dermatol</u>., 118:143 (1982).
2. N. Crutcher, E.W. Rosenberg et al., Oral nystatin in the treatment of psoriasis, <u>Arch. Dermatol</u>., 120:435 (1984).
3. E.W. Rosenberg and P.W. Belew, Improvement of psoriasis of the scalp with ketoconazole, <u>Arch. Dermatol</u>., 118:370 (1982).
4. Ü. Soyuer, Ö. Aşçıoğlu, and E. Aktaş, The pathogenetic role of <u>Candida</u> in psoriasis, in:"FEMS-Symposium on <u>Candida</u> and Candida-mycosis-Abstracts", p.74, Bilgehan Publishing House, Izmir (1989).
5. T. Frediksson, and V. Petterson, Severe psoriasis oral therapy with a new retinoid, <u>Dermatologica</u>, 157:238 (1978).

SINGLE-DOSE FLUCONAZOLE TREATMENT OF VAGINAL CANDIDOSIS

E. Tümbay[1]   K. Soy[2]   R. İnci[3]   G. Karakartal[4]   S. Ural[5]
A. Karaman[6] A. Zeytinoğlu[4]   T. Özacar[4]   M. Otkun[4]   O. Demir[1]

[1]Department of Microbiology, Faculty of Medicine, Ege
University, Bornova, Izmir
[2]The First Clinic of Obstetrics and Gynecology,
Izmir State Hospital, Yeşilyurt, Izmir
[3]Specialist of Bacteriology and Infectious Diseases,
Cumhuriyet Meydanı, Batman
[4]Department of Clinical Bacteriology and Infectious
Diseases, Faculty of Medicine, Ege University, Bornova,
Izmir
[5]Microbiology/Infectious Disease Section, Izmir State
Hospital, Yeşilyurt, Izmir
[6]Dermatology Clinic, Izmir State Hospital, Yeşilyurt,
Izmir, Turkey

INTRODUCTION

Fluconazole is a novel, systemically administered triazole
antifungal agent, used primarily in the treatment of mucosal and
systemic <u>Candida</u> infections and cryptococcosis[1].

Shortly after oral intake, fluconazole is almost completely
absorbed in the intestinal tract and diffuses well into body
fluids[2-4]. The binding of the drug to plasma proteins is rather
low – about 11-12%[2,5,6].

The plasma half-life is 25-29 hours[2,7], and 75-90% of fluconazole
is excreted by kidneys in bio-active form[3,6,8]. The drug attains and
maintains high levels in cerebro-spinal fluid, other body fluids and
-related to the subject matter of this paper- in the vaginal tissue[3].

In view of its long plasma half-life and attained high tissue
concentrations in the vaginal tissue, fluconazole has been tried in
the treatment of vulvo-vaginal candidosis. It has been observed that
single-dose fluconazole administered to mice with experimental
<u>Candida</u> vaginitis gave results similar to those obtained by drug
administration on days 3,4 and 5 following infection[9].

Single-dose, 150 mg P.O. fluconazole treatment of vulvo-vaginal
candidosis has also been clinically tried[10-13].

This is a preliminary report on the clinical and mycological
efficacy of single-dose fluconazole in vulvo-vaginal candidosis.

Candida and Candidamycosis, Edited by E. Tümbay *et al.*
Plenum Press, New York, 1991

MATERIAL AND METHOD

Twenty-four females, diagnosed both clinically and mycologically[14] to have vulvo-vaginal candidosis, were taken into trial with fluconazole. All patients were in child-bearing age, non-pregnant, non-lactating and non-diabetic (Table 1).

All patients received 150 mg single-dose fluconazole P.O. and were clinically and mycologically controlled twice - approximately 1 week and 4 weeks following drug administration.

RESULTS

Out of 24 patients, 19 could be evaluated in the first control (~after 1 week) and 14 in the second control (~after 4 weeks) visit.

The patients and clinical and mycological results during controls are presented in Tables 1, 2 and 3.

Side-effects due to fluconazole were not encountered.

All 14 patients (100%) who came to the second control found single-dose oral fluconazole a highly practical application, preferring it to other forms of treatment; and 12 (85%) were satisfied with the therapy results.

Table 1.   Presentation of 19 cases of candidal vulvo-vaginitis evaluated in the study.

Mean age	30
Duration of infection	
1 month	5
>1 month	14
Previous antifungal therapy	
Yes	8
No	11
Number of exacerbations in the last year (mean)	4
Agent of vulvo-vaginal candidosis	
Candida albicans	16
Candida (Torulopsis) glabrata	3

Table 2. Clinical effect of fluconazole.

Effect	After 1 week (%)	After 4 weeks (%)
Care	63	72
Improvement	37	14
Ineffective	–	14*

* Two cases; agents, C.albicans and C.glabrata.

Table 3. **Candida** eradication by fluconazole.

Mycology	After 1 week (%)	After 4 weeks (%)
Culture, negative	86	72
Culture, positive	14	14*
Relaps/Reinfection	–	14**

* Two cases; agent, **C.albicans**.
** Two cases; agent, **C.glabrata**.

Table 4. Comparison of clotrimazole, ketoconazole and fluconazole in the treatment of vulvo-vaginal candidosis.

Chemotherapeutic	Clinical efficacy		Mycological efficacy	
	Early Control (1-2 weeks) %	Late Control (4-9 weeks) %	Early Control (1-2 weeks) %	Late Control (4-9 weeks) %
Clotrimazole vag.tab.; 200 mg/day-3 days [10,11,13]	80-97	80-90	64-81	62-64
Ketoconazole oral tab.; 400 mg/day-5 day [10]	88	88	82	77
Fluconazole cap.;[10-13] 150 mg-single dose	80-99	73-93	72-94	72-77
Fluconazole cap.; 150 mg-single dose (Tümbay et al-present study)	100	84	86	72

DISCUSSION

Single dose, 150 mg oral fluconazole is a new application in the treatment of vulvo-vaginal candidosis, but some pioneer multi-center clinical trials have given promising results [10-13,15].

The studies have shown that single-dose oral fluconazole in candidal vulvo-vaginitis is clinically effective in 80-89% and 73-93% of patients as determined in controls after 1-2 weeks and 4-9 weeks, respectively; the drug eradicates the **Candida** in the vagina in 72-94% and 72-77% of cases as found in early and late control visits, respectively [10-13].

In the present study, the authors found clinical cure in 63% and clinical improvement in 37% (clinical efficacy, 100%) and **Candida** eradication in 84% of the patient approximately one week after the drug administration. In late controls, approximately 4 weeks after the drug intake, clinical cure was 72%, clinical improvement 14% (clinical efficacy, 86%) and fungal eradication 72% (Tables 2 and 3). The obtained results were similar to those of former trials[10-13] (Table 4).

The efficacy of single-dose oral fluconazale has been compared with those of clotrimazole vaginal tablet and ketoconazole oral tablet [10,11,13]. Clotrimazole vaginal tablet 200 mg/day for three days has yielded 80-97% and 80-90% clinical cure and 64-81% and 62-64% fungal eradication rates in early (after 1-2 weeks) and late (after 5-9 weeks) controls, respectively [10,11,13]. With ketoconazole tablet 400 mg/day for five days the clinical care rates were 88% and 88%, and fungal eradication rates 82% and 77% in early and late controls, respectively [10].

The comparative investigations have shown that 150 mg single-dose oral fluconazole was as effective in vaginal candidosis as a total of 600 mg vaginal clotrimazole tablet and also as a total of 2000 mg oral ketoconazole [10,11,13] (Table 4).

As already shown in previous studies [10,13], fluconazole was well tolerated by all patients and no side effects were recorded. Also all patients expressed their preference of single-dose oral fluconazole to other drug administrations in vulvo-vaginal candidosis.

Single-dose oral fluconazole -due to its efficacy, safety and ease of administration- seems to be a promising drug in the treatment of Candida vulvo-vaginitis.

REFERENCES

1. E.Tümbay, Sistematik etkili yeni bir antifungal kemoterapötik: Flukonazal (Fluconazole : a novel antifungal chemotherapeutic with systemic effect), İnfeks. Derg. 3 (1), Ek Baskı Seri No 1 (Suppl. 1) 27 pp. (1989).
2. K.W.Brammer and M.H. Tarbit, A review of the pharmakokinetics of fluconazole (UK-49,858) in laboratory animals and man, in : "Recent Trends in the Discovery, Development and Evaluation of Antifungal Agents", R.A. Fromtling, ed., pp. 141-150, Prous Science Publishers, Barcelona (1987).
3. P.R. Farrow, J.K.Faulkner, K.W.Brammer, The pharmacokinetics of fluconazole, in "Fluconazole Symposium Abstracts, 8-9 October 1988, Dorado, Puerto Rico", p. 18 (1988).
4. R.A.Fromtling : Overview of medically important antifungal azole derivatives, Clin. Microbiol.Rev., 1 : 187 (1988).
5. K.W.Brammer, Fluconazole : Profile of a new systemically-effective azole antifungal, in "Abstracts of Biennial Conference on Chemotherapy of Infectious Diseases and Malignancies, 5-8 March 1989, Montreux, Switzerland", Abstract No 73, Zbl. Antimikrob.Antineoplas.Chemother., Suppl 1 (1989).
6. W.E. Dismukes, Azole antifungal drugs : Old and new, Ann. Intern. Med., 109 : 177 (1988).
7. G.Foulds, C.Brennan, D.J.Weidler, D.C.Garg, P.Gibson, Steady state parenteral kinetics of fluconazole in man, in :"Abstracts of the First International Conference on Drug Research in Immuno-logic and Infectious Diseases. Antifungal Synthesis, Preclinical Evaluation, October 1987, New York" (1987).
8. P.F.Troke, R.J.Andrews, G.W.Pye, K.Richardson, Fluconazole and other azoles : Translation of in vitro activity to in vivo and clinical efficacy, in : "Fluconazole Symposium Abstracts, 5-9 October 1988, Dorado, Puerto Rico", p. 7 (1988).
9. P.F.Troke, R.I.Andrews, K.W.Brammer, M.S.Marriott, K.Richardson, Efficacy of UK-49,858 (Fluconazole) against Candida albicans experimental infection in mice, Antimicrob. Agents Chemother. 28 : 815 (1985).
10. K.W.Brammer and J.M.Feczko, Oral fluconazole in the treatment of acute vaginal candidiasis - A review of the phase III clinical program in: "Fluconazole Symposium Abstracts, 8-9 October 1988, Dorado, Puerto Rico" p. 21 (1988).

11. K.W.Brammer and L.J.Lees, Single dose of oral fluconazole in the treatment of vulvovaginal candidosis : An interim analysis of comparative study versus three-day intra-vaginal clotrimazole tablets, in: "Recent Trends in the Discovery, Development and Evaluation of Antifungal Agents", R.A. Fromtling, ed., pp.151-156, Barcelona, Prous Science Publishers, Barcelona (1987).

12. K.W.Brammer et al (Multicenter Study Group), Treatment of vaginal candidiasis with a single dose of fluconazole, Eur. J. Clin. Microbiol., 7 : 364 (1988).

13. A.Lassus, E.Kaappa, T.Virrankoski, A.Eskelinen, An open study to compare efficacy, safety, and toleration of a single dose of fluconazole (UK-49,858) and clotrimazole intravaginal tablets given once daily for 3 days in the treatment of patients with vaginal candidosis, in : "Abstracts of the Second World Congress for Sexually Transmitted Diseases, June 1986, Paris" (1986).

14. E.Tümbay "Pratik Tıp Mikolojisi (Practical Medical Mycology)" Bilgehan Basımevi, Izmir (1983).

15. E.Tümbay, K.Soy, G.Gürel, Bir çiftte saptanan genital kandidozun flukonazol ile sağaltımı (Fluconazole treatment of genital candidosis in sexual partners), İnfeks. Derg. 3 : 265 (1989).

FLUCONAZOLE IN MUCOSAL CANDIDOSIS:

PRELIMINARY RESULTS IN OROPHARYNGEAL CANDIDOSIS

E.Tümbay[1]   B.Üçer[2]   G.Karakartal[3]   A.Bilgiç[1]   R.İnci[4]
M.A.Özinel[1]   A.Havuk[5]   M.Tüker[6]   O.Demir[1]

[1] Department of Microbiology, Faculty of Medicine,
Ege University, Bornova, Izmir
[2] Anesthesiology/Intensive Care Section,
SSK Buca Hospital, Buca, Izmir
[3] Department of Clinical Bacteriology and Infectious
Diseases, Faculty of Medicine, Ege University,
Bornova, Izmir
[4] Specialist of Bacteriology and Infectious Diseases,
Cumhuriyet Meydanı, Batman
[5] Microbiology/Infectious Disease Section,
SSK Buca Hospital, Buca, Izmir
[6] Microbiology/Infectious Disease Section,
Izmir State Hospital, Yeşilyurt, Izmir, Turkey

INTRODUCTION

Fluconazole, (Flu), a novel triazole derivative administered both orally and intravenously, has been used particulary in the treatment of superficial and systemic Candida infections and crypto-coccosis. The drug has good pharmacokinetic properties: It is promptly absorbed shortly after oral intake, is diffused extensively into cerebrospinal fluid and other body fluids, shows low binding to plasma proteins, has a plasma half-life of 25-29 hours, and is excreted by the kidneys 75-90 % in bioactive form. Its toxicity is low and side effects are encountered in less than 1 % of the patients[1].

Flu has been tried in the treatment of oro-pharyngeal candidosis[2-13,15].

This paper is on clinical and mycological effect of Flu in Candida oropharyngitis.

MATERIAL AND METHOD

Flu was administered to 11 patients with oropharyngeal candidosis-clinically suspected and mycologically proven. Eight of the patients were in immunosuppression due to their primary malignant diseases and therapy (Table 1).

In all 11 patients, the causative agent of oropharyngeal infection was <u>Candida albicans</u>. The agent was seen in direct smears and grown in cultures from oropharyngeal swabs. Cultures on Sabouraud-dextrose-agar plates were semi-quantitatively evaluated : +(1-5 CFU), ++ (6-10 CFU), +++ (11-20 CFU) and ++++ (over 20 CFU).

Flu was administered in a daily dose of 100-300 mg/day for 8-42 days. The total dose for patients varied between 1200 mg and 4200 mg (Table 1). The daily dose and route of administration (PO or IV) were adjusted according to each patients's general condition, primary disease, clinical severity of the oropharyngeal infection and suspected oesophagitis.

All patients were controlled clinically every day and mycologically once a week until they were discharged from the clinic.

During observation period; each patient's liver functions, blood urea nitrogen and blood creatinine were controlled once or twice every week.

RESULTS

With Flu oropharyngeal candidosis was cured in all 11 patients. Clinical cure was obtained in 7 cases within 1 week and in 4 within 2 weeks (Table 1).

<u>Candida</u> was eradicated from the oropharynx of 4 patients by the end of first week, of 3 patients by the end of the second week, and of 1 patient by the end of the fourth week; in 3 patients the presence of the yeast continued in decreased numbers (Table 1).

Two patients (patient no. 5 with AIDS and patient no. 8 with larynx carcinoma) returned with relaps of oropharyngeal candidosis after a few months. These two patients had been previously clinically cured and eradicated from <u>Candida</u> (Table 1).

In all patients no toxicity/side effects were recorded.

DISCUSSION

Studies on the effect of Flu on oropharyngeal candidosis have been, in general, promising[2-15].

In this trial, Flu was administered to 11 patients with various diseased - generally malignant like leukemia, Hodgkin's Disease, carcinoma, AIDS, etc. The drug was given at a dose of 100-300 mg/day for 8-42 days orally or/and intravenously - the dose and route of administration being determined and adjusted according to the clinical picture and response of the patients to the drug (Table 1).

In previous studies 50 mg/day Flu administered to patients of oropharyngeal candidosis, most in immunosuppression due to primary disease, for at least seven days resulted in clinical cure rates ranging from 60% to 90% [11,12,13]. Meunier [10] obtained clinical cure or improvement in 90% of her patients who received 100 mg/day Flu for 10 days-19 weeks. In this present study, the authors have obtained clinical cure in all 11 patients - most of them immunosuppressed - in 1-2 weeks with,in general, 100-200 mg/day Flu (Table 1).

Table 1. Results of fluconazole administration in oropharyngeal candidosis (agent in all cases, C.albicans).

Case No.	Case	Administration	Clinical Result	Mycological Result
1	ÜZ, 62, male AML + Diabetes mellitus	100 mg/day IV 3 days 300 mg/day IV 4 days 200 mg/day IV 7 days 100 mg/day IV 13 days	Cure (in 1 week)	No eradication; Candida continued (++++)
2	DB, 26, male ALL	200 mg/day PO 8 days	Cure (in 1 week)	Eradication of Candida (in 4 weeks)
3	SM, 63, male ALL	200 mg/day PO 12 days	Cure (in 1 week)	Eradication of Candida (in 1 week)
4	MA, 57, male Lymphoma	100 mg/day PO 24 days	Cure (in 2 weeks)	No eradication; Candida continued in decreased numbers (++)
5	MI, 19, male AIDS	300 mg/day IV 1 day 200 mg/day IV 1 day 100 mg/day IV 2 days 100 mg/day PO 20 days	Cure (in 1 week)	Eradication of Candida (in 1 week) (later, relaps)
6	CZ, 58 male Hodgkin Dis. + Diabetes mellitus	200 mg/day IV 3 days 200 mg/day PO 10 days 100 mg/day PO 13 days	Cure (in 2 weeks)	Eradication of Candida (in 2 weeks)
7	ZE, 75, male Pemphigus	100 mg/day PO 10 days 100 mg/48h PO 21 days	Cure (in 2 weeks)	No eradication; Candida continued in decreased numbers (++)
8	ME, 58, male Larynx carcinoma	100 mg/day IV 9 days 100 mg/day PO 33 days	Cure (in 2 weeks)	Eradication of Candida (in 2 weeks) (later, relaps)
9	FY, 68, male Coronary by-pass operation	100 mg/day IV 8 days 100 mg/48h IV 8 days	Cure (in 1 week)	Eradication of Candida (in 1 week)
10	CŞ, 18, male Cerebral operation	200 mg/day IV 7 days 100 mg/day IV 7 days	Cure (in 1 week)	Eradication of Candida (in 1 week)
11	HÖ, 36, female Antibiotic therapy	100 mg/day PO 14 days	Cure (in 1 week)	Eradication of Candida (in 1 week)

Previous studies [11,12,13] and the results of this study indicate that it is necessary to administer Flu against oropharyngeal candidosis longer than two weeks - particularly in case of immuno-suppressed patients.

The dose of Flu used in Candida oropharyngitis is usually 50-100 mg/day [5,6,10,12,13]. Although clinical cure can be obtained by 50 mg/day dose, higher doses are recommended for the eradication of the fungus - particularly for patients with malignancies [11,12]. In the present study; in some cases the starting dose was 200-300 mg/day which was then lowered to 100 mg/day - in view of the general condition of the patient, degree of oropharyngeal inflammation and probable oesophagitis (Table 1).

Therapy with Flu gives different rates of clinical cure and eradication of the agent [3,6,10]. In a group of AIDS patients who received Flu 50 mg/day for 4 weeks, clinical cure was obtained in

100%; almost in all [except in a case with Candida (Torulopsis) infection] the fungus was eradicated [6]. In another study, 100 mg/day Flu given for 2-19 weeks resulted in 90% cure and 50% fungus eradication [10]. In another clinical trial, although patients showed clinical cure, Candida eradication rate was 48% [3]. Also in this investigation, along with clinical cure in 1-2 weeks, Candida was eradicated from throats of 8 patients in 1-4 weeks, and in the rest 3 patients Candida presence continued in lower numbers than the start (Table 1).

Relaps of orapharyngeal candidosis is seen particularly in the immunosuppressed [5,6,10]. Relaps was observed in all patients of an AIDS group treated and cured by Flu, 10-30 days following therapy [6].

Comparative studies with fluconazole and ketoconazole as therapeutic agents in oropharyngeal candidosis showed no difference in relaps rates with both drugs [5,10]. Relaps is due to the immune state of the patient. In the series presented, two immunocompromised patients (patients no.5 and no.8) came back with relaps a few months following therapy (Table 1).

In view of relaps, prophylactic use of Flu in oropharyngeal candidosis was considered [2]. A prophylaxis study showed that 50 mg/day Flu could protect patients from infection; Candida oropharyngitis developed in only 2-3% of patients under Flu prophylaxis and in 28% of the placebo patients [2]. The authors of this paper did no prophylactic study due to limited Flu at hand.

The patients of this study received totaly 1200-4200% mg of Flu All patients tolerated the drug very well, and no toxicity/side effects were encountered. Other investigators have also found Flu to be safe [6,8,11,12].

Fluconazole seems to be an efficacious and safe drug for oropharyngeal candidosis. Particularly in cases of immunosuppression, it should be administered for long periods and also its prophylactic administration be considered.

REFERENCES

1. E.Tümbay, Sistemik etkili yeni bir antifungal kemoterapötik : (Fluconazole, a novel antifungal agent with systemic effect), İnfeks. Derg., 3 (1), Ek Baskı Seri No. 1 (Suppl. 1), 27 pp. (1989).
2. J.P.Bodey, G.Samonis, K.Rolston, Prophylaxis of oropharyngeal candidiasis with fluconazole, in : "Abstracts of the Fifth International Symposium on Infections in the Immunocompromised Host, June 1988, Noordwijkerhout, The Netherlands" (1988).
3. K.W.Brammer, P.R.Farrow, J.M.Feczko, Fluconazole, a new systemically effective antifungal agent for use in hospitalized patients, in : "Abstracts of the Fifth International Conference of the Hospital Infection Society, August 1987, London" (1987).
4. K.W.Brammer, J.M.Feczko, R.R.G.Leeming, Fluconazole in the treatment of systemic candidosis and cryptococcal infections, in : "Abstracts of the Tenth Congress of the International Society for Human and Animal Mycology, June 1988, Barcelona" (1988).
5. N.Clumeck, S.De Wits, D.Weerts, P.Hermans, F.De Cock, Double-blind randomized study of oral fluconazole (F) versus ketoconazole (K) in 40 episodes of oropharyngeal candidiasis (DC) in ARC/AIDS

patients (P), in : "Abstracts of Biennial Conference on Chemotherapy of Infectious Diseases and Malignancies (2nd BICON)", Abstract
No. 83 Zbl. Antineoplas. Chemother.,Suppl. 1 (1989).

6. P.Dellamonica, E.Bernard, Y.Lefichoux, M.Carles, S.Politano,
   Fluconazole treatment of amphotericin B resistant mucosal
   candidiasis infection in AIDS patients, in : "Abstract of the
   Fifth International Symposium on Infections in the Immunocompromised Host, June 1988, Noordwijkerhout, The Netherlands"
   (1988).

7. B.Dupont and E.Drouhet, Fluconazole in the management of oropharyngeal candidiasis in predominantly HIV antibody-positive
   group of patients, in : "Recent Trends in the Discovery, Development and Management of Antifungal Agents", R.A. Fromtling, ed.,
   pp. 163-168, Prous Science Publishers, Barcelona (1987).

8. A.Gil, P.Lavillame, M.E.Valencia, V.Pintado, J.M.L.Dupla,
   J.Lahuerta, Fluconazole treatment of esophageal candidiasis in
   AIDS patients, in : "Abstracts of Biennial Conference on Chemo-
   therapy of Infectious Diseases and Malignancies (2nd BICON), 5-8
   March 1989, Montreux, Switzerland", Abstract No. 79,
   Zbl. Antimikrob. Antineoplas. Chemother., Suppl. 1 (1989).

9. R.J.Hay, Overview of studies of fluconazole in oropharyngeal
   candidiasis, in :"Fluconazole Symposium Abstracts, 8-9 October
   1988, Dorado, Puerto Rico", p. 20 (1988).

10. F.Meunier, Therapy of candidiasis in the immunocompromised host,
    in : "Fluconazole Symposium Abstracts, 8-9 October 1988, Dorado,
    Puerto Rico", p. 26 (1988).

11. F.Meunier, M.Aoun, M.Gerhard, N.Delhaye, Therapy of oro-
    pharyngeal candidiasis in the immunocompromised host, in :
    "Abstracts of Biennial Conference on Chemotherapy of Infectious
    Diseases and Malignancies (2nd BICON), 5-8 March 1989, Montreux,
    Switzerland", Abstract No. 70, Zbl. Antimikrob. Antineoplas.
    Chemother., Suppl. 1 (1989).

12. F.Meunier, J.Gerain, R.Snoek, F.Libotte, C.Lambert,
    A.M.Cueppens, Fluconazole therapy of oropharyngeal candidiosis
    in cancer patients, in : "Recent Trends in the Discovery, Devel-
    opment and Management of Antifungal Agents", R.A. Fromtling, ed.,
    pp. 169-174, Prous Science Publishers, Barcelona (1987).

13. P.A.Robinson, A.K.Knirsch, J.A.Joseph, Fluconazole treatment of
    serious fungal infections in patients who have failed conventional
    antifungal therapy, in : "Fluconazole Symposium Abstracts, 8-9
    October 1988, Dorado, Puerto Rico", p. 25 (1988).

14. D.A.Stevens, E.Brummer, J.G.McEwen, A.Perlman, Efficacy of
    fluconazole, a new oral triazole, in blastomycosis and in
    comparison with ketoconazole, in : "Fluconazole Symposium
    Abstracts, 8-9 October 1988, Dorado, Puerto Rico, p. 16 (1988).

15. P.F.Wragg, R.A.Howell, P.Hardy, P.J.Farrelly, M.V.Martin, An
    open non-comparative trial of fluconazole, a new systemically
    administered antifungal, J.Dent.Res., 65 : 515 (1986).

CONCLUSIONS AND PERSPECTIVES OF THE FEMS-SYMPOSIUM ON <u>CANDIDA</u> AND

CANDIDAMYCOSIS

Ch. De Vroey

Laboratory for Mycology, Institute of Tropical
Medicine, Antwerp, Belgium

It is my privilege to have been asked to present the conclusions
and perspectives of this interesting and pleasant symposium on
<u>Candida</u> and Candidamycoses.

There have been about 22 invited and over 50 free papers and
posters covering all aspects of <u>Candida</u> and candidamycosis.

It is, therefore, not possible to cite all contributions and
contributors of this symposium; it is only possible to thank them all
together.

I will limit myself to just commenting on different topics.

First, I would like to make a comment on the title of this
symposium : candidamycosis and candidosis are all right for me, but I
would prefer "Candida and candidamycoses". As was shown in the
various papers dealing with pathological, clinical, epidemiological
and mycological aspects, <u>Candida</u> infections cannot be treated as one
entity : there are several different forms of candidamycosis and the
clinical manifestations, the etiology and sources of infection depend
completely on the type of host in which the infection develops. This
is clearly shown by candidosis in neutropenia, in non-neutropenic
surgical patients, in heroin addicts, in AIDS, etc...

It is with this in mind that all aspects of <u>Candida</u> and candidamycoses have to be considered.

The first series of papers was devoted to fundamental aspects of
<u>Candida</u> molecular biology : each of these studies has already or will
lead to a better understanding of pathogenesis, to the development
of serological diagnostic procedures, to the discovery of new targets
for new antifungal drugs.

Much emphasis has been made on pathogenesis : clearly not a
single but several factors may play a role. However, the role of each
of these factors must be analysed for each type of candidosis. In
other words, in different hosts and for different <u>Candida</u> species
different factors are involved.

Candida and Candidamycosis, Edited by E. Tümbay *et al.*
Plenum Press, New York, 1991

I will give a very simple example. Much importance is given to the role of invasive mycelium of <u>Candida albicans</u> in the vaginal mucosa. How then can we explain that up to 8% of cases of vaginitis are due to <u>C. glabrata</u> - which as a good "<u>Torulopsis</u>" species is unable to produce hyphae?

From the many reports on chemotherapy and prevention and from the satellite symposium on therapy of candidosis, it appears once more that to demonstrate not only <u>in vitro</u> but also <u>in vivo</u> efficacy of useful antifungal drugs, the techniques and the animal models have to be "adapted" to each molecule.

This is a reflection of what happens in a clinical context :the therapeutic decision has to be made, mainly by the clinician, depending on the type of patient, the severity of the disease, the localisation of the lesions and the etiological agent. The pharmaceutical companies provide us with more and more new antifungal drugs. This precise cellular/molecular mode of action is sometimes completely understood before their clinical efficacy has been demonstrated!

It is noteworthy that the value of "old" drugs, the polyenes- whose exact mode of action is still not entirely clear - has been often referred to during this symposium. For certain compromised patients amphotericin B remains the first choice. It is a challenge for all the pharmaceutical companies : they still have to find a less toxic drug with the same or higher efficacy, optimally a drug with <u>in vivo</u> fungicidal effect.

A limited number of papers have been devoted to the diagnosis, identification and particularly to serology. The diagnosis of invasive <u>Candida</u> is still a problem but there are so many papers and publications on this topic that it was, I think, agreed by the organising committee to restrict the number of papers on this aspect.

I would like to stress the fact that especially in the field of serological diagnosis it is very important to take into account the type of patient when new techniques are evaluated.

At the end of this remarkable symposium I would like to make a suggestion : as this symposium was so successful and since so many aspects of <u>Candida</u> and candidamycoses have still to be clarified, I would suggest that Prof. E. Tümbay should organize (with the help of the Turkish Microbiological Society and the generosity of FEMS and the pharmaceutical companies) another symposium; since it was such a success why not follow the example of certain TV or movie series with <u>Candida</u> and candidamycoses n°2, n°3, etc. ?!!!

Congratulations and many thanks to the organizers!

Teşekkür ederim ve allahaısmarladık! (Thank you and good-bye!).

296

AUTHOR INDEX

SUBJECT INDEX